Rapid Assessment Program
Programme d'Évaluation Rapide

RAP
Bulletin
of Biological
Assessment

Bulletin RAP
*d'*Évaluation
Rapide
35

A Rapid Biological Assessment of the Forêt Classée du Pic de Fon, Simandou Range, South-eastern Republic of Guinea

Une Évaluation Biologique Rapide de la Forêt Classée du Pic de Fon, Chaîne du Simandou, Guinée

Editor
Jennifer McCullough

Center for Applied Biodiversity
Science (CABS)

Conservation International

Rio Tinto

Le *Bulletin RAP d' Évaluation Rapide* est publié par:
The *RAP Bulletin of Biological Assessment* is published by:
Conservation International
Center for Applied Biodiversity Science
Department of Conservation Biology
1919 M St. NW, Suite 600
Washington, DC 20036
USA

202-912-1000 telephone
202-912-0773 fax
www.conservation.org
www.biodiversityscience.org

Editor/*Editeur:* Jennifer McCullough
Design/production: Glenda Fábregas
Map/*Carte:* Mark Denil
Photography/*Photo couverture:* Piotr Naskrecki, Jan Decher et Andulai Barrie
French Translation/*Traduction française:* Sylvain Dufour, Eleana Gomez et Fanja Andriamialisoa

RAP Bulletin of Biological Assessment Series Editors:
Terrestrial and AquaRAP: Leeanne E. Alonso et Jennifer McCullough
Marine RAP: Sheila A. McKenna

Conservation International est un organisme privé à but non lucratif exonéré selon la section 501 c(3) du Internal Revenue Code de tout impôt sur les bénéfices.
Conservation International is a private, non-profit organization exempt from federal income tax under section 501 c(3) of the Internal Revenue Code.

ISBN: 1-881173-75-5
© 2004 by Conservation International
Tous droits de reproduction, de traduction et d'adaptation réservés. All rights reserved.

*Numéro au catalogue de la Bibliothèque du Congrès/*Library of Congress Catalog Card Number: 2003116214

Les désignations des entités géographiques dans cette publication et la presentation du matériel ne sous-entendent pas l'expression de toute opinion quelle qu'elle soit de Conservation International ou des organisations qui la soutiennent concernant le statut légal de tout pays, territoire ou région ou encore de ses autorités ou concernant la délimitation de ses frontières ou limites.
The designations of geographical entities in this publication, and the presentation of the material, do not imply the expression of any opinion whatsoever on the part of Conservation International or its supporting organizations concerning the legal status of any country, territory, or area, or of its authorities, or concerning the delimitation of its frontiers or boundaries.

Les opinions exprimées dans les Bulletins RAP d'Évaluation Rapide sont celles des auteurs et ne reflètent pas nécessairement celles de Conservation International ou de ses coéditeurs.
Any opinions expressed in the *RAP Bulletin of Biological Assessment* are those of the writers and do not necessarily reflect those of CI.

Les Bulletins RAP d'Évaluation Rapide étaient precedemment publiés sous le nom RAP Working Papers. Les numéros 1-13 de cette série ont été publiés sous l'ancien nom.
RAP Bulletin of Biological Assessment was formerly *RAP Working Papers.* Numbers 1–13 of this series were published under the previous title.

Citation Proposée:
McCullough, J. (éd.). 2004. Une Evaluation Biologique Rapide de la Forêt Classée du Pic de Fon, Chaîne du Simandou, Guinée. Bulletin RAP d'Evaluation Rapide No. 35. Conservation International, Washington, DC.
Suggested citation:
McCullough, J. (ed.). 2004. A Rapid Biological Assessment of the Forêt Classée du Pic de Fon, Simandou Range, Southeastern Republic of Guinea. RAP Bulletin of Biological Assessment 35. Conservation International, Washington, DC.

Table of Contents

Preface

In many regions of the globe, natural resource development overlaps with competing environmental values, social values and economic considerations. As natural resource development expands into new areas, it is expected that this overlap will occur with increasing frequency. To ensure that resource development does not unnecessarily compromise these other factors, there is an urgent need for better mechanisms of dialogue and cooperation among key stakeholders and the values they strive to uphold. It was in this spirit that Rio Tinto Mining and Exploration Limited (RTM&E) and Conservation International (CI) negotiated a memorandum of agreement to initiate activities in and around the Pic de Fon classified forest in the Simandou Range in November of 2002.

There has been very little published data available on the Simandou Range, and more specifically on the Pic de Fon. However, based on its unique habitat mosaic, including rare montane habitats, the Pic de Fon was thought likely to contain high biological diversity and was defined as a priority area for biodiversity assessment during the West Africa priority-setting workshop in Elmina, Ghana in 1999. In addition to its biological wealth, the Simandou Range also harbors potential mineral wealth, and in particular extensive iron ore deposits. Since 1997, RTM&E has been conducting exploration activities on contiguous exploration licenses covering the Simandou Range.

A Rio Tinto-funded Rapid Assessment Program (RAP) biological survey was conducted in November and December 2002 by a multi-disciplinary team of 13 international and regional scientists specializing in West African terrestrial biodiversity. This was done in conjunction with a socio-economic threats and opportunities assessment, comprised of a literature review and a stakeholder workshop with 33 participants from 17 organizations including national government, multilateral and bilateral donors, NGOs, scientific research centers and RTM&E. The workshop identified threats to biodiversity in the region, as well as opportunities for conservation.

In addition to expanding scientific knowledge of the region, this initial work will also greatly assist RTM&E's ongoing environmental management at the site. The data will contribute to more detailed baseline studies and to the Environmental and Social Impact Assessment (ESIA) that will run parallel to overall feasibility studies, if the project proceeds beyond the current exploration phase.

These activities have also formed the basis for an Initial Biodiversity Action Plan (IBAP) for the Pic de Fon and West African Greater Nimba Highlands Region, which includes both the Simandou and Nimba Ranges. The overall conservation objective identified in the IBAP is to improve natural resource management in the Greater Nimba Highlands through addressing threats to biodiversity, promoting sound agricultural production practices and developing sustainable income alternatives for communities that rely heavily on forests for their livelihoods.

Both RTM&E and CI are hopeful that the results of this work will not only promote improved resource management and conservation efforts in the Pic de Fon, but in other areas in the Greater Nimba Highlands. As such, this report should not be seen as an end product, but rather a starting point for generating knowledge, interest and support for promoting regional conservation and sustainable development through greater cooperation and dialogue with key local and regional stakeholders.

Colin Harris
Africa/Europe Region
Rio Tinto Mining and Exploration Limited

Glenn Prickett
The Center for Environmental Leadership in Business
Conservation International

Participants and Authors

Mohamed Alhassane Bangoura (reptiles and amphibians)
Chercheur Independent affilié au Centre de Gestion de
l'Environnement des Monts Nimba
BP 1869, Conakry
Guinea
Email. Mohamed_alhassane@yahoo.fr

Abdulai Barrie (large mammals)
Department of Biological Sciences
Faculty of Environmental Sciences
Njala University College
University of Sierra Leone
PMB, Freetown
Sierra Leone
Email. ahbarrie@yahoo.com

Jan Decher (small mammals)
Department of Biology
University of Vermont
Burlington, VT 05405
USA
Email. Jan.Decher@uvm.edu

Nicolas Londiah Delamou (plants)
Chef d'Antenne Forêt Classée Mont Béro
P.G.R.R./Centre Forestier N'Zérékoré
BP 171, N'Zérékoré
Guinea
Email. cfzpgrr@sotelgui.net.gn

Ron Demey (birds)
Van der Heimstraat 52
2582 SB Den Haag
Netherlands
Email. rondemey@compuserve.com

Mamadou Saliou Diallo (coordination)
Guinée-Ecologie
210, rue DI 501, Dixinn
B.P. 3266, Conakry
Guinea
Email. dmsaliou@mirinet.com

Njikoha Ebigbo (bats)
Department of Experimental Ecology
University of Ulm
Albert-Einstein-Allee 11
89069 Ulm
Germany
Fax. +49(731) 5022683
Email. njikoha.ebigbo@biologie.uni-ulm.de

Jakob Fahr (bats)
Department of Experimental Ecology
University of Ulm
Albert-Einstein-Allee 11
89069 Ulm
Germany
Fax. +49(731) 5022683
Email. jakob.fahr@biologie.uni-ulm.de

Ilka Herbinger (primates)
Max Planck Institute for Evolutionary Anthropology
Department of Primatology
Deutscher Platz 6
04103 Leipzig
Germany
Email. herbinger@eva.mpg.de

Jean-Louis Holié (plants)
Consultant botaniste
P.G.R.R./Centre Forestier N'Zérékoré
BP 624, Conakry / BP 171, N'Zérékoré
Guinea

Soumaoro Kante (large mammals)
Division Faune et Protection de la Nature
Direction Nationale des Eaux et Forêts
BP 624, Conakry
Guinea
Email. dfpn@sotelgui.net.gn

Jennifer McCullough (coordination, editor)
Conservation International
1919 M Street NW, Suite 600
Washington, DC 20036
USA
Email. j.mccullough@conservation.org

Piotr Naskrecki (invertebrates)
Director, Invertebrate Diversity Initiative
Conservation International
Museum of Comparative Zoology
Harvard University
26 Oxford St.
Cambridge, MA 02138
USA
Email. p.naskrecki@conservation.org

Hugo Rainey (birds)
School of Biology
Bute Medical Building
University of St. Andrews
St. Andrews, Fife KY16 9TS
United Kingdom
Email. hjr3@st-andrews.ac.uk

Mark-Oliver Rödel (reptiles and amphibians)
Department of Animal Ecology and Tropical Biology
Biocenter
Am Hubland, D-97074 Würzburg
Germany
Email. roedel@biozentrum.uni-wuerzburg.de

Elhadj Ousmane Tounkara (primates)
ENRM Project
Winrock International
BP 26, Conakry
Guinea
Email. ourypdiallo@yahoo.com
jfischer@sotelgui.net.gn

Organizational Profiles

CONSERVATION INTERNATIONAL

Conservation International (CI) is an international, nonprofit organization based in Washington, DC. CI believes that the Earth's natural heritage must be maintained if future generations are to thrive spiritually, culturally and economically. Our mission is to conserve the Earth's living heritage, our global biodiversity, and to demonstrate that human societies are able to live harmoniously with nature.

Conservation International
1919 M Street NW, Suite 600
Washington, DC 20036
USA
tel.   800 406 2306
fax.   202 912 0772
web.   www.conservation.org
         www.biodiversityscience.org

RIO TINTO MINING AND EXPLORATION LIMITED

Rio Tinto Mining and Exploration Limited is the exploration division of Rio Tinto p.l.c., a global mining company based in the United Kingdom. Rio Tinto's worldwide operations supply essential minerals and metals that help to meet global needs and contribute to improvements in living standards. The company recognizes that excellence in managing Rio Tinto's health, safety, environment and community responsibilities is essential.

Rio Tinto Mining and Exploration
6 St James Square
London
SW1Y 4LD
UK
tel.   0207 9302399
fax.   0207 9303249
web.   www.riotinto.com

CENTRE FORESTIER DE N'ZÉRÉKORÉ

Centre Forestier de N'Zérékoré (CFN) is the field implementation office of the National Directorate of Water and Forests (DNEF) in Guinea's forest region. Through the Projet de Gestion des Ressources Rurales (PGRR), CFN is currently conducting forestry management work in the Pic de Fon region, including delineation and demarcation of forest boundaries.

PGRR/Centre Forestier de N'Zérékoré
BP 171, N'Zérékoré
GUINEA
tel.   (224) 91 15 03
email.   cfzpgrr@sotelgui.net.gn

GUINÉE ECOLOGIE

Guinée Ecologie is a Guinean registered (1990) non-governmental, volunteer and nonprofit organization. The vision of the organization is based upon global concerns regarding the state of the earth under the high pressure of a number of human activities which are not sustainable. The mission of Guinée Ecologie is to contribute through research, education, information, communication and management to the protection of the environment and the conservation of biodiversity in Guinea following the principles of sustainable development.

Guinée Ecologie
210, rue DI 501, Dixinn
BP 3266, Conakry
GUINEA
tel.   (224) 46 24 96
email.   dmsaliou@afribone.net.gn

Acknowledgments

The success of this RAP survey was due to the support of many people. The participants thank Mamadou Saliou Diallo, Kolon Diallo, and Guinée-Ecologie for the superb organization of logistics in Guinea. Without their thoughtful assistance the whole tour would have been almost impossible. In the Pic de Fon area we received all possible support from the Centre Forestier de N'Zérékoré and their Conservateur, Cécé Papa Condé. Thanks also to the Rio Tinto Mining and Exploration Limited (RTM&E) staff, represented by John Merry and Luc Stévenin, for providing fuel, food and all help needed on the mountain. Work on the second site would not have been possible without the assistance and support of the people from Banko village. Our local assistants, Bernard Dore and Kaman Camara among many others, were of invaluable help during field work, as was Direction Nationale des Eaux et Forêts (DNEF) agent Ouo Ouo Bamanou. We wish also to thank all the RAP participants for their pleasant and inspiring companionship. The Ministère de l'Agriculture des Eaux et Forêts, Republic of Guinea, Mathias Rodolphe HABA, issued the collecting and export permissions (Code 100, Nr. 000741-742). The RAP participants are indebted to Conservation International (CI) in general and Jennifer McCullough and Leeanne E. Alonso in particular for the invitation to participate in and organization of this Rapid Assessment Program survey. Thanks also to Francis Lauginie and Guy Rondeau for their assistance in preparing the RAP.

We wish to recognize the contributions of John Merry, Colin Harris, and Sally Johnson from RTM&E and David Richards of Rio Tinto p.l.c. Without their substantive and financial support, this project would not have been possible, and their feedback on this report has been a valuable addition. We wish to thank CI's Center for Environmental Leadership in Business (CELB), and particularly Assheton Carter, Marielle Canter, and Greg Love for facilitating the collaboration with RTM&E and for guiding the Initial Biodiversity Action Plan. Leonie Bonnehin, of CI-Côte d'Ivoire, was an invaluable asset to this project both before the survey and afterwards, in supporting logistical arrangements, beautifully organizing and running the threats and opportunities workshop (in cooperation with Mamadou Saliou Diallo), and conducting follow-up work in Guinea as a result of this survey. Thanks is also due to other members of CI's West Africa program including Jessica Donovan, Mohamed Bakarr, Olivier Langrand, and Tyler Christie for input and guidance both before and after the expedition. Additionally, without the help of Rebecca Kormos this expedition may never have happened. We wish to sincerely thank her for all of her enthusiasm and for sharing her many contacts within and extensive knowledge of Guinea. We also owe thanks to Karl Morrison of CI's Regional Conservation Strategies Group for his participation and assistance with the workshop and IBAP, to Eduard Niesten for his assistance in formulation of the threats and opportunities assessment methodology, to Mark Denil of CI's Conservation Mapping Program, and to Glenda Fábregas for her patience in designing RAP publications.

The bat team is very grateful to the following curators for providing data on specimens in the collections under their care: Peter Taylor (Durban), William Stanley (Chicago), David Harrison (Sevenoaks), Georges Lenglet (Bruxelles), Manuel Ruedi (Genève), Wim van Neer (Tervuren), Nico Avenant (Bloemfontein), Susan Woodward (Toronto), Dieter Kock (Frankfurt/M.), Fritz Dieterlen (Stuttgart), and Teresa Kearney (Pretoria). Meredith Happold (Canberra), Ernest Seamark (Pretoria), Paul van Daele (Livingstone, Zambia) and P. Taylor kindly made available unpublished distribution records of *Myotis welwitschii*. The present work was partially funded through the BIOLOG-program of the German Ministry of Education and Research (BMBF; Project W09 BIOTA-West, 01LC0017). The non-volant small mammal team wishes to thank Dr. R. Hutterer at the Museum Alexander Koenig, Bonn, Germany; L. Gordon and Dr. M. Carleton at the United States National Museum, Washington; and Dr. C. W. Kilpatrick at the University of Vermont for assistance with voucher specimen preservation and identification.

Report at a Glance

A RAPID BIOLOGICAL ASSESSMENT OF THE FORÊT CLASSÉE DU PIC DE FON, SIMANDOU RANGE, SOUTH-EASTERN REPUBLIC OF GUINEA

Expedition Dates
27 November-7 December 2002

Area Description
The Simandou Mountain Range is located in south-eastern Guinea and extends for 100 km from Komodou in the north to Kouankan in the south. At the southern end of the Siman-dou Range lies the Pic de Fon classified forest. Approximately 25,600 ha of this forest were given limited protection in 1953 for the purpose of conserving forest and soil. The Simandou Range also acts as a natural barrier against savannah fires and regulates the flow of the water sources that originate from within the range (including the Diani, Loffa, and Milo rivers). The location of this reserve in the transition between forest and savannah zones offers habitat types ranging from rainforest to humid Guinea savannah to montane gallery and ravine for-est. The Pic de Fon classified forest covers an altitudinal range from about 600 to more than 1,600 m asl (containing the Pic de Fon, the highest point on the range at 1,656 m asl and second highest point in Guinea) resulting in the additional habitat type of montane grass-land, a rare and threatened habitat in West Africa. Additionally, surrounding the Pic de Fon classified forest are 24 villages whose residents currently rely on the forest as a source of food, water, fuelwood, and medicine. Two sites were surveyed during the RAP expedition: one close to the summit of the Pic de Fon and the other at lower elevation and near to Banko vil-lage.

Reason for the Expedition
There is very little published data available on the Simandou Range, and more specifically the Pic de Fon. However, based on its unique habitat mosaic, including rare montane habi-tats, the Pic de Fon was thought likely to contain high biotic diversity and was defined as a priority area for biodiversity assessment during Conservation International's (CI's) prior-ity-setting workshop in Elmina, Ghana in 1999. In addition to the unique assemblages of ecosystems, the Simandou Range also harbors potential mineralogical wealth, and in par-ticular extensive iron ore deposits. Rio Tinto Mining and Exploration Limited (RTM&E) is currently conducting exploration activities on four contiguous exploration licenses within the Simandou Range.

This RAP was instigated by Rio Tinto as a contribution to their initial environmental and social assessment studies. It will serve as a contribution to the baseline studies and more detailed Environmental and Social Impact Assessment (ESIA) that will run parallel to overall feasibility studies, if the project proceeds beyond the current exploration phase. Given the prediction of high biodiversity within the Pic de Fon classified forest, RTM&E entered into an agreement with CI, facilitated by their Center for Environmental Leadership in Business

(CELB), to assess the region's biodiversity, as well as potential threats to and opportunities for conservation in the Pic de Fon classified forest. This partnership was formed in the spirit of providing significant gains for biodiversity conservation, as well as for the communities of people who rely on resources within the Simandou Range.

Major Results

A variety of terrestrial habitats and taxa were observed during the RAP survey, including montane grassland, montane forest (both gallery and ravine forest), semi-evergreen lowland forest (both primary and secondary), savannah, mountain streams, shrubby edge habitats, perennial plantations (coffee, cocoa, banana), and farmbush. Habitats were evaluated with respect to plants (both woody and herbaceous), invertebrates (katydids), amphibians, reptiles, birds, small mammals (including bats), and large mammals (with special attention given to primates).

In total, the RAP team documented 797 species. Several of these species are new to science, including at least five invertebrates (of which two are also new genera) and three amphibians. The RAP team recorded range extensions for a number of species and added as new records for Guinea: 11 invertebrates, 3 amphibians, 7 birds, 3 bats and 1 shrew. The Pic de Fon classified forest harbors a number of species of international conservation concern, including one tree (*Neolemonniera clitandrifolia*) listed by IUCN as "Endangered", as well as 15 trees listed as "Vulnerable" and one listed as "Near Threatened"; one amphibian (*Bufo superciliaris*) and four reptile species (*Chamaeleo senegalensis, Varanus ornatus, Python sebae,* and *Osteolaemus tetraspis*) that are protected by international law (CITES); two primate species listed by IUCN as "Endangered": *Pan troglodytes verus* (West African Chimpanzee) and *Cercopithecus diana diana* (Diana Monkey), and two listed as "Near Threatened": *Cercocebus atys atys* (Sooty Mangabey) and *Procolobus verus* (Olive Colobus); eight birds of global conservation concern including *Schistolais leontica* (Sierra Leone Prinia), *Criniger olivaceus* (Yellow-bearded Greenbul) and *Lobotos lobatus* (Western Wattled Cuckoo-shrike) listed by IUCN as "Vulnerable"; and seven bat species listed by IUCN, one species as "Vulnerable" (*Epomops buettikoferi*) and six species as "Near Threatened" (*Rhinolophus alcyone, R. guineensis, Hipposideros jonesi, H. fuliginosus, Kerivoula cuprosa, Miniopterus schreibersii*). Local hunters and guides confirmed the presence of the Nimba Otter Shrew (*Micropotamogale lamottei*), listed by IUCN as "Endangered."

Number of Species Recorded

Plants	409
Katydids (Orthoptera)	40
Amphibians	32
Reptiles	12
Birds	233
Bats (Chiroptera)	21
Small mammals (non-volant)	17
Large mammals	39
Primates	13

New Species Discovered

Katydids (Orthoptera)	At least 4 spp.: *Thyridorhoptrum* sp. *Afrophisis* sp. 2 spp. of new genera of Meconematinae possibly *Ruspolia* sp.
Cave cockroach (Blaberidae: Oxyhaloinae)	*Simandoa conserfariam* Gen. and sp. n
Amphibians	*Arthroleptis/Schoutedenella* sp. *Amnirana* sp. *Bufo* ? sp.

New Records for Guinea

Katydids (Orthoptera)	*Phaneroptera minima* *Phaneroptera* ? *maxima* *Zeuneria melanopeza* *Weissenbornia praestantissima* *Ruspolia jaegeri* *Thyridorhoptrum* sp. n. Meconematinae Gen. n., sp. n. 1 Meconematinae Gen. n., sp. n. 2 *Afrophisis* sp. n. *Anodeopoda* cf. *lamellate*
Cave cockroach (Blaberidae: Oxyhaloinae)	*Simandoa conserfariam* Gen. et sp. n
Amphibians	*Arthroleptis/Schoutedenella* sp. *Amnirana* sp. *Bufo* ? sp.
Birds	*Bubo poensis* *Neafrapus cassini* *Indicator willcocksi* *Smithornis capensis* *Phyllastrephus baumanni* *Cercotrichas leucosticta* *Vidua camerunensis*
Bats	*Myotis welwitschii* *Kerivoula phalaena* *Kerivoula cuprosa*
Non-volant small mammals	*Crocidura grandiceps* (shrew)

Conservation Recommendations

The results of this RAP survey show that the Pic de Fon classified forest contains unique terrestrial and aquatic species; intact connections between habitat, flora and fauna; and one of the last remaining, intact montane habitats, both within Guinea and the Upper Guinea Highlands. The primary current threats to the biodiversity of the area include bushmeat hunting, conversion of forest to agricultural land, and bushfires. Other threats include logging, artisanal mining, road development and unsustainable collection of non-timber forest products. As a result of this study, we strongly recommend that the Pic de Fon classified forest receive increased protection status and that current regulations with regards to its classified status be monitored, enforced and strengthened, as necessary. The limits of the protected area should be defined and enforced in collaboration with local communities and should be accompanied by education and outreach programs, as well as additional scientific and socio-economic studies. If the Pic de Fon classified forest is incorporated into an effectively managed protected area system that includes both forests and grasslands, and representatives of the many microhabitat variations, with special attention paid to critical habitats and endemic and threatened species, a high proportion of plant and animal species of this region would be protected. The RAP team found few invasive species during this survey and steps should be taken to prevent the introduction of such species.

Our results also point to the fact that any potential exploitation of mineral resources in the Pic de Fon area, if undertaken, will require further investigation into potential hydrological impacts and possible impacts to species relying on high altitude habitats (eg. West African Chimpanzee) as part of an Environmental and Social Impacts Assessment (ESIA). As montane grassland habitats are extremely rare in West Africa, any activities that will destroy or irreversibly damage this habitat should be carefully considered.

The 24 surrounding villages rely on this forest for a number of ecosystem services (and for food, water and medicines) and every effort should be made to ensure that the forest will continue to provide these services. Along these lines, community needs must be addressed in order for any protected area to remain viable, while illegal encroachment (and subsequent destruction of both flora and fauna within the forest) must be inhibited to ensure long-term viability, both in the interest of local communities and to protect some of the last intact forest left in West Africa. Land tenure is an extremely sensitive issue, and while it is critical to place the Pic de Fon region under some form of increased legal protection as soon as possible, the success of this protected area depends on open discussion and close collaboration with communities and other stakeholders in the region. These dialogues will facilitate mutual understanding between the government, conservation groups, and communities, with the intention of achieving conservation through a variety of measures that target the need for productive agricultural land, fuelwood supplies, sustainable wildlife management, and other services provided by the forest.

Maps and Photos

Cartes et photos

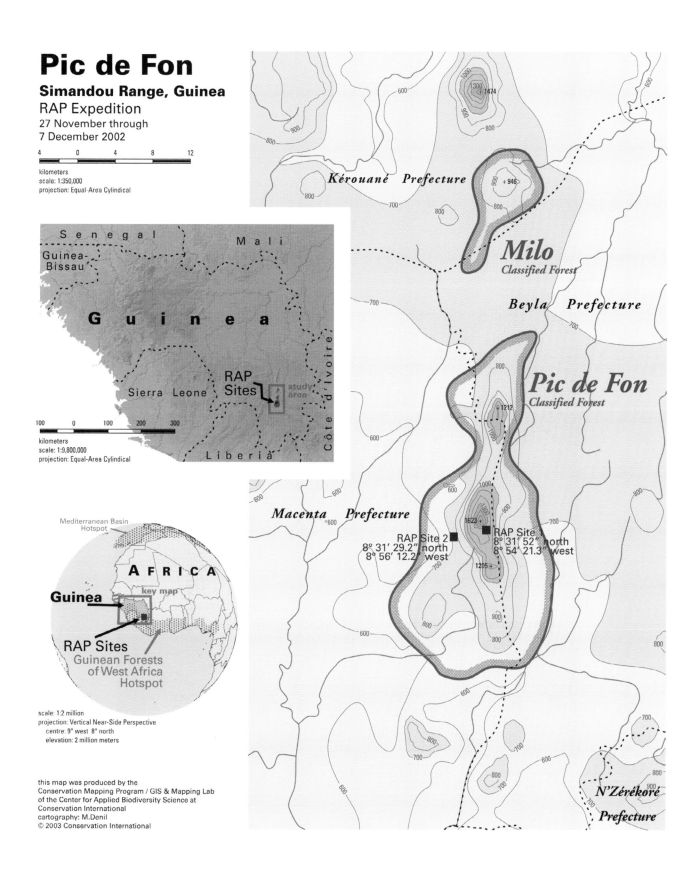

Pic de Fon

Simandou Range, Guinea

RAP Expedition
27 November through
7 December 2002

kilometers
scale: 1:350,000
projection: Equal-Area Cylindical

kilometers
scale: 1:9,800,000
projection: Equal-Area Cylindical

scale: 1:2 million
projection: Vertical Near-Side Perspective
centre: 9° west 8° north
elevation: 2 million meters

this map was produced by the
Conservation Mapping Program / GIS & Mapping Lab
of the Center for Applied Biodiversity Science at
Conservation International
cartography: M.Denil
© 2003 Conservation International

Banko village
Village de Banko

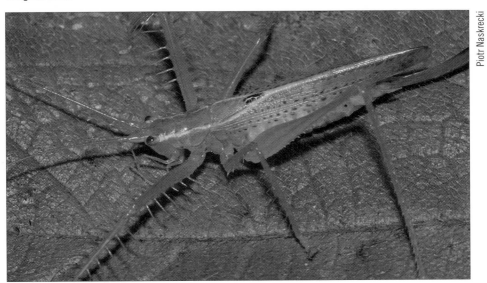

A new species of katydid discovered in the Pic de Fon forest, *Afrophisis* sp. nov.
Une nouvelle espèce de sauterelle découverte dans la Forêt Classée du Pic de Fon, *Afrophisis* sp. nov.

Blue-headed Bee-eater, *Merops muelleri*
Guêpier à tête bleue, *Merops muelleri*

A new katydid species, *Anoedopoda* sp. nov.
Une nouvelle espèce de sauterrelle, *Anoedopoda* sp. nov.

Arthroleptis sp., an undescribed frog species discovered during the RAP survey
Arthroleptis sp., une espèce de grenouille non décrite découverte pendant le RAP

Another new katydid species of an undescribed genus
Une autre nouvelle espèce de sauterrelle d'un genre non décrit

Camera trap photo of a brush-tailed porcupine, *Atherurus africanus*
Photo d'une Athérur africain, *Atherurus africanus*, prise avec un piège photographique

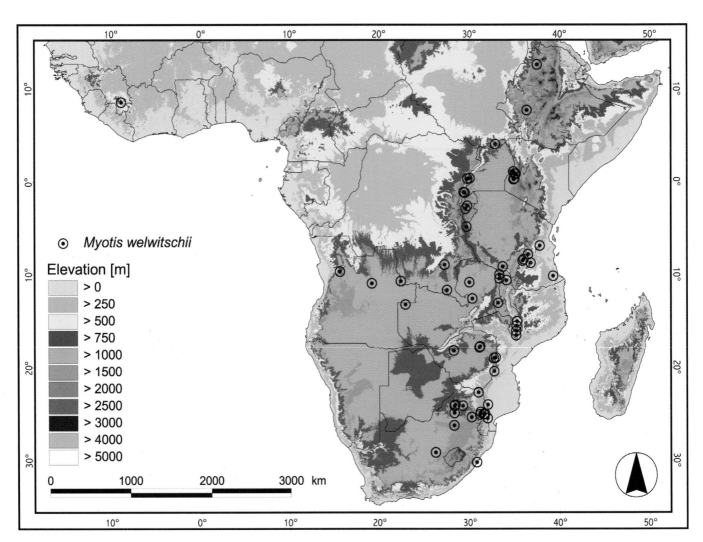

Distribution of *M. welwitschii* plotted on the 30-arc second digital elevation model (USGS; modified from Fahr and Ebigbo 2003)
Distribution de *M. welwitschii* (USGS ; d'après Fahr et Ebigbo 2003)

Piotr Naskrecki

A column of driver ants
Une colonne de fourmis magnans

Piotr Naskrecki

New genus and species of katydid Meconematinae
Nouveau genre et espèce de sauterrelle, Meconematinae

Jan Decher

RAP Campsite 1 near the summit of the Pic de Fon
Emplacement du camp 1 près du sommet du Pic de Fon

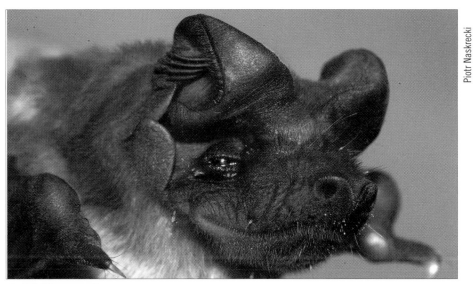

Piotr Naskrecki

The bat, *Mops spurelli*
La chauve-souris, *Mops spurelli*

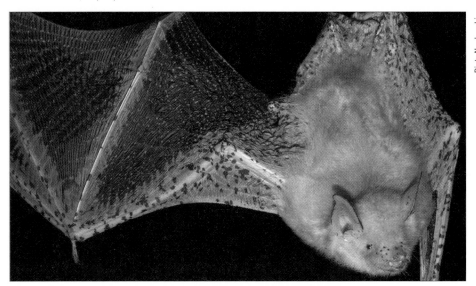

Piotr Naskrecki

Myotis welwitschii, recorded for the first time in West Africa during the RAP survey
Myotis welwitschii, enregistré pour la première fois en Afrique de l'Ouest pendant le RAP

Jan Decher

Praomys rostratus, eating a palm nut
Praomys rostratus, mangeant une noix de palme

A small stream running through typical habitat of RAP site 2
Un petit cours d'eau dans un habitat typique du site 2

Rousettus aegyptiacus, giving birth inside a cave
Rousettus aegyptiacus, donnant naissance à l'intérieur d'une grotte

Sierra Leone Prinia, *Schistolais leontica*, listed by IUCN as Vulnerable, found in higher elevation forest at Pic de Fon
La Prinia de Sierra Leone, *Schistolais leontica*, classée par l'UICN comme Vulnérable, a été trouvée à 1300-1350 m d'altitude

Piotr Naskrecki

Afronatrix anoscopus

Piotr Naskrecki

Woman from Banko village
Femme du village de Banko

Piotr Naskrecki

The Pic de Fon RAP Team
L'équipe RAP du Pic de Fon

Executive Summary

INTRODUCTION

Located in West Africa, the Guinean Forest is one of the world's 25 biologically richest and most endangered terrestrial ecosystems. The Upper Guinea Forest, one of two main blocks in the Guinean Forest ecosystem, extends across six West African countries, from southern Guinea to Sierra Leone and eastward through Liberia, Côte d'Ivoire, Ghana and Togo. Originally, the Upper Guinea Forest ecosystem is estimated to have covered as much as 420,000 km², but centuries of human activity have resulted in the loss of nearly 70 % of the original forest cover (Bakarr et al. 2001). The remaining Upper Guinea Forest is restricted to a number of isolated patches acting as refugia for the region's unique flora and fauna. These remaining forests contain exceptionally diverse ecological communities, distinctive flora and fauna, and a mosaic of forest types providing refuge to a number of endemic species.

The topography of West Africa is generally quite uniform and flat, with only a few mountainous regions, i.e. the Upper Guinea Highlands, the Togo/Volta Highlands, the Jos Plateau in Nigeria and the Cameroon Mountains. Accordingly, West African montane habitats are very limited in extent and consequently especially endangered. The Simandou Mountain Range, a part of the Upper Guinea Highlands, is located in south-eastern Guinea and extends for 100 km from Komodou in the north to Kouankan in the south. The southern portion of the Simandou Range lies within the Upper Guinea Forest ecosystem and at the southern end of the Simandou Range lies the Pic de Fon classified forest.

The Pic de Fon classified forest, created in 1953, is the third largest in the forested region of Guinea (Guinée Forestière) and covers approximately 25,600 ha. (In Guinea, classified forest status provides some protection as government property, though not necessarily for biodiversity conservation.) Its location in the transition between forest and savannah zones offers habitat types ranging from rainforest to humid Guinea savannah. Also, as the Pic de Fon classified forest covers an altitudinal range from about 600 to more than 1,600 m asl (containing the Pic de Fon, the highest point on the range at 1,656 m asl and second highest point in Guinea) the forest harbors a significant and increasingly rare habitat type for the Upper Guinea Forest zone, montane grassland.

Very little published data are available for the Simandou Range, and more specifically for the Pic de Fon classified forest. However, based on its montane habitats, altitudinal range, and relatively intact forest, the Pic de Fon was thought likely to contain high biotic diversity and the area was defined as a priority area for biodiversity assessment during Conservation International's (CI's) priority-setting workshop in Elmina, Ghana in 1999 (Bakarr et al. 2001).

Scope of Project

In addition to a unique assemblage of ecosystems, the Simandou Range also harbors potential mineralogical wealth and, in particular, extensive iron ore deposits. Rio Tinto Mining and Exploration Limited (RTM&E) is currently conducting exploration activities on four contiguous exploration licenses within the Simandou Range. This RAP survey, instigated by Rio Tinto as a contribution to their initial prefeasibility environmental and social assessment studies, will serve as a contribution to the baseline studies and more detailed Environmental and Social Impact Assessment (ESIA) that will run parallel to overall feasibility studies, if the project proceeds beyond the current exploration phase.

Given the potential of high biodiversity within the Pic de Fon classified forest, Rio Tinto entered into an agreement with Conservation International to assess the region's biodiversity, as well as potential threats to and opportunities for conservation in the Pic de Fon classified forest. This partnership was formed in the spirit of providing significant gains for biodiversity conservation and industry, as well as for the government and people of Guinea.

The collaboration originally focused on a set of activities of mutual interest: a RAP biological survey; a socio-economic threats and opportunities assessment; and building from these first two activities, development of an Initial Biodiversity Action Plan (IBAP) for the Pic de Fon. Subsequently, CI and RTM&E, along with other local and regional stakeholders, are forming a more comprehensive alliance in the region to carry forward the initial recommendations presented in this report.

The alliance seeks to strengthen the capacity of local communities and regional authorities to foster sustainable economic development and to effectively manage regional biodiversity and natural resources. Partner organizations will collaborate with local stakeholders to understand underlying drivers of threats to biodiversity while addressing the need for alternative livelihoods and locally-managed natural resource management. Although Pic de Fon classified forest is the priority target, partners will work to leverage benefits across the Greater Nimba Highlands, including Mount Nimba Nature Reserve, Diécké classified forest, Parc National Ziama, Béro-Tetini classified forest, Tibé classified forest, and Milo classified forest. Regional conservation planning will be facilitated through an integrated landscape design process that considers underlying biodiversity patterns, ecosystem processes (watersheds and resource availability) and land use.

RAP Expedition Overview and Objectives

Conservation International's Rapid Assessment Program (RAP), a department within the Center for Applied Biodiversity Science (CABS), was founded in 1990 in response to the increasing loss of biodiversity in tropical ecosystems. It is an innovative biological inventory program designed to generate scientific information to catalyze conservation action in tropical areas that are under imminent threat of habitat conversion. Together with the Center for Environmental Leadership in Business (CELB) and CI's West Africa program, RAP organized an expedition to the Pic de Fon classified forest in November 2002 to better understand the biological diversity of this area. The primary objective of the RAP expedition was to collect scientific data on the diversity and status of species of the Pic de Fon classified forest in order to make recommendations regarding their conservation and management. In addition, data collected by the RAP team was used to develop an Initial Biodiversity Action Plan for the Pic de Fon area (see pg. 34).

The specific aims of the expedition were to:

- Derive a brief but thorough overview of the existing diversity and integrity within the Pic de Fon ecosystem and evaluate its relative conservation importance;

- Undertake an evaluation of threats to the biodiversity within the area surveyed;

- Provide further "on-site" training for Guinean biologists under the guidance and mentorship of experienced field ecologists;

- Provide preliminary management and research recommendations for the Pic de Fon ecosystem together with recommendations for conservation priorities with a view to influencing local, national and international conservation policies relating to the Pic de Fon/Simandou Range in particular and the Upper Guinea Highlands in general; and

- To make RAP data publicly available to identify and describe the biodiversity and unique features of the Pic de Fon ecosystem for decision-makers as well as members of the general public in Guinea and elsewhere, with a view to increasing awareness of this ecosystem and promoting its conservation.

The expedition's multi-disciplinary team of 13 scientists included representatives of two Guinean government departments, the Direction Nationale des Eaux et Forets (DNEF) and the Centre Forestier (see participants section), and two local organizations, the Centre de Gestion de l'Environnement des Monts Nimba and Winrock International. The scientific team comprised international and national scientists specializing in West African terrestrial ecosystems and biodiversity.

The RAP team examined selected taxonomic groups to determine the area's biological diversity, its degree of endemism, and the uniqueness of the ecosystem. RAP expeditions survey focal taxonomic groups as well as indicator species, with the aim of choosing taxa whose presence can help identify a habitat type or its condition. We chose to survey plants, katydids (Orthoptera), amphibians and reptiles, birds, small mammals including bats, and large mammals.

Study Area

The RAP survey took place from 26 November to 7 December 2002, at the beginning of the dry season, and covered two sites. The first RAP survey site (Site 1) was situated between montane grasslands and montane forests that were directly connected to larger forests at a lower altitude. This site covered an altitudinal range of between 1,000 and 1,600 m asl. The campsite was located at 1,350 m asl (08°31'52"N, 08°54'21"W) and was 3.5 km (3,461 m) from the Pic de Fon.

The second site (Site 2) comprised secondary and primary forests and some small savannah patches at around 600 m asl. The campsite (08°31'29"N, 08°56'12"W) was located approximately 6 km from the village of Banko (in the direction of the Simandou Ridge) and was approximately 1.5 km inside the boundary of the classified forest. The camp was situated near lowland forest with many streams of different size exiting the mountain range and some hills with derived savannah vegetation. On more level ground and toward the village, most forest has been converted to cocoa and coffee plantations, and this area also contains some secondary forest.

SUMMARY OF RESULTS

Criteria generally considered during RAP surveys in order to identify priority areas for conservation across taxonomic groups include species richness, species endemism, rare and/or endangered species, and critical habitats. Measurements of species richness can be used to compare the number of species per area between areas within a given region. Measurements of species endemism indicate the number of species endemic to some defined area and give an indication of both the uniqueness of the area and the species that will be threatened by alteration of that area's habitat (or conversely, the species that will likely be conserved through protected areas). Assessment of rare and/or endangered species within a given area indicates the proportion of richness of IUCN Red Data species (IUCN 2002) that are known or suspected to occur within the area. Many of these Red-listed species carry increased legal protection thus giving greater importance and weight to conservation decisions. Describing the number of critical habitats or subhabitats within an area identifies sparse or poorly known habitats within a region that contribute to habitat variety and therefore to species diversity. The following is a summary, based on these criteria, of the key findings from our two-week field study.

Overall Significance
Species of Conservation Concern:

- We recorded 17 plant species listed by IUCN – one species listed as "Endangered" (*Neolemonniera clitandrifolia*), fifteen listed as "Vulnerable" and one listed as "Near Threatened".

- We identified one amphibian (*Bufo superciliaris*) and four reptile species (*Chamaeleo senegalensis, Varanus ornatus, Python sebae*, and *Osteolaemus tetraspis*) that are protected by international law (CITES species).

- We recorded eight birds of global conservation concern – two in the highlands, seven in the lowlands. The most important of these was *Schistolais leontica* (Sierra Leone Prinia), listed by IUCN as "Vulnerable", a species with a very limited distribution in the West African highlands and known from only three other sites in the world.

- We recorded seven bat species, 33.3 % out of the total of 21 species recorded during this survey, listed by IUCN (Hutson et al. 2001): one species as "Vulnerable" (*Epomops buettikoferi*) and six species as "Near Threatened" (*Rhinolophus alcyone, R. guineensis, Hipposideros jonesi, H. fuliginosus, Kerivoula cuprosa, Miniopterus schreibersii*).

- Local hunters and guides confirmed the presence of the Nimba Otter Shrew (*Micropotamogale lamottei*), listed by IUCN as "Endangered."

- We observed numerous tracks of several large mammal species listed by IUCN: one species, *Syncerus caffer* (African buffalo) listed as "Conservation Dependent", and five species, *Cephalophus maxwelli* (Maxwell's duiker), *Cephalophus niger* (Black duiker), *Cephalophus dorsalis* (Bay duiker), *Cephalophus silvicultor* (Yellow-backed duiker), and *Neotragus pygmaeus* (Royal antelope), listed as "Near Threatened".

- We directly observed two primate species listed on the most recent IUCN Red List of Threatened Species (IUCN 2002) as "Endangered": *Pan troglodytes verus* (West African Chimpanzee) and *Cercopithecus diana diana* (Diana Monkey), and one listed by IUCN as "Near Threatened": *Cercocebus atys atys* (Sooty Mangabey). Of the indirectly confirmed species, one, *Procolobus verus* (Olive Colobus), is listed as "Near Threatened".

New or Endemic Species:

- A new toad is most likely an endemic species to the Simandou Range.

- We recorded at least two additional new, and possibly endemic, amphibian species: an *Arthroleptis* species with warty skin and an *Amnirana* species.

- We collected four new, and likely endemic, species of katydids, as well as a new genus of cockroach found in the one bat cave surveyed.

In terms of the criteria generally considered during RAP surveys in order to identify priority areas for conservation (as described above) the Pic de Fon classified forest qualified as an important site in all regards. In all groups of organisms surveyed we found a rich and unique composition of forest and grassland species, confirming the Pic de Fon classified forest as one of the more diverse of the studied regions in Africa.

Concerning endemism and rare and/or threatened species, this site qualifies as an Important Bird Area (IBA) as defined by BirdLife International on the basis of the number of threatened species and presence of large numbers of both restricted-range and biome-restricted species (Fishpool and Evans 2001). Most of the amphibian species found have a very restricted distribution in the forested and mountainous habitats of Upper Guinea and are known only from a few localities. As another note-worthy find during this survey, we recorded the bat *Myotis welwitschii* for the first time in West Africa, extending its distribution range by more than 4,400 km. The list of threatened and endemic species recorded during this brief survey points to the need for both increased protection for the forest and additional surveys.

In particular, the chimpanzee population warrants additional study. Chimpanzees are listed as "Endangered" under the IUCN Red List and are listed under Appendix I of CITES (Most Critically Endangered Species), and, out of the three subspecies the West African Chimpanzee, *Pan troglodytes verus* is the most threatened by habitat destruction, hunting pressure, bushmeat and pet trade, as well as disease transmission. Currently there are thought to be between 25,000 to 58,000 chimpanzees left in all of West Africa, with the majority living in unprotected areas (Kormos and Boesch 2003). It is therefore of high importance to protect the remaining chimpanzee populations and to establish additional protected areas. It is likely that the Pic de Fon classified forest holds between one and four distinct community groups of *Pan troglodytes verus*. Chimpanzee density in this forest is higher than average density for degraded forest suggesting that this area still holds an important number of chimpanzees, and is an important habitat in regards to chimpanzee conservation in Guinea. We note that during our survey we only found chimpanzee nests above 900 m altitude, indicating that in this forest chimpanzees presumably have their range center or core area in the higher elevations.

Finally, with regard to critical habitats, much of the forest surveyed during this RAP was high-quality undisturbed primary forest, and as only 4.1 % of the original closed canopy forest remains in Guinea (Sayer et al. 1992), protecting what does remain is critical. Furthermore, the montane grassland of Site 1 represents a rare and endangered habitat within West Africa (Fahr et al. 2002).

Plants

During the RAP survey we recorded 409 plant species. The montane grasslands of the Pic de Fon result from underlying geology, altitude, and climate-related factors and, with the exception of areas adjacent to roads, the vegetation here is not, for the most part, a result of anthropogenic activities. The species composition of the ravine and gallery forests at Site 1 and some classified forest at Site 2 also represents high-quality mostly undisturbed forest. In the montane forests we recorded 18 tree species listed by IUCN – 1 species listed as "Endangered" (*Neolemonniera clitandrifolia*), 16 species listed as "Vulnerable" (*Albizia ferruginea, Antrocaryon micraster, Cordia platythyrsa, Cryptosepalum tetraphyllum, Drypetes afzelii, D. singroboensis, Entandrophragma candollei,*

E. utile, Garcinia kola, Guarea cedrata, Khaya grandifoliola, Lophira alata, Milicia regia, Nauclea diderrichii, Neostenanthera hamata, and *Terminalia ivorensis*) and 1 species listed as "Near Threatened" (*Milicia excelsa*). Agriculture is encroaching into this forest, particularly at Site 2, and we observed the use of many unsustainable farming practices (such as cutting down trees to harvest fruit versus just taking the fruit) in the secondary forests and plantations.

Invertebrates

This study is the first survey of the katydids (Orthoptera: Tettigoniidae) of the Pic de Fon area. Katydids, insects related to grasshoppers and crickets, are important in forest and savannah ecosystems as they are key herbivores and thus shape an area's vegetation. We collected 40 species, of which at least four are new to science, and ten are new to Guinea (a 16 % increase). These 40 species represent at least 24 genera and 6 subfamilies. The *Anoedopoda lamellata* savannah population may be distinct from the remainder of the African population. The Pic de Fon katydid fauna has a very high potential for endemicity and should be further investigated. Additionally a new genus of cockroach was found in the one bat cave surveyed.

Amphibians

We recorded at least 32 amphibian species with at least three species that are new to science and presumably endemic to the Simandou Range. We continuously added new species to the list and, because our survey took place at the beginning of the dry season, we believe that in total more than double the number of recorded species may occur in this area. Of the species found, only 10 (28.6 %) range outside of West Africa; 71.4 % of the species found are restricted to West Africa. More than half of the species found (19, 54.3 %) are endemic to the Upper Guinea rainforest. Eleven species found are endemic to the western part of the Upper Guinea Forest zone. Many of the latter species have a very restricted distribution in the mountainous area of Upper Guinea and are known only from a few localities. The amphibian fauna of the whole area was made up of a very unique mixture of savannah and forest amphibians, ranging from species that require undisturbed primary rainforest to those requiring drier savannah habitats. A new toad is most likely a species endemic to the Simandou Range. In addition, we recorded an *Arthroleptis* species with warty skin (like the Mount Nimba endemic *A. crusculus*) and a new *Amnirana* species. One amphibian species recorded (*Bufo superciliaris*) is protected by international law (CITES species) and is of conservation concern. Based on the results of this survey, we believe the Pic de Fon to be one of the most diverse places for amphibians in West Africa.

Reptiles

During our survey we found or obtained evidence for 12 species of reptiles. However, the snake fauna alone is estimated to contain over 60 species in this area, based on surveys in neighboring areas of similar or less diverse habitat types (Angel et al. 1954a, b; Barbault 1975; Rödel et al. 1995, 1997, 1999; Böhme 1999; Rödel and Mahsberg 2000; Ernst and Rödel 2002; Ineich 2002; Branch and Rödel 2003). Of the species we did record, four reptiles (*Chamaeleo senegalensis, Varanus ornatus, Python sebae*, and *Osteolaemus tetraspis*) are of conservation concern and protected by international laws (CITES species).

Birds

The total number of bird species recorded in all of Guinea to date is around 625. We recorded 233 bird species – 131 in the highland area (Site 1), 198 in the lowland area (Site 2). In comparison, 287 have been found in Ziama Forest Reserve and 141 in Diécké Forest Reserve – the two other sites in south-eastern Guinea that have been studied; 383 are known from the Liberian side of Mount Nimba, but this is the result of many years of intensive study. We recorded eight birds of conservation concern – two in the highlands, seven in the lowlands. The most important of these was *Schistolais leontica* (Sierra Leone Prinia), a bird listed as "Vulnerable" by IUCN that has a very limited distribution in the West African highlands. This species seems to be found only in dense vegetation at forest edge and along streams over 700 m and could be particularly vulnerable to alteration of the higher altitude habitats. We observed 6 of the 15 species of birds that make up the Upper Guinea Forests Endemic Bird Area – these are land bird species having a global breeding range of less than 50,000 km². Of the 153 species of the Guinea-Congo Forests biome that occur in Guinea, we recorded 104 (68 %) in this study. Additionally, we recorded seven species for the first time in Guinea. This site qualifies as an Important Bird Area (IBA) as defined by BirdLife International on the basis of the number of threatened species and presence of large numbers of both restricted-range and biome-restricted species. We note that the Pic de Fon classified forest is similar in many respects to Mount Nimba because of its altitudinal range and different habitats in close proximity, and believe many further species may be found in the Pic de Fon forest reserve.

Bats

We recorded 21 species of bats across 6 families and 14 genera, representing one-third of the bat species currently known from Guinea. We recorded seven bat species, 33.3 % out of the total recorded during this survey, listed by IUCN: one species as "Vulnerable" (*Epomops buettikoferi*) and six species as "Near Threatened" (*Rhinolophus alcyone, R. guineensis, Hipposideros jonesi, H. fuliginosus, Kerivoula cuprosa, Miniopterus schreibersii*). We surveyed only one bat cave, inhabited by *Rousettus aegyptiacus,* but we expect that many more caves occur in the area. *Hipposideros jonesi* and *Rhinolophus guineensis* sampled during this survey are endemic to the Upper Guinea Region and dependent on caves. We recorded *Miniopterus schreibersii,* a species with isolated occurrence in the Upper Guinea Region and likewise dependent

on caves. Both *H. jonesi* and *M. schreibersii* are not known from Mount Nimba. We recorded *Myotis welwitschii* for the first time not only for Guinea and the Upper Guinea Forest region but for all of West Africa, extending the documented distribution range by 4,400 km to the west. This species is tied to mountainous regions in its distribution range and is extremely rare at all localities where it is known to occur (Fahr and Ebigbo 2003). We recorded three species of *Kerivoula*, a genus of very rare forest bats. Two of these, *K. phalaena* and *K. cuprosa*, are the first records for Guinea. During the survey, species were continuously added to the total species number, indicating that the species list is far from complete. The high topodiversity of this area and its proximity to Mount Nimba (where 49 species have been recorded) and Massif du Ziama suggests that the Simandou Range is likely to support 50 to 65 species, potentially including threatened species that have been documented for areas nearby.

Non-volant small mammals

We recorded three shrew species and seven rodent species and observed four additional squirrel species. The shrew *Crocidura grandiceps* appears to be the first record from Guinea. However, we suspect that higher species numbers will be found with longer sampling time and more sample sites in forest and savannah. Results were characteristic of montane semi-evergreen forest with higher levels of small mammal biomass at the lower elevation suggesting increased forest productivity due to an abundance of water, high plant and microhabitat diversity, and in core areas, relatively low levels of disturbance. Local hunters and guides confirmed the presence of *Micropotamogale lamottei* (the Endangered Nimba Otter Shrew) based on pictures and descriptions of the animal, but we did not observe this species during the RAP.

Large mammals

We observed, identified by sound or photographed 31 species of large mammals in Site 1 and 34 species in Site 2 for a combined total of 39 species of large mammals. Using camera traps we obtained four photographs of two large mammal species, three photographs of *Atherurus africanus* (Brush-tailed Porcupine) and one of the Endangered *Pan troglodytes verus* (West African Chimpanzee). No leopards or aardvark were observed but local hunters reported that these species still occur in the Pic de Fon forest. We confirmed through tracks, dung and observations of local hunters the presence of *Syncerus caffer* (African Buffalo), listed by IUCN as "Conservation Dependent", as well as a number of duikers, *Cephalophus maxwelli* (Maxwell's Duiker), *Cephalophus niger* (Black Duiker), *Cephalophus dorsalis* (Bay Duiker), *Cephalophus silvicultor* (Yellow-backed Duiker), and *Neotragus pygmaeus* (Royal antelope) listed by IUCN as "Near Threatened".

Primates

Of the fifteen species of primates that could be expected to live in the Pic de Fon classified forest, we were able to directly confirm the presence of eight species (through visual or acoustic observations) and indirectly of an additional five species. Out of the eight directly observed, two species are listed by IUCN as "Endangered": *Pan troglodytes verus* (West African Chimpanzee), *Cercopithecus diana diana* (Diana Monkey), and one is listed as "Near Threatened": *Cercocebus atys atys* (Sooty Mangabey). Of the indirectly confirmed species, one, *Procolobus verus* (Olive Colobus), is listed as "Near Threatened". While the diversity of primates is quite high in the area, abundance is low as most species were encountered infrequently, with the exception of *Galagoides demidoff* (Demidoff's Galago) that we heard regularly and chimpanzees that we heard daily at Site 2.

We estimate the population of chimpanzees in the Pic de Fon classified forest (based on an estimate of 10,000 ha of forested area out of the 25,600 ha of the classified forest) to be around 75 (within a range of 21-246). Community sizes for chimpanzees are usually between 10 to 100 individuals, so it is likely that the Pic de Fon classified forest holds between one and four distinct community groups. Chimpanzee density in this forest is higher than average density for degraded forest suggesting that this area still holds an important number of individuals, and is an important habitat in regard to chimpanzee conservation in Guinea.

SOCIO-ECONOMIC THREATS AND OPPORTUNITIES ASSESSMENT

Overview and Objectives

Conservation International (CI) conducted a socio-economic threats and opportunities assessment of the Pic de Fon classified forest during November and December 2002. The objectives of this assessment were:

- To identify and confirm the principal stakeholders in the region;

- To identify and characterize threats to biodiversity and obstacles confronting conservation in the region;

- To develop an initial prioritization of these biodiversity threats; and

- To identify opportunities and possible solutions to these biodiversity threats.

As a first step in this assessment, CI oversaw a thorough search of published and grey literature including two preliminary studies commissioned by Rio Tinto. The literature review generated very little socio-economic data specific to the Pic de Fon area and revealed a general lack of historic records and time-series studies, making an analysis of socio-economic trends almost impossible.

Following this literature review, CI convened a workshop in Conakry on 12-13 December 2002, at the conclusion of the RAP survey, to elicit information on threats to and opportunities for biodiversity conservation in the Pic de Fon area from a broad spectrum of stakeholders and experts. This workshop sought to complement literature-based research with information from parties directly engaged in conservation in the immediate project area, Guinea, and the broader region. The exercise also afforded an opportunity to solicit additional documentation that may have been unobtainable through the initial literature searches.

Participants were selected to represent a broad range of perspectives and expertise regarding conservation efforts throughout Guinea, the Greater Nimba Highlands, and the Pic de Fon area. In total, 33 participants from 17 organizations including national government, multilateral and bilateral donors, NGOs, scientific research centers and Rio Tinto were present.

SUMMARY OF CURRENT AND POTENTIAL THREATS TO BIODIVERSITY

Hunting

Bushmeat hunting impacts species of large mammals—including primates—most immediately. RAP findings confirm that wildlife in the forest is subject to intensive hunting pressure. The RAP team found 11 shotgun shells, two snares, and a number of hunting camps near Site 1 and found 36 shotgun shells, 38 snares or traplines, and heard gunshots daily and at least four times at night near Site 2, despite reduced hunting levels during our visit as (a) we were employing local hunters as guides and (b) the Banko village elders had banned hunting during the duration of our survey out of consideration for our safety. The threats and opportunities assessment corroborated bushmeat hunting as a primary threat through the literature review and workshop.

Agricultural encroachment

PGRR, a field implementation project of Guinea's National Directorate of Water and Forests (DNEF), estimates that 35 to 40 % of the Pic de Fon classified forest has been impacted by agricultural encroachment from surrounding communities for both subsistence and cash crops. The planting of perennial tree crops such as coffee requires a substantial investment on the part of farmers, indicating a perception that long-term cultivation within the Pic de Fon boundaries is unlikely to provoke a reaction from authorities.

Bushfires

Uncontrolled bushfires, whether natural or set to clear land, prepare pasture, or drive game for hunting, are an ever-present threat. The use of fire to prepare pasture for livestock was identified as a regional threat in the literature and workshop and while we found no specific references to livestock encroachment in the Pic de Fon, each of the 24 villages surrounding the Pic de Fon engages in animal husbandry.

Logging

Materials reviewed in the literature survey and workshop mentioned timber exploitation as a current threat to biodiversity in the Pic de Fon classified forest. Commercial logging is expanding in Guinea and demand for fuelwood is rapidly outpacing the regenerative capacity of the resource base. Additionally, the RAP team noted that the area from the entrance of the classified forest up to the foot of the mountain near Camp 2 is stripped of trees (a few skeletons stand out), likely due to local wood use for houses, furniture, firewood, etc. in addition to clearance for farmland. The scale and impact of collection of wood for fuel, construction, and other uses was not directly observed during the RAP survey. We thus identify logging as a threat that may be significant now, and, if not, is likely to become increasingly acute over time.

Overfishing and Collection of non-timber forest products

Unregulated extraction of resources such as fish and non-timber forest products (NTFPs) from the Pic de Fon classified forest was identified in the literature survey and workshop. However, neither exercise yielded information regarding fish species typically caught or the range of NTFPs typically collected, not to mention extraction rates or overall impact. The RAP team did observe some unsustainable practices of NTFP collection, in particular fruit collection methods that included chopping down entire trees to collect fruit rather than harvesting fruit from trees left standing. However, the nature and scale of extraction of fish and NTFPs has yet to be determined.

Artisanal mining

Several sources mention illegal artisanal mining inside the Pic de Fon classified forest boundaries, and Conakry workshop participants also identified such mining as a threat. In particular, the literature and participants note the destructive impact of artisanal or small-scale mining for gold and diamonds on streambeds and waterways. The RAP team observed this practice in at least one instance and we identify artisanal mining as a likely threat. The exact nature and scale of this threat has yet to be determined and we strongly recommend bringing the level of information on such activities up to date.

Road development

Recently introduced roads, built to support mineral exploration activities, will likely have potential impacts including exacerbation of the threat of bushmeat hunting as roads potentially provide easier access to formerly remote areas of the Pic de Fon. Also, improvements to existing roads may provide better access to more distant markets for bushmeat. Another potential threat from road development is the introduction or proliferation of non-native and invasive species. Additionally, potential impacts on species due to habitat alteration or noise disturbance associated with roads warrant consideration. The RAP team observed sediment erosion along roads; erosion can lead to increased siltation in local waterways, which could have an adverse effect on aquatic and riparian communities. However, what impact road erosion is currently having on species of the Pic de Fon has yet to be explored.

Habitat fragmentation

Habitat fragmentation resulting from continued encroachment on and conversion of the Pic de Fon classified forest is a potential future threat to the region and was observed in the lowland forest near Site 2. There are few sites in West Africa that stretch over such a wide range of altitudes in close proximity and such sites are important for a number of species. Birds, for example, sometimes require habitat at different altitudes between seasons. Certain endemic species may be widespread throughout the Pic de Fon classified forest and require only more general habitat-type protection while others, for example, may occur in only one particular cave.

Commercial mining operations

Preliminary results from an RTM&E exploration program indicate an exceptionally high-grade iron ore deposit in the Simandou Range. Possible impacts from commercial mining operations within the Pic de Fon classified forest are of significant concern. Alteration of the topography at higher elevations of the Pic de Fon classified forest could have a significant effect on the rare montane grassland habitat and may also affect endemic and threatened species (such as the chimpanzees nesting above 900 m, threatened bat species that are dependent on caves, endemic amphibian species, and *Schistolais leontica* (Sierra Leone Prinia)). Topographic alteration could also potentially have a broader impact on the climatic and hydrological functions of the Pic de Fon and surrounding areas within dependant watersheds. Mine development can also potentially result in an increase in the rate of forest clearance as workers migrate into an area.

Open cast mining is of particular concern. When employed in the past, particularly by companies giving little consideration to environmental and social impacts, this method has resulted in destruction of rare habitats, as well as severe impacts to villages in the immediate vicinity of mines (Colston and Curry Lindahl 1986, Gatter 1997, Toure and Suter 2001, 2002). While practices of international mining companies often differ significantly from those of most national mining companies, the potential damage to the Pic de Fon habitat from open cast mining should be carefully considered.

Many of these sources of pressure on biodiversity will most likely intensify as human population growth and migration increase demand for bushmeat, agricultural land, and other resources. Therefore, the fact that this region features the highest population growth rates in Guinea lends urgency to the need for concerted conservation efforts. While not examined in depth, it was also noted that there are interrelated drivers underpinning these threats, including a lack of capacity among regional actors (community, government, NGOs, and the scientific community), civil unrest, and land tenure conflict that lend an added dimension of complexity to the aforementioned threats.

Perhaps the most daunting fact confronting conservation planning in this region is that Guinea is considered to be one of the poorest countries in the world, though one of the richest in terms of minerals. The United Nations Development Program's (UNDP's) Report on Human Development has for several years ranked Guinea among the last of some 170 countries (UNDP 2003). The Guinean economy is not highly diversified and its potential development is constrained by a widespread lack of basic infrastructure.

Inhabitants of the 24 villages surrounding the Pic de Fon rely on this forest for food, water, fuel, medicine and a number of ecosystem services and every effort should be made to ensure that the forest will continue to provide these services, or that appropriate substitutes can be found where current practices are unsustainable (as in the case of bushmeat hunting or conversion of forest to farmland). Many of the threats detailed above can only be addressed through careful planning and cooperation with these communities and assistance from national and international partners. Along these lines, community needs must be addressed in order for any protected area to remain viable, while illegal encroachment (and subsequent destruction of both flora and fauna within the forest) must be controlled to ensure long-term viability, both in the interest of local communities and for protecting some of the last intact forest left in West Africa.

Substantial further fieldwork is required to develop a robust, quantitative understanding of the current and future threats to and opportunities within the Pic de Fon area, beyond the scope of the current effort with respect to both budget and timeline. This additional information will be critical to the design of an appropriate conservation strategy for the region.

RECOMMENDED CONSERVATION ACTIONS

The results of this RAP survey show that the Pic de Fon ecosystem contains unique terrestrial and aquatic species; intact connections between habitat, flora, and fauna; and one of the last remaining, relatively intact montane habitats, both within Guinea and the Upper Guinea Highlands. As a result of this study, we strongly recommend that the Pic de Fon classified forest receive increased protection status and that current regulations with regards to its classified status be monitored, enforced and strengthened, as necessary. The limits of the protected area should be defined and enforced in collaboration with local communities and should be accompanied by education and outreach programs, as well as additional scientific and socio-economic studies. Biodiversity conservation will require substantial investment in capacity building.

The implications of intensified enforcement to control access by local communities to forest resources must be addressed as well. Land tenure is an extremely sensitive issue, and while it is critical to place the Pic de Fon region under some form of increased legal protection as soon as possible, the success of this protected area depends on open discussion and close collaboration with communities in the region. These dialogues will facilitate mutual understanding between the government, conservation groups and communities, with the intention of achieving conservation through a variety of measures that target the need for productive agricultural land, fuelwood supplies, sustainable wildlife management and other services provided by the forest.

If the Pic de Fon classified forest is incorporated into such a protected area system that includes both forests and savannahs and representatives of the many microhabitat variations, a high proportion of plant and animal species of this region would be protected.

Moreover, the Simandou Range forms a component of the Greater Nimba Highlands, which includes Mount Nimba, Diécké, Wonegizi-Ziama Range and Mount Béro-Tetini. Planning of conservation investments therefore must proceed at a regional level, since many of the most pressing threats also comprise regional dynamics. In the immediate term, a large obstacle confronting a conservation strategy for the region is the paucity of concrete data, particularly on socio-economic trends over time, so that additional research with a focus on field-level primary data collection features among the highest priorities.

Specific recommended actions include:

Immediate actions:

- Collect GIS satellite imagery and remote sensing data to monitor forest cover of the Pic de Fon, in order to assess changes over time.

- Clarify existing laws governing hunting and land use in the Pic de Fon classified forest through conducting and publishing a study on existing laws that effect protection of "Endangered" and "Threatened" species from hunting and other threats in the Pic de Fon, identifying any gaps or inconsistencies.

- Develop and implement an awareness campaign to counter unsustainable hunting practices.

- Conduct socio-economic baseline studies.

- Conduct futher baseline biological surveys throughout the Simandou Range to determine conservation priorities and possible future threats, including surveys of montane grasslands and of the large forest block to the south of the current RAP study sites.

- Monitor/prevent the introduction of invasive species (noting that very few were found during this survey).

Long-term actions:

- Increase capacity of national, regional and local actors to enforce existing laws governing hunting and other exploitation of resources in the Pic de Fon; continue to delimit Pic de Fon forest boundaries and establish and build capacity for a monitoring and enforcement regime.

- Increase the capacity of local and regional actors to research and manage the Pic de Fon classified forest.

- Establish partnerships between NGOs, government agencies and local research institutions to develop better long-term management plans and carry out monitoring protocols for the Pic de Fon.

- Develop incentive-based conservation agreements. These are agreements that offset the opportunity cost of conserving the resources in question and provide economic stimuli for alleviating poverty and improving social welfare through providing income-generating activities.

- Assist with capacity building to strengthen Guinean hunting and wildlife protection laws and enforcement.

Recommended Scientific Studies
See individual chapters for more details.

Hydrology
We strongly recommend an investigation into the hydrology of the Simandou Range and the Pic de Fon and the possible climatological outcomes of potential mining activities that will have the effect of lowering the mountain ridge. The ridge of the Simandou Range most likely acts as a barrier to southwestern winds, resulting in higher precipitation amounts than usual for areas of this latitude. Lowering of the ridge would probably result in decreased precipitation rates. In addition, the whole mountain possibly acts as a water reservoir. A reduction of this reservoir in connection with reduced rainfall would almost certainly result in the disappearance of the forest along the mountain range. This would not only impact aquatic and forest faunas, but could also have serious consequences for the local human population and requires further evaluation. We also recommend monitoring water quality and mineral levels in streams to assess effects of erosion and mining exploration activities.

Herpetofauna
Three new species found only along the mountain ridge during our survey, a dwarf toad and two frogs (*Arthroleptis*

sp. and *Amnirana* sp.), are presumably endemic to this area. These species and their respective population sizes and distribution need more investigation in order to define restricted nature preserves along the ridge. Further surveys during the wet season and in additional areas along the Simandou Range should also be undertaken to ascertain a more complete knowledge of the herpetological diversity in the area.

Primates
In order to provide adequate protection to the primates of Pic de Fon, we recommend further surveys during different seasons and covering longer time spans to confirm the distribution, ranging patterns, and abundance of the primate population. Data from such surveys are needed to further inform adequate protection measures, especially for the four threatened species of the Pic de Fon.

Large mammals
We recommend that an inexpensive camera-trapping monitoring protocol be implemented throughout the forest. A sustained long-term camera photo-trapping research program can provide scientists and managers with continuous information on local wildlife resources, thus enabling more informed conservation actions.

Bats
There is good preliminary evidence that the mixture of habitat types in the Pic de Fon and the elevational gradient of the area support an exceptionally rich bat assemblage that requires documentation through additional surveys. A striking feature of the bat fauna in the Upper Guinea Highlands is the disproportionally high number of species that are either endemic to Upper Guinea or which are represented in this region with isolated populations. As some species are not only endemic to Upper Guinea but also have astonishingly small distribution ranges, we emphasize that further surveys are urgently needed to assess the presence of globally threatened species that have been documented from similar habitats and in proximity to the Pic de Fon classified forest (Fahr et al. 2002).

Non-volant small mammals
A continued survey effort is recommended in particular to verify the presence of additional shrew species, such as the endemic Nimba shrew (*Crocidura nimbae*) and the Nimba otter shrew (*Micropotamogale lamottei*; see Nicoll et al. 1990) found at both Mount Nimba and Ziama, by employing a variety of survey methods for extended periods and at various sites around Pic de Fon.

Birds
Further surveys should be conducted at different sites throughout the forest reserve and should be carried out in different seasons (e.g. at the start of the rainy season, in April-May, when more species may be breeding and are vocally most active) and in habitats that we were unable to

cover. It is likely that other threatened species will be found in future surveys. It is also possible that different species of migratory birds from Europe may be encountered and that altitudinal migration of birds within the forest may be observed.

Plants

We recommend using GIS imagery and remote sensing data to monitor the forest cover of the Pic de Fon, assessing any changes over time. In addition, surveys of the flora on the drier, eastern side of the Pic de Fon are needed to investigate potential endemic species.

Invertebrates

Further studies along the entire Simandou Range are needed in order to estimate the distribution and population sizes of the newly discovered species found during this survey. Also, additional collecting is needed at the beginning of the rainy season when most species appear in high numbers and as adults. The cryptic complex of species of the katydid genus *Ruspolia* may include several undescribed species, and a genetic and bioacoustic study should be conducted among the species of this genus in the Simandou Range.

Fishes

We recommend a survey of the area's freshwater fishes to complement terrestrial data collected during the RAP and to address fisheries issues brought up during the threats and opportunities assessment. Another potential component of a fish survey could be to assess water quality and mineral levels in streams to help in determining any significant impacts of erosion or other anthropogenic activities.

REFERENCES

Angel, F., J. Guibe and M. Lamotte. 1954a. La réserve naturelle intégrale du mont Nimba. Fascicule II. XXXI. Lézards. Memoirs de l'Institute fondamental d'Afrique noire, sér. A, 40: 371-379.

Angel, F., J. Guibe, M. Lamotte and R. Roy. 1954b. La réserve naturelle intégrale du mont Nimba. Fasicule II. XXXII. Serpentes. Bulletin de l'Institute fondamental d'Afrique noire, sér. A, 40: 381-402.

Bakarr, M., B. Bailey, D. Byler, R. Ham, S. Olivieri and M. Omland (eds.). 2001. From the Forest to the Sea: Biodiversity Connections from Guinea to Togo. Washington DC: Conservation International.

Barbault, R. 1975. Les peuplements de lézards des savane de Lamto (Côte d'Ivoire). Annales de l'Université d'Abidjan, sér. E, 8: 147-221.

Böhme, W. 1999. Diversity of a snake community in a Guinean rain forest (Reptilia, Serpentes). In: Rheinwald, G. (ed.): Isolated Vertebrate Communities in the Tropics. Pp. 69-78. Proceedings of the 4th International Symposium, Bonner zoologische Monographien, 46.

Branch, W.R. and M.-O. Rödel. 2003. Herpetological survey of the Haute Dodo and Cavally forests, western Ivory Coast, Part II: Trapping results and reptiles. Salamandra, Rheinbach 39(1): 21-38.

Colston, P.R. and K. Curry-Lindahl. 1986. The Birds of Mount Nimba, Liberia. British Museum (Natural History). London. 129 pp.

Ernst, R. and M.-O. Rödel. 2002. A new *Atheris* species (Serpentes: Viperidae) from Taï National Park, Ivory Coast. Herpetological Journal 12: 55-61.

Fahr, J. and N.M. Ebigbo. 2003. A conservation assessment of the bats of the Simandou Range, Guinea with the first record of *Myotis welwitschii* (GRAY, 1866) from West Africa. Acta Chiropterologica 5(1): 125-141.

Fahr, J., H. Vierhaus, R. Hutterer and D. Kock. 2002. A revision of the *Rhinolophus maclaudi* species group with the description of a new species from West Africa (Chiroptera: Rhinolophidae). Myotis 40: 95-126.

Fishpool, L.D.C. and M.I. Evans (eds.). 2001. Important Bird Areas in Africa and Associated Islands: Priority sites for conservation. BirdLife Conservation Series No. 11. Pisces Publications and BirdLife International. Newbury and Cambridge, UK.

Gatter, W. 1997. Birds of Liberia. Pica Press, The Banks, Mountfield. 320 pp.

Hatch & Associates, Inc. 1998. "Preliminary Environmental Characterisation Study: Simandou Iron Ore Project Exploration Programme, South-Eastern Guinea". Montreal, Canada.

Hutson, A.M., S.P. Mickleburgh and P.A. Racey (comp.). 2001. Microchiropteran Bats: Global Status Survey and Conservation Action Plan. IUCN/SSC Chiroptera Specialist Group. Gland, Switzerland: IUCN. x + 258 pp.

Ineich, I. 2002. Diversité spécifique des reptiles du Mont Nimba. Unpublished manuscript.

IUCN. 2002. 2002 IUCN Red List of Threatened Species. www.redlist.org.

Kormos, R. and C. Boesch. 2003. Status survey and conservation action plan. The West African Chimpanzee. IUCN/SSC Action Plan.

Nicoll, M.E., G.B. Rathbun and IUCN, SSC Insectivore Tree-Shrew and Elephant-Shrew Specialist Group. 1990. African insectivora and elephant-shrews: an action plan for their conservation. IUCN, Gland, Switzerland, iv+53 pp.

Rödel, M.-O. and D. Mahsberg. 2000. Vorläufige Liste der Schlangen des Tai-Nationalparks / Elfenbeinküste und angrenzender Gebiete. Salamandra 36: 25-38.

Rödel, M.-O., K. Grabow, C. Böckheler and D. Mahsberg. 1995. Die Schlangen des Comoé-Nationalparks, Elfenbeinküste (Reptilia: Squamata: Serpentes). Stuttgarter Beiträge zur Naturkunde, Stuttgart, Serie A, Nr. 528: 1-18.

Rödel, M.-O., K. Grabow, J. Hallermann and C. Böckheler. 1997. Die Echsen des Comoé-Nationalparks, Elfenbeinküste. Salamandra, 33: 225-240.

Rödel, M.-O., K. Kouadio and D. Mahsberg. 1999. Die Schlangenfauna des Comoé-Nationalparks, Elfenbeinküste: Ergänzungen und Ausblick. Salamandra 35: 165-180.

Sayer, J.A., C.S. Harcourt and N.M. Collins (eds.). 1992. The Conservation Atlas of Tropical Dorests: Africa. Macmillan. 288 pp.

Toure, M. and J. Suter. 2001. Workshop report of the 1st trinational meeting (Côte d'Ivoire, Guinea, Liberia), 12-14 September 2001, Man, Côte d'Ivoire. Initiating a Tri-national Programme for the Integrated Conservation of the Mount Nimba Massif. Fauna & Flora Int., Conservation International & BirdLife Int., Abidjan. 56 pp. http://www.fauna-flora.org/around_the_world/africa/mount_nimba.htm

Toure, M. and J. Suter. 2002. Workshop report of the 2nd trinational meeting (Côte d'Ivoire, Guinea, Liberia), 12-15 February 2002, N'Zérékoré, Guinea. Initiating a Tri-national Programme for the Integrated Conservation of the Nimba Mountains. Fauna & Flora Int., Conservation International & BirdLife Int., Abidjan. 82 pp.

[UNDP] United Nations Development Programme. 2003. Human Development Report 2003. Oxford University Press, Oxford. 375 pp.

Initial Biodiversity Action Plan for the
Forêt Classée du Pic de Fon, Guinea

This Initial Biodiversity Action Plan (IBAP) is based on data collected during the terrestrial biological survey by the Rapid Assessment Program (RAP) and the Socio-economic Threats and Opportunities Assessment for the Pic de Fon classified forest. The intention of this IBAP is to catalyze biodiversity conservation efforts in the Pic de Fon classified forest, and also to promote conservation action in the Greater Nimba Highlands, as many of the biodiversity threats noted for the Pic de Fon classified forest relate to dynamics operating at the regional level within the Greater Nimba Highlands. It is expected that this plan will evolve and strengthen as the biological and socio-economic dynamics of the Pic de Fon and Greater Nimba Highlands are better understood, and to this end, Conservation International (CI) and Rio Tinto Mining and Exploration (RTM&E), along with other local and regional stakeholders, are now forming a more comprehensive alliance in the region to carry the initial recommendations presented in this report forward. Feedback on the plan is welcome.

The Regional Context

The Pic de Fon classified forest forms a component of the Greater Nimba Highlands, which includes Mount Nimba, Diécké, Wonegizi-Ziama Range and Mount Béro-Tetini. Within the Pic de Fon classified forest, several threats to biodiversity currently exist; as discussed in the Socio-economic Threats and Opportunities Assessment chapter of this report, hunting, agricultural encroachment, bushfire, logging, overfishing and collection of non-timber forest products, artisanal mining and road development were all identified as existing threats. Additionally, habitat fragmentation and large-scale commercial mining operations, if they do commence, were identified as significant future threats to biodiversity.

Many of these threats to biodiversity are likely to intensify as human population growth and migration increase demand for bushmeat, agricultural land and other resources. Therefore, the fact that this region features the highest population growth rates in Guinea lends urgency to the need for concerted conservation efforts. While not examined in depth, it was also noted that there are interrelated drivers that operate at a regional level and which underpin these threats, including a lack of capacity among regional actors (community, government, non-governmental organizations and the scientific community), civil unrest, and land tenure conflict that lend an added dimension of complexity to the aforementioned threats.

Since many of the threats can be exacerbated by these regional dynamics, planning of conservation investments must proceed at a regional level. In the immediate term, a challenge to identifying conservation outcomes and building a conservation strategy for the region is the paucity of concrete data, particularly on socio-economic trends over time, so that additional research with a focus on field-level primary data collection features among the highest priorities.

Key Conservation Objective, Milestones and Outputs

The results of this RAP survey show that the Pic de Fon classified forest ecosystem contains unique terrestrial and aquatic species; intact connections between habitat, flora, and fauna; and one of the last remaining, relatively intact montane habitats, both within Guinea and the Upper Guinea Highlands. Furthermore, the Socio-Economic Threats and Opportunities Assessment suggests that the Pic de Fon classified forest, and the Greater Nimba Highlands, face a number of threats to biodiversity conservation. Based on the results of this study, the overall conservation objective of this Initial Biodiversity Action Plan is to to improve natural resource management in the Greater Nimba Highlands through addressing threats to biodiversity, promoting sound production practices and developing sustainable income alternatives for communities that rely heavily on the forests and natural resources of the region. We identified five key conservation milestones to achieving the conservation objective, and several outputs for each:

1. *Biodiversity data and analyses for the Greater Nimba Highlands and, in particular, the Pic de Fon classified forest, are made publicly available.*

- An analysis of existing satellite imagery and remote sensing data covering the Upper Guinea Highlands (including the Pic de Fon classified forest);

- Biological surveys of taxa identified in the RAP covering different seasons, including surveys in the large forest block to the south of the RAP study sites;

- A scientific assessment of the significance of the biodiversity of the Pic de Fon classified forest relative to comparable surrounding areas in the Greater Nimba Highlands; and

- A study of the hydrology of the Pic de Fon classified forest.

2. *Biodiversity monitoring protocol established for the Pic de Fon classified forest.*

- Camera traps put in place in the Pic de Fon classified forest to monitor distribution and ranging patterns of large mammals, including primate species, and hunting activities;

- Baseline data acquired from satellite imagery and change detection used to monitor changes over time and to update the biodiversity strategy for the Pic de Fon classified forest;

- A protocol established to monitor water quality and mineral levels in streams to determine impact of human activities in the Pic de Fon classified forest; and

- Partnerships between NGOs, government agencies, industry and local research institutions established to develop long-term management plans and carry out monitoring protocols in the Pic de Fon classified forest.

3. *A community-based sustainable resource use program developed for the Pic de Fon classified forest.*

- Socio-economic baseline studies of communities surrounding the Pic de Fon classified forest conducted and made publicly available, with particular attention to livelihood strategies and how those impact biodiversity;

- An analysis of roads adjacent to or creating access to the Pic de Fon classified forest to better understand their impact on biodiversity; and

- Based upon socio-economic baseline studies, a sustainable resource use plan developed and implemented with communities surrounding the Pic de Fon classified forest. Such a plan should include a variety of measures that target the need for productive agricultural land, fuelwood supplies, sustainable wildlife management, and other services provided by the forest, and may include incentive-based conservation agreements that offset the opportunity cost of conserving the resources in question and that provide economic stimuli for alleviating poverty and improving social welfare through providing income-generating activities.

4. *Awareness about and capacity to address biodiversity conservation issues is built in key government agencies, national and regional scientific institutions, and local communities with an interest in the conservation of biodiversity in the Pic de Fon classified forest.*

- A local cadre of scientists and technicians trained to research and monitor biodiversity in the Pic de Fon classified forest and sufficient funding available to support this capacity; and

- A community-based awareness campaign implemented to counter destructive hunting practices in the Pic de Fon classified forest.

5. *Legal and physical framework is in place for protection of biodiversity in the Pic de Fon classified forest and Greater Nimba Highlands.*

- A review of existing laws regarding hunting and wildlife protection in Guinea, and in particular the Pic de Fon classified forest;

- A strategy developed with Pic de Fon classified forest communities and the Guinean government to strengthen and enforce Guinean hunting and wildlife protection laws; and

- A network of protected areas in the Greater Nimba Highlands developed that formally protects the region's biodiversity over the full altitudinal range from the peaks to the lowland forests, based upon results of the scientific assessment of the significance of the biodiversity of the Pic de Fon classified forest relative to comparable surrounding areas in the Greater Nimba Highlands.

Recommended Scientific Studies – see Executive Summary p. 33

Chapter 1

An ecological and conservation overview
of the Forêt Classée du Pic de Fon, Simandou Range, Guinea

INTRODUCTION

Current biodiversity patterns and plant and animal endemism in the Upper Guinea Forest
date to the Pleistocene epoch, 15,000-250,000 B.P. The ensuing dry conditions in the tropics
created isolated refugia, and with repeated expansions and contractions in the original forest,
the resulting flora and fauna living in new habitats experienced considerable speciation (Lebbie
2001). Originally, the Upper Guinea Forest ecosystem is estimated to have covered as much as
420,000 km^2, but centuries of human activity have resulted in the loss of nearly 70 % of the
original forest cover (Bakarr et al. 2001). The remaining Upper Guinea Forest is restricted to a
number of isolated patches acting as refugia for the region's unique species of flora and fauna.
These remaining forests contain exceptionally diverse ecological communities, and a mosaic of
forest habitat providing refuge to numerous endemic species.

Site-specific studies conducted in different countries within the Upper Guinea Forest show
high levels of local endemism in plant species, which is likely reflective of the region as a whole
(Bakarr et al. 2001). Of 1,300 species known from Taï National Park in Côte d'Ivoire (4,550
km^2), nearly 700 are confined to the Upper Guinea Forest ecosystem and 150 are endemic to
Taï (Davis et al. 1994). On Mount Nimba, bordering Guinea, Côte d'Ivoire and Liberia, 2,000
plant species are estimated to occur in an area of only 480 km^2, with 13 being endemic to that
site.

While the Upper Guinea Forest is home to a unique assemblage of ecosystems, it also
houses substantial mineralogical wealth, and all within a region facing extreme poverty, growing human population densities, weak environmental governance, and periodic, but persistent,
civil unrest. Determining how to best safeguard the biological and geological resources, to
the advantage of local communities and regional biodiversity, is a complex challenge faced by
extractive industries, West African governments, development organizations, NGOs and local
inhabitants.

Geography

The Simandou Range forms one component of the Upper Guinea Highlands, which also
includes Mount Nimba, Diécké, Wonegisi-Ziama Range, Mount Béro-Tetini, Loma-Tingi
Mountains and the Fouta Djallon. The Simandou Range consists of high altitude peaks and
plateaus extending through south-eastern Guinea from Komodou in the north to Kouankan
and Boola in the south. The highest point, the Pic de Fon, is 1,656 m asl. As the Simandou
Range is situated in the transition between savannah and forest zones, it offers habitat types
ranging from humid Guinea savannah, to Western Guinean lowland rain forest, to Guinean
montane and gallery forests, and to the rare and exceptional habitat type of montane grassland.
West African montane habitats are very limited in extent and consequently especially endangered, and the Simandou Range is one of few places in all of West Africa harboring this habitat
type. In places, the Simandou Range is steeply sloping covering an altitudinal range from about
500 to over 1,600 m, and as a result placing all of these different habitat types within close
proximity.

At the southern end of the Simandou Range lies the Pic de Fon classified forest (containing the Pic de Fon, the highest point on the range and second highest point in Guinea). The Republic of Guinea has 113 national classified forests, most of which were classified in the mid 1900's by the French colonial government for commercial and environmental purposes. The Pic de Fon classified forest, created in 1953, is the third largest in the Guinée Forestière region and covers approximately 25,600 ha. The original reasons for the classification of forests in Guinea included conservation of forest and soil, which, because of the steepness of the slopes, in the case of the Pic de Fon, become unstable when cultivated. Also, it was recognized that the Simandou Range acts as a natural barrier against savannah fires and regulates the flow of the water sources that originate from within the range (including the Diani, Loffa and Milo rivers) (Hatch 1998).

Climate

Compared to many major rain forest areas around the world, the Guinean Rain Forest is drier and seasonality is more pronounced (WWF and IUCN 1994). Most of the Guineo-Congolian Region receives between 1,600 and 2,000 mm of rainfall per year and this rainfall is less evenly distributed than that of other rain forests (White 1983). In the main eastern block of the Guineo-Congolian Region, in general, rainfall shows two peaks separated by one relatively severe and one less severe dry period (White 1983). Throughout the Guineo-Congolian Region, average monthly temperature is almost constant throughout the year (White 1983).

The Simandou Range crosses two of Guinea's four natural regions. The northern part of the range is located in the Upper Guinea or Mandigue Plateau Region characterized by a hot, dry "soudano-guinéen" climate, while the southern portion of the range lies in the Forested Guinea or Southern Dorsal Guinea Region, characterized by a "guinéen forestier" climate with higher annual rainfall and a short three-month dry season.

Vegetation and Habitat Diversity

While previous biological studies are mostly unavailable, diversity and endemism within this classified forest have been suspected to be quite high. The Pic de Fon is steep and densely vegetated on the southwestern side, making access to the high-altitude forests quite difficult. The high-altitude areas within the Pic de Fon classified forest contain two distinct habitats: montane forest restricted at higher elevation to gallery and ravine forests running in narrow strips along streambeds (these streambeds are usually at the bottom of ravines running between slopes) and montane grassland. Both the montane forests and grasslands are typical native vegetation of the Simandou Mountains, yet, as there are few mountains of this elevation within West Africa, these habitats are rare and exceptional within the region. Another factor increasing the potential for high diversity and endemicity is that these different habitats occur in close proximity to

one another as well as to the Guinean lowland rain forest found at lower elevations in the classified forest.

Because the Pic de Fon acts as a physical barrier between the harmattan (dry and dust-charged northeastern winds) and the more moisture-laden southwestern trade winds, the steep western slopes of the Pic de Fon are more humid than the eastern slopes and therefore much more densely forested. On the eastern side of the Pic de Fon, the forest is restricted to gallery forests as thin strips running along streams, descending from just below the ridge. The western side of the ridge has wider, denser strips of gallery forests and large patches of ravine forests that descend from the ridge and then connect to the lowland forests in the west. This barrier effect also results in the Pic de Fon receiving higher precipitation than usual compared to lowland areas of similar latitude. The Simandou Mountain Range appears to be a major water recharge area for the whole surrounding region and the source of a complex hydrographic system feeding the Niger, Loffa, Makona and Diani drainage systems (Hatch 1998).

Geology

In West Africa, almost the whole of the Guineo-Congolian Region is underlain by Precambrian rock. The landscape is formed of relatively low plateaux and plains interrupted by residual inselbergs and small higher plateaux. The most important of the latter are the Fouta Djalon, the Upper Guinea Highlands, and the Togo-Atacora range (White 1983). The Guinea Highlands attain 1,752 m in Mount Nimba and 1,947 m in the Loma Mountains. In contrast to the Fouta Djalon, the Guinea Highlands have few level surfaces and the hills are rounded (White 1983).

Almost all land areas in the Guineo-Congolian Region have an altitude lower than 1,000 m asl. In contrast to the vast majority of West Africa, the southern part of the Simandou Range rises above the basement plain by some 600 to 800 m with the main ridge ranging in elevation from 1,250 to 1,650 m. On the eastern side of the Simandou main ridge, at least in the southern portion, the more gentle lower slopes correspond to canga talus that occurs as outwash fans, the largest being 1 km² in area. In the same area, the upper east facing slopes are steep and entirely covered by mineralized outcrops, scree material made up of similar rocks and canga-like breccias. It is common that summit areas are crowned with more or less scattered blocks of semi-massive or massive mineralization. This feature is particularly developed at the top of the Pic de Fon where massive iron ore blocks are more or less closely packed over a wooded area (Hatch 1998).

Hydrological System

The mountainous areas of Guinea within the Upper Guinea Forest ecosystem, which overlap with the Pic de Fon and its classified forest, have been described as the "Water Tower of West Africa" (MMGE 2002) due to their critical role in the hydrology of the entire region. The Simandou mountain range appears to be a major water recharge area for the whole surrounding region and the source of a complex hydrographic system feeding the Niger, Loffa, Makona and Diani drainage systems (Hatch 1998).

Description of the RAP Study Sites

The RAP expedition took place from 26 November to 7 December 2002 at the beginning of the dry season and surveys were based from two sites, one close to the summit of the Pic de Fon and the other at lower elevation and near to Banko village (see Map). Both camp sites were chosen based on proximity to the desired sampling sites, but selection of camps was restricted based on accessibility. The camps allowed entry to all of the main habitat types of the region investigated, granting the team a thorough first impression, however they are likely not wholly representative of the entire forest that potentially contains richer, more diverse areas that are less easily accessible.

The first site (Site 1) was situated between montane grassland and montane forest (including both ravine and gallery forest) pockets that were directly connected to larger lowland rain forests at lower altitude. This site covered an altitudinal range of between 1,000 and 1,600 m asl. The camp site was located at 1,350 m asl (08°31'52"N, 08°54'21"W) and was 3.5 km (3,461 m) from the Pic de Fon (see Map).

The second site (Site 2) comprised secondary and primary forests and some small savannah patches at around 600 m asl. The camp site (08°31'29"N, 08°56'12"W) was located approximately 6 km from the village of Banko (in the direction of the Simandou Ridge) and was approximately 1.5 km inside the boundary of the classified forest (see Map). The camp was situated near lowland rain forest with many mountain streams of different size exiting the mountain range and some hills with derived savannah vegetation. This lowland forest is characterized by *Parinari excelsa*, a plant whose seeds are fruit bat dispersed, which is a remarkable feature of montane forests west of the Dahomey Gap. On more level ground and toward the direction of the village, most forest has been converted to cocoa and coffee plantations, and this area also contains some secondary forest. Banana plantations and farm-bush, intermingled with cultivated spices, such as pepper, were located within the boundary of the classified forest with a clear and sudden transition to forest beginning just at the foot of the mountain.

Population Profile

At present, the population density in the south-eastern quarter of Guinea, the Guinée Forestière region, remains relatively low, documented at 9 to 16 inhabitants per km². However, population density may be higher in the southern foothills of the Pic de Fon region (Konomou and Zoumanigui 2000).

There has been a massive influx of refugees into Guinée Forestière since the wars in Sierra Leone and Liberia. There are more refugees seeking refuge in Guinea than in any other African country. At the end of 1996 it was estimated that there were about 650,000 refugees in Guinea from Liberia and Sierra Leone combined (UNHCR 1997). Konomou and Zoumanigui (2000) report a refugee population of 629,275, or about 40 % of the population in Guinée Forestière. Dramatic influxes of refugees from Liberia and Sierra Leone (and recently from Côte d'Ivoire) may account for the high population growth rates recorded in the Kérouané, Beyla, and Macenta prefectures (See chapter 10 for more).

Legal Protection Status

Guinea is one of 150 member countries of CITES and has ratified the Convention Concerning the Protection of World Culture and Natural Heritage (WHC, Paris, 1972) and the Convention for the Cooperation in the Protection and Development of the Marine and Coastal Environment of the Western and Central African Region (Abidjan, 1981). Guinea has signed but not ratified The African Convention for the Conservation of Nature and Natural Resources (ACCN) (Barnett and Prangley 1997).

In Guinea, the government body responsible for wildlife is the Ministry of Agriculture and the Direction Nationale des Eaux et Forêt (DNEF). The law governing the use of wildlife is the "Code de la Protection de la Faune Sauvage et Réglementation de la chasse" (République de Guinée, 1988). Drafted in 1988, adopted in 1990 and amended in 1997, in this code species are listed as either (1) integrally protected, (2) partially protected, or (3) other species. Species that are integrally protected cannot be hunted, captured, detained or exported (except if a scientific permit is obtained from the government). The penalty for hunting, capturing or detaining an integrally protected species is between six months to one year in prison and a fine of 40,000 to 80,000 FG, or one of these two penalties. For species that are not specially protected, hunters must obey the "Réglementation de la chasse" and must have a permit to hunt, can only hunt between 13 December and 30 April, and only between sunrise and sunset (Kormos et al. 2003).

In the Pic de Fon area a forest block of about 25,600 ha has been designated a Forêt Classée (classified forest) – the Forêt Classée du Pic de Fon. The classified forests of Guinée Forestière total nearly 323,000 ha, with Ziama and Diécké being the two largest at 116,170 ha and 59,000 ha respectively (Robertson 2001). Pic de Fon classified forest ranks third in size (Konomou and Zoumanigui 2000). Forêt Classée status is supposed to provide some protection as government property, though not necessarily for biodiversity conservation. There are six types of protected area in Guinea: Parc national (National Park), Réserve naturelle intégrale (Strict Nature Reserve), Réserve naturelle gérée (Managed Nature Reserve), Réserve spéciale (Special Reserve) or Sanctuaire de Faune (Faunal Sanctuary), Zone d'Intérêt cynégétique (Trophy Hunting Zone) and Zone de Chasse (Hunting Zone). A classified forest may receive status as one of these types of protected area, but in and of itself does not denote conservation status.

REFERENCES

Bakarr, M., B. Bailey, D. Byler, R. Ham, S. Olivieri and M. Omland (eds.). 2001. From the Forest to the Sea: Biodiversity Connections from Guinea to Togo. Washington DC: Conservation International.

Camara, W. and K. Guilavogui. 2001. "Identification des villages riverains autour des Forêts classées du Pic de Fon et Pic de Tibé". PGRR/Centre Forestier N'Zérékoré, Mesures Riveraines.

Davis, S.D., V.H. Heywood and A.C. Hamilton (eds.). 1994. Centers of Plant Diversity: A Guide and Strategy for Their Conservation (Volume 1). Gland, Switzerland: World Wide Fund for Nature and IUCN-The World Conservation Union.

Fahr, J. and N.M. Ebigbo. 2003. A conservation assessment of the bats of the Simandou Range, Guinea, with the first record of *Myotis welwitschii* (GRAY, 1866) from West Africa. Acta Chiropterologica 5(1): 125-141.

Fahr, J., H. Vierhaus, R. Hutterer and D. Kock. 2002. A revision of the *Rhinolophus maclaudi* species group with the description of a new species from West Africa (Chiroptera: Rhinolophidae). Myotis 40: 95-126.

Hatch & Associates, Inc. 1998. "Preliminary Environmental Characterisation Study: Simandou Iron Ore Project Exploration Programme, South-Eastern Guinea." Montreal, Canada.

Konomou, M. and K. Zoumanigui. 2000. "Zonage Agro-Ecologique de la Guinée Forestière". Centre de Recherche Agronomique de Sérédou. Ministère de l'Agriculture et des Eaux et Forêts: Guinée.

Kormos, R., Boesch, C., Bakarr, M.I. and Butynski, T. (eds.). 2003. West African Chimpanzees. Status Survey and Conservation Action Plan. IUCN/SSC Primate Specialist Group. IUCN, Gland, Switzerland and Cambridge, UK.

Lebbie, A. 2001. Distribution, Exploitation and Valuation of Non-Timber Forest Products from a Forest Reserve in Sierra Leone. PhD Dissertation, University of Wisconsin-Madison, USA.

Mittermeier, R.A., N. Myers, C.G. Mittermeier and P.R. Gil. 1999. Hotspots: Earth's Biologically Richest and Most Endangered Terrestrial Ecoregions. CEMEX.

MMGE [Ministry of Mines, Geology, and the Environment]. 2002 (January). National Strategy and Action Plan for Biological Diversity; Volume 1: National Strategy for Conservation Regarding Biodiversity and the Sustainable Use of these Resources. Guinea/UNDP/GEF: Conakry.

[PGRR] Projet de Gestion des Ressources Rurales. 2001. "Pic de Fon – Situation Actuelle, Analyse et Recommandations (Proposition d'axes de collaboration)". Centre Forestier N'Zérékoré.

[PGRR] Projet de Gestion des Ressources Rurales. 2002a. "Proposition de stratégie pour la surveillance du Pic de Fon – Attente PGRF". Centre Forestier N'Zérékoré, Section Conservation – Biodiversité.

[PGRR] Projet de Gestion des Ressources Rurales. 2002b. "Rapport d'Activites – Délimitation du versant sud-ouest de la Forêt Classée du Pic de Fon". Centre Forestier N'Zérékoré (Antenne Béro).

Robertson. 2001. *In:* Fishpool, L.D.C. and M.I. Evans (eds.). Important Bird Areas in Africa and Associated Islands. Pisces Publications. 1162 pp.

White, F. 1983. The vegetation of Africa: A descriptive memoir to accompany the Unesco/AETFAT/UNSO vegetation map of Africa. Paris, Unesco. 356 pp.

WWF and IUCN. 1994. Centers of plant diversity. A guide and strategy for their conservation. Vol 1. Europe, Africa, South West Asia and the Middle East. 3 Volumes. IUCN Publications Unit, Cambridge, U.K.

Chapter 2

A rapid botanical study of the Forêt Classée du Pic de Fon, Guinea

Jean-Louis Holié and Nicolas Londiah Delamou

SUMMARY

A rapid assessment of the flora of the Pic de Fon classified forest was conducted from 27 November to 7 December 2002. Data were collected from two sites using line transects within different habitat types to create a list of species present within the classified forest and to describe habitats. The species recorded from the two sites suggested that both the gallery and ravine forests and montane grassland at Site 1 were relatively undisturbed, while the semi-evergreen lowland rain forest at Site 2 faced more pressure from agriculture by local residents. We recorded 409 species including 17 tree species listed by IUCN, one as "Endangered", 15 as "Vulnerable", and one as "Near Threatened".

INTRODUCTION

Including lowland, montane, swamp and mangrove forest, the approximate original extent of closed canopy tropical moist forest cover in Guinea was 185,800 km². Today, an estimated 7,655 km² of this forest cover remains (or 4.1 % of the original closed canopy forest; Sayer et al. 1992) with an average annual loss of closed canopy forest in Guinea between 1981 and 1985 of 1.8 % (WRI 1992).

The south-eastern forested areas of Guinea, including the Pic de Fon classified forest, are included in the Guineo-Congolian Regional Center of Endemism (RCE), an area of about 2.8 million km² in West and Central Africa. Within the Guineo-Congolian RCE, an estimated 12,000 species occur (WWF and IUCN 1994), with around 80% of these species and 20% of the genera endemic to the region (White 1983). The western portion of the Guineo-Congolian RCE, the Guinean Rain Forest, covers about 420,000 km² and is centered around the forest areas of Guinea, Sierra Leone, Liberia, southern Côte d'Ivoire and Ghana, with a ridge of uplands to the North (WWF and IUCN 1994).

The flora of Guinea is comparatively poorly known overall and large areas of the country have yet to be surveyed. Within Guinea, only the flora of Mount Nimba is fairly well studied. Major works on the flora of Nimba are by Schnell (1952) and Leclerc et al. (1955). More than 2,000 plant species have been described from Mt. Nimba and about 16 of these are thought to be endemic (Adam 1971-1983). The area has been identified as a center of plant diversity under the IUCN-WWF Plants Conservation Programme (IUCN/WWF 1988). The vegetation of the Pic de Fon classified forest, including the highest point of the Simandou Range, can be expected to resemble that of Mount Nimba, based on similar geology, climate and altitude. However, prior to this survey, published vegetation studies of the Simandou Range consisted of only one study of the Pic de Fon by Schnell (1961).

STUDY SITE AND METHODS

As a result of the southwest orientation of the mountain range, Pic de Fon classified forest covers three primary habitat types. The mountains act as a physical barrier between dry northeastern winds and humid winds from the southwest. This wind pattern results in the western slopes of the Pic de Fon being more humid, and hence more densely forested, than the eastern slopes. While forest on the eastern slopes is mostly restricted to small patches of gallery and ravine forest running along numerous small rivers and creeks, on the western slopes high altitude montane forest patches are directly connected to an intact continuous belt of lowland evergreen rain forest. Residents of the 24 villages surrounding the classified forest have actively used this forest for at least the past 25 years and at the lower elevation near villages little primary forest remains. On the outer edges of the classified forest, particularly near to Banko village, much area has been converted to perennial plantations or rice fields, while other areas have been fallow for between one and five years. The lowlands adjacent to the east of the classified forest are dominated by bush-tree savanna, with small gallery forests along creeks exiting the Simandou Range.

We carried out 11 days of field work, four days in the highlands at altitudes of 900 to 1,500 m (Site 1, 27-30 November) and seven days in the lowlands at 550-800 m (Site 2, 1-7 December). Our initial focus was on identifying habitat types. We used transects to conduct the species inventory, collecting specimens for plants that were impossible to identify in the field. In the grasslands, for the inventory of lianas and grasses, we recorded plants up to 2 m on either side of the transect. For the inventory of shrubs, we recorded plants up to 20 m on either side of the transect. For the gallery and ravine forests of Site 1, which were often wedged into the valleys, we adopted the method of crossing these forests while following the course of the streams which generally take their source from within the mountain. We then followed a transect that crossed perpendicular to the water's course. For the semi-evergreen lowland rain forest, which is more dense, we used the same method as in the montane forests.

At each site inventoried we completed a description of the vegetation. The determination of species was done progressively. Species not identified on the spot were collected and identified later with the help of a key. Specimens are deposited at the Centre Forestier de N'Zérékoré, Guinea. Parallel to making inventories for each site, we made observations on the health, history and impacts to the sites.

Accessing the forest of Site 1 was more difficult as a result of no access by road and few existing foot paths. Sampling at Site 2, near the village of Banko and in the interior of the classified forest, was more easily accessible by foot due to a well established footpath leading from the village to the entrance of the classified forest as well as a number of well established hunting trails within the classified forest.

RESULTS

Pic de Fon classified forest contains three major habitat types: high altitude montane grassland (Site 1); montane forest, consisting of both ravine and gallery forest in the hollows and stream beds of the Pic de Fon (Site 1); and semi-evergreen lowland rain forest (Site 2). In our inventory of these three habitat types within the classified forest, we catalogued 409 species of plants (see Appendix 1).

The vegetation of the montane grassland is related to the edaphic conditions (lateritic ground with a high iron content) and is dominated mostly by Fabaceae and Poaceae. Montane grassland covers most of the land of the eastern and northern parts of the Pic de Fon. The most typical species of grasses found in the montane grassland of Site 1 include *Bouteloua gracilis* (Poaceae), *Dolichos nimbaensis* (Fabaceae), *Droogmansia scaettaiana* (Fabaceae) and *Protea angolensis* (Proteaceae), among others. The species of shrubs such as *Craterispermum laurinum* and *C. caudatum* (Rubiaceae) and especially *Ficus eriobotryoides* (Moraceae) stand out among the grasses here.

A number of grasses and shrubs typical of secondary grassland were also found including *Pennisetum purpureum* (Poaceae), *Imperata cylindrica* (Poaceae) and *Hyparrhenia rufa* (Poaceae) and shrubs including *Crossopteryx febrifuga* (Rubiaceae) and *Parkia biglobosa* (Mimosaceae). However, as Fabaceae such as *Dolichos nimbaensis* (the grass endemic to Mt. Nimba's montane grasslands) and *Rhynchospora corymbosa* (Cyperaceae) are associated with edaphic grassland and are more typical of the Pic de Fon montane grassland, we believe this habitat represents an edaphic climax.

The montane forest of Site 1 along the streambed is dominated by the riparian species *Cryptosepalum tetraphyllum* (Caesalpiniaceae), *Teclea verdoorniana* (Rutaceae), *Garcinia polyantha* (Clusiaceae), and *Tarenna vignei subglabrata* (Rubiaceae). At higher altitudes we noted the presence of *Strychnos spinosa* (Loganiaceae), *Balanites wilsoniana* (Balanitaceae), *Morus mesozygia* (Moraceae) and *Stereospermum acuminatissimum* (Bignoniaceae). The inventoried zones for the most part are narrow bands along the courses of water running between the mountains with steep sides that are difficult to access and navigate. These ravine and gallery forests constitute an intact habitat with very few traces of disturbance.

The semi-evergreen lowland rain forest of Site 2 is dominated by a great number of trees, lianas and grasses not found in other sites within the area. The families most represented are the Rubiaceae, Sapotaceae, Clusiaceae, Ulmaceae, Moraceae, Euphorbiaceae and Caesalpiniaceae. The species of large trees recorded were *Neolemonniera clitandrifolia* (Sapotaceae), *Xylia evansii* (Mimosaceae), *Sterculia oblonga* (Sterculiaceae), *Guarea cedrata* (Meliaceae), *Khaya grandifoliola* (Meliaceae), *Celtis adolfi-friderici* (Ulmaceae), *Mansonia altissima* (Sterculiaceae), *Parkia bicolo*r (Mimosaceae) and other species found in Appendix 1. Lianas inventoried included *Strychnos aculeatea* (Loganiaceae), *S.*

spinosa, *Dalbergia hostilis* (Papilionaceae), *Hippocratea* spp. (Hippocrateaceae), and *Calycobolus africanus* (Convolvulaceae). For grasslands, we noted the presence of *Geophila* spp. (Rubiaceae), *Solanum nigrum* (Solanaceae), *Trema orientalis* (Ulmaceae), *Marantochloa* spp. (Marantaceae), *Leptaspis cochleata* (Poaceae), *Olyra latifolia* (Poaceae) and *Cyathula prostrata* (Amaranthaceae).

The classified forest of Pic de Fon contains both primary unexploited and heavily exploited forest units. In less disturbed areas trees such as *Cryptocepalum tetraphyllum* (Caesalpiniaceae), *Parinari excelsa* (Chrysobalanaceae), *Teclea verdoorniana* (Rutaceae), *Morus mesozygia* (Moraceae), *Celtis* spp. (Ulmaceae), *Funtumia* spp. (Apocynaceae) and *Guarea cedrata* (Meliaceae) were typical. In exploited areas, early colonizing trees, grasses and herbs such as *Solanum incanum* (Solanaceae), *S. verbascifolium*, *Ageratum conyzoides* (Asteraceae), *Physalis angulata* (Solanaceae), *Trema orientalis* (Ulmaceae), *Anthocleista nobilis* (Loganiaceae) and *Harungana madagascariensis* (Clusiaceae) were present.

DISCUSSION

Based on species composition, we believe the montane grassland of Site 1, prior to the recent mineral exploration activity, has had little human interference and, other than disturbance created by roads, the montane grassland species are mostly intact. We note the presence of a number of species associated with secondary grassland, but due to the predominance of species associated with edaphic grassland, believe the vegetation here represents climax vegetation for this habitat.

Likewise, the ravine and gallery forests of Site 1 also constitute an intact habitat with very few traces of disturbance, with a species composition resembling both upland *Parinari excelsa* forest and drier peripheral semi-evergreen Guineo-Congolian rain forest as described by White (1983).

We wish to note that the topography of West Africa is generally quite uniform and flat with only a few mountain regions making West African montane habitats very limited in extent. The Simandou Range is one of the few locations in all of West Africa harboring montane grasslands, and thus one of the only places where species adapted to this particular habitat can live. The distribution map of *Myotis welwitschii* (pg. 18) indicates the rarity of high elevation habitats within all of West Africa.

A factor increasing the potential for high diversity and endemicity within the Pic de Fon classified forest is that these rare habitat types occur in close proximity to one another as well as to the lowland forest found at lower elevations within the boundaries of the classified forest.

Species of Conservation significance
Among the 409 species recorded, we documented 18 species of trees listed on IUCN's Red List, one as Endangered, 16 as Vulnerable, and one as Near Threatened. See Table 2.1

for list and explanations of threat categories. Of these tree species, all 18 were found in the lowland forest, and only one (*Milicia excelsa* – nt) was also found in the montane forest. In addition, for the montane forests, the species *Teclea verdoorniana*, *Schefflera barteri* and *Balanites wilsoniana* constitute a rare group that ought to be protected.

Species used for food, medicine and international trade
Among the species recorded, we identified 98 species with known value for food, medicine or international trade. Of these, 76 species are used for food, 24 are commonly used medicinally in Africa, and 16 species are traded internationally and economically important (including *Elaeis guineensis*, oil palm, a species native to West Africa). Some of the most important species, such as *Rauvolfia vomitoria*, are used locally as food and for medicine, while they are also exported internationally, in the case of *R. vomitoria* for the pharmaceutical industry. For a complete list of plants identified during our survey used for food, medicine and in trade, see Appendix 2.

Local use of forest resources
The residents of villages around the Pic de Fon classified forest have actively used this forest for at least the past 25 years. Hence, at the lower elevations near villages only a small percentage of primary forest remains. We noted that on the outer edges of the classified forest, particularly near to Banko village (and presumably near to other villages as well) many areas have been converted to perennial plantations or rice fields, while other areas are fallow (most fallow plots observed had been fallow for between one and five years). We noted a number of newly planted plots of perennial trees, such as coffee and bananas, indicating a perception by villagers that long-term cultivation within the boundaries of the classified forest is unlikely to provoke a reaction from authorities. Previous studies by PGRR (the Projet de Gestion des Ressources Rurales), a project of the Centre Forestier de N'Zérékoré, have estimated that between 35 and 40 % of the classified forest has been impacted in some way by agricultural encroachment (PGRR 2001). We also observed a number of unsustainable harvesting practices, in particular, fruit collecting methods that included chopping down entire trees to collect fruit from the trees, rather than harvesting fruit from trees and leaving trees standing to fruit again in another season.

Table 2.1. Plant species of global conservation concern recorded during this survey.

Family	Genus/species	IUCN Status
Anacardiaceae	*Antrocaryon micraster*	VU A1cd
Annonaceae	*Neostenanthera hamata*	VU A1c, B1+2c
Boraginaceae	*Cordia platythyrsa*	VU A1d
Caesalpiniaceae	*Cryptosepalum tetraphyllum*	VU A1c, B1+2c
Clusiaceae	*Garcinia kola*	VU A1cd
Combretaceae	*Terminalia ivorensis*	VU A1cd
Euphorbiaceae	*Drypetes afzelii*	VU A1c, B1+2c
Euphorbiaceae	*Drypetes singroboensis*	VU A1c
Meliaceae	*Entandrophragma candollei*	VU A1cd
Meliaceae	*Entandrophragma utile*	VU A1cd
Meliaceae	*Guarea cedrata*	VU A1c
Meliaceae	*Khaya grandifoliola*	VU A1cd
Mimosaceae	*Albizia ferruginea*	VU A1cd
Moraceae	*Milicia excelsa*	nt
Moraceae	*Milicia regia*	VU A1cd
Ochnaceae	*Lophira alata*	VU A1cd
Rubiaceae	*Nauclea diderrichii*	VU A1cd
Sapotaceae	*Neolemonniera clitandrifolia*	EN B1+2c

Threat Status (2002 IUCN Red List of Threatened Species) (IUCN 2002)

EN: Endangered

  B) Extent of occurrence estimated to be less than 5,000 km^2 or area of occupancy estimated to be less than 500 km^2, and estimates indicating any two of the following:

    1) Severely fragmented or known to exist at no more than five locations

    2) Continuing decline, inferred, observed or projected, in any of the following:

      c) Area, extent and/or quality of habitat

VU: Vulnerable

  A) Population reduction in the form of either of the following:

    1) An observed, estimated, inferred or suspected reduction of at least 20% over the last 10 years or three generations, whichever is the longer, based on (and specifying) any of the following:

      c) A decline in area of occupancy, extent of occurrence and/or quality of habitat

      d) Actual or potential levels of exploitation

  B) Extent of occurrence estimated to be less than 20,000 km^2 or area of occupancy estimated to be less than 2000 km^2, and estimates indicating any two of the following:

    1) Severely fragmented or known to exist at no more than ten locations

    2) Continuing decline, inferred, observed or projected, in any of the following:

      c) Area, extent and/or quality of habitat

nt : Near Threatened

Taxa which are close to qualifying for Vulnerable.

REFERENCES

Adam, J.G. 1971-1983. Flore descriptive des Monts Nimba. Vols. 1 – 6. Mémoires du Muséum d'Histoire Naturelle B.20: 1-527; 22: 529-908.

Cunningham, A.B. 1993. African Medicinal Plants: setting priorities at the interface between conservation and primary health care. Working paper 1. UNESCO, Paris.

[FAO] Food and Agriculture Organization of the United Nations. 1990. Appendix 3 - Commonly Consumed Forest and Farm Tree Foods in the West African Humid Zone. The Major Significance of 'Minor' Forest Products: The Local Use and Value of Forests in the West African Humid Forest Zone. FAO, Rome.

[FAO] Food and Agriculture Organization of the United Nations and the US Department of Health, Education and Welfare. 1968. Appendix 4 - Index of Scientific Names of Edible Plants. Food Composition Table for Use in Africa.

IUCN. 2002. 2002 IUCN Red List of Threatened Species. Downloaded on 13 October 2003. www.redlist.org.

IUCN/WWF. 1988. Centres of Plant Diversity: a guide and strategy for their conservation. IUCN-WWF Plants Conservation Programme/IUCN Threatened Plants Unit. 40 pp.

Leclerc, J.C., M. Lamotte, J. Richard-Molard, G. Rougerie and P. Porteres. 1955. La Réserve Naturelle Intégrale du Mont Nimba. La chaine du Nimba: essai géographique. Mémoires de l'Institut Française d'Afrique Noire 43: 1-256.

MacKinnon, J. and K. MacKinnon. 1986. Review of the proteced areas system in the Afrotropical Realm. IUCN, Gland, Switzerland, in collaboration with UNEP. 259 pp.

Marshall, N.T. 1998. Searching for a Cure: Conservation of Medicinal Wildlife Resources in East and Southern Africa, TRAFFIC International.

PGRR. 2001. Pic de Fon – Situation Actuelle, Analyse et Recommandations (Proposition d'axes de collaboration). Centre Forestier N'Zérékoré.

Safowora, A. 1982. Medicinal Plants and Traditional Medicine in Africa. John Wiley and Sons Limited, Chichester.

Sayer, J.A., C.S. Harcourt and N.M. Collins (eds.). 1992. The Conservation Atlas of Tropical Forests: Africa. Macmillan. 288 pp.

Schnell, R. 1952. Végétation et flore de la region montagneuse du Nimba. Mémoires de l'Institut Française d'Afrique Noire, 22, p. 1-604.

Schnell, R. 1961. Contribution à l'étude botanique de la chaîne de Fon (Guinée). Bull. Jard. Bot. État Brux., 31, p. 15-54.

White, F. 1983. The vegetation of Africa: a descriptive memoir to accompany the Unesco/AETFAT/UNSO vegetation map of Africa. Natural Resources Research XX. Unesco, Paris. 356 pp.

[WRI] World Resources Institute. 1992. World Resources 1992-93: a guide to the global environment. Oxford University Press, New York. 385 pp. (Prepared in collaboration with UNEP and UNDP.)

WWF and IUCN. 1994. Centres of plant diversity. A guide and strategy for their conservation. 3 volumes. IUCN Publications Unit, Cambridge, U.K. Vol. 1, p. 119.

Chapter 3

A rapid survey of katydids (Insecta: Orthoptera: Tettigoniidae) of the Forêt Classée du Pic de Fon, Guinea

Piotr Naskrecki

SUMMARY

This study is the first survey of the katydids (Tettigoniidae) of the Pic de Fon area. 40 species were collected, of which at least 4 are new to science, and 10 are new to Guinea (a 16 % increase). *Anoedopoda* cf. *lamellata* savanna population may be distinct from the remainder of the African population. The Pic de Fon katydid fauna has a very high potential for endemicity and should be further investigated.

INTRODUCTION

The species level diversity taken into consideration in tropical conservation decision-making is by and large restricted to large, well-studied ("charismatic") vertebrates, such as primates, elephants or large predators. Less frequently, the bird, reptile, or amphibian diversity are factors in designating protected areas. But so far there has not been a single case in which invertebrate data alone influenced a decision to award the protected status to a tropical area or site (although such cases are becoming more frequent in Europe, New Zealand, and North America; Murphy and Weiss 1988; Sherley and Hayes 1993; Ortiz 2001). Several factors are responsible for this situation. Most importantly, the state of our knowledge of tropical invertebrate communities is rudimentary, or non-existent for many areas, and as a result it is virtually impossible to monitor the abundance or even presence or absence of individual invertebrate species.

This overwhelming lack of data on invertebrate diversity in the tropics has the tragic result of species becoming extinct or seriously endangered without anybody even realizing their existence. There are very few documented cases of invertebrate species extinction (e. g., St. Helena earwig, California dune neduba; Rentz 1977), and virtually all of them come to light only when a species from a group known for its high degree of endemicity and habitat fidelity is discovered by a taxonomist in a museum collection, collected from a site already altered or destroyed by human actions.

While there are thousands of vertebrate specialists working on tropical faunas, and identification of species is not a problem, in the case of the great majority of invertebrate groups identification is problematic and often impossible under the field conditions. In addition, the high number of yet undescribed species (in some invertebrate groups over 90 % of species are still waiting to be formally recognized and named) makes collecting and comparing data from tropical locations very difficult.

Ironically, invertebrates could not be better suited to serve as conservation tools and indicators of habitat health. Many species are highly sensitive to even minor changes in humidity, temperature, concentration of heavy metals in the soil and water, or composition of plant communities, thus potentially serving as an early warning system of changes that will eventually affect vertebrate communities. Also, many species of invertebrates are the primary food source of other animals and perform immeasurable ecological services, such as pollination, soil production or decomposition. Many small species, due to their oligotypic requirements and limited dispersal abilities, are more vulnerable than larger animals. Yet they routinely slip under the radar of conservation authorities and are irreversibly lost.

During the Rapid Assessment Program (RAP) survey of the southern part of the Simandou Range in eastern Guinea one group of invertebrates, the katydids (Tettigoniidae), were sampled in order to determine their diversity and potential as indicators of the uniqueness of two sites. Members of this group exhibit remarkable potential as environmental indicators (Samways 1997) and many species have remarkably small distribution ranges, resulting in a high degree of endemicity within the group.

With the exception of a single study based on the material collected at Mount Nimba (Chopard 1954) and a few individual records scattered in taxonomic revisions (Brunner von Wattenwyl 1895; Beier 1962, 1965) the fauna of katydids of Guinea is virtually unknown. There exists a high potential, partially demonstrated by this study, to discover in this country new taxa, both at generic and specific level. Chopard (1954) reported 60 species from the Mount Nimba area and Beier (1965) added one additional species to the list.

METHODS

Three collecting methods have been proven highly effective in collecting katydids in mixed habitats, such as the area under investigation: the black light (UV light), visual search, and net sweeping. Unfortunately, the black light method was not available during the current survey, thus potentially reducing the chance to collect flying, nocturnal species, such as many members of the Phaneropterinae. However, the availability of other light sources (fluorescent light at Site 1, petroleum lamps at Site 2) helped to determine that virtually no night flying katydids were present in the area, thus the impact of not using the UV light as one of the collecting method had probably only a minimal effect on the final count of collected species.

Net sweeping was employed in savannah habitats, the forest understory, and bushy edge habitats adjacent to the forest. This method was highly effective in collecting seed feeding katydids in tall grasses as well as a number of arboreal katydids that cling upside-down to the lower surface of leaves. Sweeping was standardized by performing six consecutive sweeps in a series before the content of the net was inspected, and four series of sweeps were performed at each 100 m transect in savannah habitats. A total of 30 transects were sampled.

By far the most effective method of collecting, in terms of the number of species collected, rather than the sheer number of collected specimens, was the visual search at night. Most of the collecting was conducted between the hours of 8 pm and 2 am when the activity of virtually all katydid species is the highest. This method was standardized only by the time spent collecting.

Representatives of all encountered species were collected and voucher specimens were preserved in 95 % alcohol and as pinned, dry specimens. These specimens will be deposited in the collections of the Museum of Comparative Zoology, Harvard University and the Academy of Natural Sciences of Philadelphia (the latter will also become the official repository of the holotypes of several new species encountered during the present survey upon their formal description.)

In addition to physical collection of specimens, stridulation of acoustic species was recorded using the Sony WM-D6C Walkman Professional tape recorder and a Sennheiser shotgun microphone. These recordings are essential to establish the identity of several cryptic species of the genera *Ruspolia* and *Thyridoropthrum*, where morphological characters alone are not sufficient for species identification.

RESULTS

Despite unfavorable weather conditions (beginning of the dry season), a relatively high number of katydids were collected. In total, 40 species were collected, representing at least 24 genera (52 % of known West African generic diversity) and 6 subfamilies. The only subfamily known from the region but not collected during the present survey was the Hetrodinae, which includes species typical of lowland, xeric habitats, and thus its absence among the collected taxa is not surprising. Among the collected species at least four are new to science, including at least one new genus, but it is possible that additional new taxa are among the yet unidentified specimens of graminicolous genus *Ruspolia*. The latter is in great need of a taxonomic revision, which should be based in large part on the acoustic properties of their species-specific calls. Figure 3.1 presents digital signatures of *Ruspolia* "acuspecies" recorded during the survey. Table 3.1 presents the complete list of species collected during the survey.

At the first, markedly dry and open site (Site 1: 08°31'52"N; 08°54'21"W) the majority of collected species were savannah elements. Most of the katydids were nocturnal seed feeders, with only two predaceous species (*Thyridorhoptrum*). Forested ravines at the site hosted remarkably few truly sylvan species. Three separate forest patches were investigated: two patches at elevations of 1,200-1,300 m, and one at an elevation of approximately 1,000 m (08°32.9'N; 008°53.9'W). The only taxa known to be associated exclusively with forest habitats were *Mormotus* sp. 1 and *Afrophisis* sp. n. The remainder of the species collected in forests were savannah elements that apparently penetrate the relatively narrow and dry forest patches. Especially surprising was the presence in one of the patches of the obligatory graminivore, *Plastocorypha nigrifrons*. Completely absent were taxa of katydids normally associated with leaf litter, such as *Afromecopoda* or *Euthypoda*. While this may be simply a sampling error resulting from the brief amount of time spent collecting at the site, the litter layer within the forest patches was very dry, with low numbers of other insects (e. g., Collembola, Dermaptera, Formicidae) normally found in such a habitat. Interestingly, the afrotropical species *Anoedopoda lamellata*, found throughout its range in forest habitats, at this site was only present in the open savanna. Further studies should be undertaken to determine if the Pic de Fon population is conspecific with the forest populations in central and southern Africa, but based on the characteristics of its call and minor morphological differences of the wing venation it is safe to assume that the Pic de Fon population represents a new species of the genus.

At least two additional species collected at the first site are new to science. *Aphrophisis* sp. n. represents a small genus of specialized insectivores, known so far only from Togo (1 species) and Cameroon (1 species). It is very likely that this species is endemic to the area as most species in its group (Phisidini) are known to have very restricted ranges (Jin and Kevan 1992). The presence of *Thyridorpthrum* sp. n. indicates the existence of a previously unknown complex of cryptic species within the genus. While morphologically very similar to the widely distributed *T. senegalense*, the new species found in the savanna and forest edges at the first site has a very distinct, unmistakable advertisement call (Fig. 3.2). However, the good flying abilities of the new species makes it unlikely that this taxon is exclusive to the area and it can probably be found at other sites in the Simondou Range. In addition to the species new to science, at least 4 species collected at the first site represent new species records for Guinea.

The second site (Site 2: 08°31'29"N; 08°56'12"W) was situated at the elevation of 590 m and exhibited much higher humidity and forest coverage. Nineteen of 29 species collected there were typical sylvan elements, found mostly in the primary forest on the western slope of the range. Due to higher humidity and a more well developed leaf litter layer of the forest, at least three ground dwelling species were present at the site (*Afromecopoda frontalis*, *Euthypoda* sp. 1 and *Conocephalus carbonarius*.) At least 10 of the collected species were exclusively arboreal, and an additional few species were heard stridulating high in the canopy (such species can only be collected by canopy fogging or after a series of heavy rains, when they are knocked to the ground by wind and water).

A striking feature of the forest katydid fauna at the second site was the presence of very low population densities of the observed species. On average, only one individual was collected per each 30 minutes of searching in the forest. This situation was most likely the effect of the beginning of the dry season, when many species undergo a short egg diapause. In addition, several genera (*Euthypoda*, *Mustius*, *Amytta*) were only present as young nymphs.

The dominant grass of the savanna, *Setaria megaphylla*, was a host to several species of *Ruspolia*, *P. nigrifrons* and *Conocephalus* spp. Remarkably absent were *P. lanceolatus* and *A.* cf. *lamellata*, two savanna species very common at the first site.

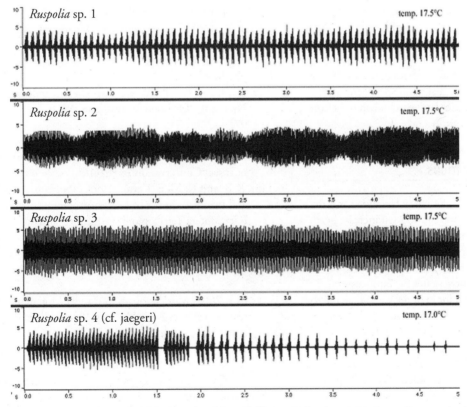

Figure 3.1. Oscillograms of 4 cryptic (sibling) species of *Ruspolia* (Conocephalinae) present in the montane grassland habitats of Site 1.

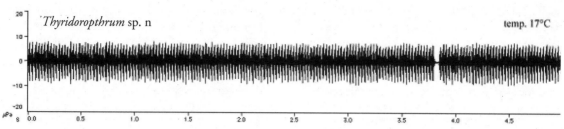

Figure 3.2. Oscillogram of a new species of *Thyridoropthrum* (Conocephalinae) from the montane grassland habitats of Site 1.

Two species of Meconematinae collected at the second site are new to science (they also represent a new genus to be described shortly) (see photos pp. 18-21), but more undescribed species may be present among the unidentified *Ruspolia*. Unfortunately, the lack of comparative sound recordings for virtually all previously described species of the genus makes it inadvisable to make taxonomic decisions based on the morphological characters alone, and an additional survey of West African species should be conducted as soon as possible. The two new species of Meconematinae are almost certainly very restricted in their distribution as both are flightless, thus of low dispersal potential. Also, at least six species collected at the second site represent new records for Guinea.

Several species collected at the second site represent the only records of the species outside of the type specimen locality record. For example, the specimen *Weissenbornia praestantissima* collected there is only the second known specimen of this species, originally described from Cameroon (Lowry-Kribi Mudung), and the large population of *Conocephalus carbonarius* found in the understory grasses of the secondary forest of the Site 2 represent the first adult specimens of this species ever collected (this species has been known only from the type nymphal specimen from Accra).

CONSERVATION RECOMMENDATIONS

While by no means exhaustive, this study has added at least 10 species to the faunal list of the country (an increase by 16 %). At the same time it found a remarkably low number of species (10) previously recorded from Mount Nimba by Chopard (1954) and Beier (1965). This appears to be indicative of high species turnover rates in south-eastern Guinea, and carries a potential for high degree of endemicity among katydids in this area. The newly discovered species of katydids should be considered restricted or endemic, at least until further studies locate additional populations of these species. It is also likely that the savanna population of *A. lamellata* is distinct from the forest populations of this species in other parts of Africa. Furthermore, the cryptic complex of species of the genus *Ruspolia* may include several undescribed species, and an extensive genetic and bioacoustic study should be conducted among the species of the genus in the Simandou Range.

Because of the high sensitivity of forest katydid species to changes in humidity and plant species composition, any activity in the area that changes the water regime and affects the climate can potentially have devastating effects on the present fauna. Further studies along the entire Simandou Range are needed in order to estimate the distribution and population sizes of the newly discovered species. Also, additional collecting is needed at the beginning of the rainy season when most species appear in high numbers and as adults.

REFERENCES

Beier, M. 1962. Pseudophyllinae. Tierreich 73.

Beier, M. 1965. Die afrikanischen Arten der Gattungsgruppe "Amytta" Karsch. Beiträge zur Entomologie 15: 203-242.

Brunner von Wattenwyl, C. 1895. Monographie der Pseudophylliden. Wien.

Chopard, L. 1954. La reserve naturelle integrale du Mont Nimba III. Orthopteres Ensiferes. Mem. IFAN 40: 25-97.

Gibbs, G.W. 1998. Why are some weta (Orthoptera: Stenopelmatidae) vulnerable yet others are common? Journal of Insect Conservation 2: 161-166.

Jin, X-B. and K. McE. Kevan. 1992. *Afrophisis*, a new genus and two new species of small orthopteroids from Africa (Grylloptera Tettigonioidea Meconematidae). Tropical Zoology 4: 317-328.

Murphy, D.D. and S.B. Weiss. 1988. A long-term monitoring plan for a threatened butterfly. Conservation Biology 2: 367-374.

Ortiz, F. 2001. Mexico moves to protect Monarch butterfly reserve. Online. Available: http://www.planetark.org/dailynewsstory.cfm/newsid/13455/story.htm. 21 Nov. 2001.

Rentz D.C.F. 1977. A new and apparently extinct katydid from Antioch Sand Dunes (Orthoptera: Tettigoniidae). Entomological News 88: 241-245.

Samways, M.J. 1997. Conservation Biology of Orthoptera.
In: Gangwere, S.K. et al. (eds). The Bionomics of Grasshoppers, Katydids and Their Kin. CAB International,
pp. 481-496.

Sherley, G.H. and L.M. Hayes. 1993. The conservation of
giant weta (*Deinacrida* n. sp. Orthoptera: Stenopelmatidae) at Mahoenui, King Country: habitat use, and other
aspects of its ecology. NZ Entomol. 16, 55–68.

Table 3.1 Complete lists of Orthoptera species collected during Pic de Fon RAP survey.

Taxon	Site 1	Site 2	Habitat	New record for Guinea
Tettigoniidae				
Phaneropterinae				
Phaneroptera minima	+	+	S/RF/BP	Yes
Phaneroptera ? maxima	+		S	Yes
Phaneroptera sp. 1	+	+	S/PF	
Eurycorypha sp. 1	+		S/RF	
Eurycorypha sp. 2		+	S	
Preussia lobatipes		+	BP	
Ducetia fuscopunctata		+	S	
Arantia ? ovalipennis		+	S	
Arantia sp. 1		+	S	
Zeuneria melanopeza		+	S	Yes
Weissenbornia praestantissima		+	PF	Yes
Conocephalinae				
Ruspolia jaegeri	+		S	Yes
Ruspolia sp. 1	+		S	?
Ruspolia sp. 2	+		S	?
Ruspolia sp. 3	+	+	S/BP	?
Ruspolia sp. 4	+	+	S/BP	?
Ruspolia sp. 5	+	+	S	?
Pseudorhynchus lanceolatus	+		S	
Plastocorypha nigrifrons	+	+	S	
Conocephalus carbonarius		+	PF	
Conocephalus maculatus	+	+	S/BP	
Conocephalus sp. 1	+		S	
Thyridorhoptrum senegalense	+	+	S/RF/BP	
Thyridorhoptrum sp. n.	+		RF/PF	Yes
Meconematinae				
Amytta sp. 1	+	+	RF/PF	
Amytta sp. 2		+	PF	
Gen. n., sp. n. 1		+	PF	Yes

S – savannah
RF – ravine forest
PF – primary forest
BP – banana plantation

Taxon	Site 1	Site 2	Habitat	New record for Guinea
Gen. n., sp. n. 2		+	PF	Yes
Listroscelidinae				
Afrophisis sp. n.	+		RF	Yes
Mecopodinae				
Anodeopoda cf. lamellata	+		S	Yes
Afromecopoda frontalis		+	PF	
Euthypoda sp. 1		+	PF	
Pseudophyllinae				
Stenampyx annulicornis		+	PF	
Mustius sp. 1	+		RF/PF	
Mustius sp. 2		+	PF	
Mormotus sp. 1		+	PF	
Mormotus clavaticercus		+	PF	
Lichenochrus cf. centralis		+	PF	
Hambrocomes sp. 1		+	PF	
Adapantus brunneus		+	PF	

Total species – 40
New to Guinea – 10 (possibly more)
New to science – 4 (possibly more)
Site 1 – 20
Site 2 – 29
Savannah – 21 (14 exclusively)
Forest – 22 (18 exclusively)
Banana plantation – 6

Chapter 4

Rapid survey of amphibians and reptiles in the Forêt Classée du Pic de Fon, Guinea

Mark-Oliver Rödel and Mohamed Alhassane Bangoura

SUMMARY

We recorded at least 32 amphibian species from the Pic de Fon area and estimate that 50-60 species likely occur. A mixture of forest and savannah species, widespread species and local endemics characterize the amphibian community of Pic de Fon, reflecting the unique habitat mosaic of the area. Several species encountered during the RAP survey are new to science and possibly endemic to the Pic de Fon area. While further surveys are necessary to confirm the presence of additional species, we believe the Pic de Fon classified forest (PFCF) is potentially one of the areas with the highest amphibian biodiversity in West Africa. The regional amphibian community currently faces a range of threats, including habitat destruction, and future habitat alteration may have severe, negative impacts. Further research is needed to assess the range and demographic characteristics of the Pic de Fon amphibians.

INTRODUCTION

Amphibians are highly diverse in West Africa (Rödel 2000b, Rödel and Schiøtz in Bakarr et al. 2001). Their taxonomy is relatively well known and most species are tightly connected to certain species-specific types of habitats. In addition, it is comparatively easy to assess a reasonable and representative part of the amphibian fauna of a given area in a short time, at least during their reproductive phase, which is nearly always the rainy season. This makes amphibians a very suitable group for rapid assessment programs focusing on the biodiversity of scientifically unknown areas (Heyer et al. 1993). In addition, the presence or absence of certain species serves as an indicator of the status (pristine versus degraded) of habitats (compare Rödel and Branch 2002).

While African savannah amphibians are mostly widespread, often having distributions that range from Senegal into southern Africa (Schiøtz 1999, Rödel 2000a), this is different in forest species. The latter are mostly restricted to West Africa, defined as an area stretching from Senegal in the northwest to Nigeria in the south-east (Schiøtz 1967, 1999; Rödel 2000b). Many of the species have even smaller ranges confined to the Upper Guinean forests (the forest zone west of the Dahomey Gap) or small parts of these forests (e.g. Rödel and Ernst 2000, 2002b; Rödel et al. 2002, 2003).

Reasonably good amphibian inventories are still scarce. So far not more than 25 have been compiled for the whole Upper Guinea region (for review see Rödel and Agyei 2003). However, those areas show a clear trend in that the amphibian diversity of forest areas is higher than that of savannah habitats, and areas that comprise a mixture of different biomes are most likely to have a higher diversity than uniform biomes.

During this RAP survey, reptiles were only encountered rarely. The time of the RAP, at the beginning of the dry season, was unfavorable for reptile records because of high grass hindering sight and a generally low activity of the species. Based on these and other survey results in other parts of West Africa (Angel et al. 1954a, b; Barbault 1975; Böhme 1994b, 1999; Rödel et al. 1995, 1997, 1999; Rödel and Mahsberg 2000; Ernst and Rödel 2002; Ineich 2002; Branch and Rödel 2003) the number of recorded reptile species may only account for 20 % of the species living in the Pic de Fon area. We therefore give only a summary of those reptiles recorded and comment on protected ones. In the analyses we focus on amphibians.

STUDY SITE AND METHODS

As a result of the southwest orientation of the mountain range, PFCF covers many different habitat types. The western slopes of PFCF are more humid than the eastern slopes and therefore more densely forested. Forest also stretches further north in the West than on the eastern slopes. Whereas forest in the East of PFCF is mostly restricted to small patches of gallery forest running along the numerous small rivers and creeks, in the West forest patches at high altitude are directly connected to a still very much intact continuous belt of lowland rainforest. The lowlands adjacent to the East of PFCF are dominated by bush-tree savanna, with small gallery forests along creeks exiting the Simandou Range. The high-altitude areas within the PFCF contain two distinct habitats: montane forest (gallery and ravine) running in narrow strips along streambeds and montane grassland, the latter being one of the most exceptional and threatened West African habitat types. Mean annual precipitation varies between 1,700 and 2,000 mm. A long dry season lasts from November to April.

The RAP survey took place from 27 November to 7 December 2002 at the beginning of the dry season. The RAP team was mainly working at two sites. The first site (Site 1; 08°31'52"N, 08°54'21"W) contained montane grassland and forest pockets along streams at an altitudinal range of about 1,000-1,600 m asl, close to the Pic de Fon. Both habitat types seemed to be in primary state. However, roads, recently built to support mineral exploration activities, have caused sediment erosion. The second site (Site 2; 08°31'29"N, 08°56'12"W) was at about 600 m asl, at a distance of 6 km from the village of Banko. It was approximately 1.5 km inside the boundary of the classified forest and comprised primary forest and some small savannah patches in a very hilly landscape. Towards the direction of the village, some forest has been converted to cocoa and coffee plantations, and this area also contains some secondary forest.

We tried to search all present habitat types equally well for amphibians. However, due to the dry weather conditions and the high grass in the montane grassland, this habitat was probably undersampled. From the highlands, most amphibian records originate from gallery forests along ravines and humid parts of the grasslands, the latter containing small patches of stagnant waters. In the lowlands we sampled with equal effort in the pristine rainforests and the degraded forests and plantations towards Banko village. Due to the hilly landscape in the primary forest, stagnant waters were nearly absent. Instead this part of the reserve offered a huge variety of slow to fast running waters of different size. Several small rivers were also running through the degraded forests and plantations. In addition these habitats also provided some stagnant ponds, puddles and swamps. Appendix 3 gives a full list of habitats investigated, including their geographic position, date of investigation, sampling effort and brief habitat characterizations. Prior to the RAP survey, the junior author did some fieldwork in the Pic de Fon area from 2-9 November 2002. Records of both field campaigns are included in this analysis.

Amphibians and reptiles were mainly located opportunistically during visual surveys of all habitats by up to four people. Surveys were undertaken during day and night. Search techniques included visual scanning of terrain and refuge examination (e.g. lifting rocks and logs, scraping through leaf litter). We also applied acoustic monitoring in all available habitat types (Heyer et al. 1993). During the whole investigation period we had rainfall only three times, twice during the first field period and on the first day of the second period (see below). Due to this fact, amphibian reproductive activity (and therefore calling males) was very limited, and most records were based on visual recordings. Many species could only be recorded as juveniles. We also applied dip netting for tadpoles in suitable waters and checked the drift fences with pitfall traps established by the small mammal group at Site 2. However, results from pitfall traps were negligible. Our methodology provided only qualitative and semi-quantitative data. For quantitative data, mark-recapture experiments along standardized transects or on plots would have been necessary. Time of the RAP survey was too limited to apply the latter methods. We measured our sampling effort in man-hours spent searching in a certain area (Appendix 3). To get comparative data we spent more time searching in complex and larger habitats than in uniform or smaller ones.

Assuming that sampling effort was comparable throughout the habitats, we calculated the total number of amphibian species occurring in PFCF. Because we had no quantitative data available, we used the Chao2 and Jack-knife 1 estimators, based on presence/absence data for all habitats (software: EstimateS, Colwell 1994-2000). Calculation bases were the daily species lists (17 days) for 32 amphibian species (see below). To prevent order effects, all calculations have been based on 500 randomized runs. For an introduction to the methods applied see Colwell (1994-2000) and literature cited therein.

Several determinations still need confirmation. Current taxonomic knowledge makes it impossible to determine *Arthroleptis/Schoutedenella* frogs to species level. Separation of species applying morphological criteria (morpho-species) is likewise difficult or impossible. The only reliable field characters are calls (Rödel and Branch 2002, Rödel and Agyei 2003). Due to the dry weather none of these frogs were ever heard calling during the RAP survey. Therefore we counted records from that group as only one species. However, it is much more realistic to assume that at least two to three species were recorded. Species richness therefore is estimated conservatively.

We comment only on species that are not well known, are of uncertain taxonomic status, and/or have not been dealt with in detail in other recent publications (e.g. Schiøtz 1999, Rödel 2000a, Rödel and Branch 2002, Rödel and Agyei 2003). A full species list including all sites where a particular species was recorded during the RAP survey is given in Table 4.1. Nomenclature mainly follows Frost (2002). For exceptions see text.

Some voucher specimens were anesthetized and killed in a chlorbutanol solution and thereafter preserved in 70 % ethanol. Vouchers were deposited in the collections of the senior author (MOR) and junior author (MAB, see Appendix 4). MOR's specimens will be transferred to collections of several natural history museums. MAB's specimens will serve as the base of a Guinean reference collection, eventually stored at the University of Conakry. Tissue samples (toe tips) of recorded species were preserved in 95 % ethanol. These samples are stored in MOR's collection and the Institute of Zoology at Mainz University, Germany.

RESULTS

Selected species list:

Bufo superciliaris Boulenger, 1888 "1887". This largest African toad could not be recorded during the RAP. However, we got such accurate descriptions by villagers concerning size, coloration and habits of a large forest toad that we have no doubt that the occurrence of this species can be taken as confirmed. According to our local guides it can be found at the feet of large buttress trees in closed forest. Local Malinké people believe that this toad gives birth to the rainbow. This belief seems to be widespread in West Africa. The senior author heard respective tales by Oubi and Guéré people in western Côte d'Ivoire. *B. superciliaris* is known to range from Côte d'Ivoire eastward to northern Democratic Republic of Congo and Gabon (Frost 2002). Recently it has been recorded from Ziama forest, south of PFCF, by Böhme (1994a). Other known West African localities are Taï National Park, Côte d'Ivoire (Rödel 2000b), Mount Nimba (Côte d'Ivoire, Liberia, Guinea; Guibé and Lamotte 1958a) and Ghana (Hughes 1988, without giving exact locality). According to Böhme (pers. comm.), Central and West African *B. superciliaris* differ considerably. *B. superciliaris* was described from "Rio del Rey, Cameroons" by Boulenger (1888 "1887"). It thus might be that *Bufo chevalieri* Mocquard, 1909 "1908", described from Côte d'Ivoire, and synonymized without discussion by Tandy and Keith (1972), has to be resurrected.

Bufo togoensis Ahl, 1924. This species was put into the synonymy of *B. latifrons* Boulenger, 1900 by Tandy and Keith (1972). *B. latifrons* was described from Gabon, *B. togoensis* was described from Togo. Our specimens have more or less smooth, straight parotids that are in contact with the eyelids (spiny and spaced from eyelid in *B. latifrons*, Perret and Amiet 1971), and a clearly discernible tympanum (indistinct in *B. latifrons*, Perret and Amiet 1971). Their warts vary in size and the skin on the top of their head is nearly smooth. Our specimens thus resemble somehow *B. camerunensis* Parker, 1936. However, their skin on flanks and back is less warty and the skin on top of the head is less smooth than in *B. camerunensis*. Furthermore, males from PFCF by far outranged males of both other species in size. Four measured, calling males had SVL of 62.5-64 mm (*B. camerunensis* males: 46-56 mm, *B. latifrons* males: 35-45 mm, Perret and Amiet 1971). We find that West African toads of this group are in need of revision (see also Rödel and Branch 2002), but because our vouchers are closest to Ahl's description of *B. togoensis,* decided to assign our specimens to this species. Toads from Mount Nimba, reported to be *B. latifrons* (Guibé and Lamotte 1958a) most likely are not *B. latifrons* but belong to the taxon dealt with herein. Another species of this group, also put into synonymy with *B. latifrons* by Tandy and Keith (1972), is *B. cristiglans* Inger and Menzies, 1961 described from the Tingi Hills in Sierra Leone.

Bufo togoensis was very abundant in the western forested parts of PFCF. Whereas most other anuran species had already stopped reproduction, we found huge aggregations of these toads along a large river (5-10 m width) near Camp 2 (> 100 males, several couples and females, one female measured 72 mm SVL). Large choruses were built up during the night with highest peaks during dusk and dawn. However, we also heard single males regularly calling during daytime. Smaller choruses and single calling males were heard at nearly all creeks within closed forest. Males were sitting along the river between 50 to 200 cm distant from each other. Calling sites were in shallow water, on stones and even on vegetation (40 cm height). At one site *B. togoensis* called in a mixed chorus with *B. maculates* Hallowell, 1854. We once observed two couples at 10 a.m, while spawning in a very shallow, slow flowing part of the river. The egg strings were glued with mud thus preventing accurate egg counts.

Calling males sometimes had nearly yellow backs but changed to other patterns within minutes when disturbed, a feature also known from other African toads (Rödel 2000a, Channing 2001) and even petropedid frogs (Rödel 2003). Color pattern was very variable. We found specimens with and without clear vertebral lines, with uniform brown backs, red heads, present or absent black spots behind neck, etc.

Bufo (?) sp. This minute, uniform brown toad was found in montane grassland in close proximity to a gallery forest at 1,350 m asl. We failed to assign this probable juvenile to a known West African toad. It is peculiar in lacking parotid glands, a feature that is clearly discernible in juveniles of all other known toads of the region (Rödel unpubl. data). Based on its general appearance (e.g. body shape, structure of feet and hands), we tentatively assign it to the genus *Bufo*, however further (i.e. adult) specimens are required to clarify the taxonomic status of this toad.

Conraua sp. *C. alleni* (Barbour and Loveridge, 1927) is the only species of that genus described from the Upper Guinea rain forest. Records under this name have been published from Liberia, Côte d'Ivoire, Guinea and Sierra Leone (Lamotte and Perret 1968). We have recorded two *Conraua* species from Côte d'Ivoire (Rödel and Branch 2002, Rödel 2003), both clearly different to *C. derooi* Hulselmans, 1972 that was described from Togo. Our records therefore comprise at least one new species. Morphologically and genetically the PFCF specimens are closer to specimens from Haute Dodo forest than to those from Mount Sangbé National Park (Rödel and Kosuch unpubl. data). However, the relationship of our records and the description of a new species require further analysis and will be dealt with in a separate publication.

Conraua frogs were present in all rivers and creeks of the region. Their calls were regularly heard during the night, but clearly peaked after dusk. Maximum calling activity was reached before dawn. Adult frogs reached up to 70 mm SVL. They were either colored brown or olive with black spots on the back. Some showed a black lateral line. Tadpoles could be observed exclusively during night. The numerous tadpoles were active in shallow, nearly stagnant parts of creeks mostly on very muddy or sandy ground. When disturbed, adults and tadpoles immediately dug themselves into the mud (compare Knoepffler 1985, Rödel and Branch 2002). Tadpoles had a reddish body, speckled with black spots. The last half of their tail including the fin was black.

Amnirana nov. sp. We found one adult *Amnirana* male (58 mm SVL) that differed considerably from other West African *Amnirana* by an extraordinary large tympanum and a skin completely covered with large spines. *Amnirana occidentalis* males have smooth skin (Rödel and Branch 2002). *A. albolabris* males have fine granular skin. Size of spines of the Simandou specimen even exceed those of the only other known *Amnirana* with a more spiny skin texture, *A. assperima* from Central Africa (Perret 1977). This new *Amnirana* species will be described elsewhere. It remains unknown if a very large female *Amnirana* (75 mm) with smooth skin and smaller tympanum, captured at the same locality, also belongs to the new species or is simply a female *A. albolabris* that was also recorded along the Simandou Range.

Ptychadena spp. In addition to *Ptychadena aequiplicata*, that is a typical forest frog ranging from Central into West Africa, we recorded two other *Ptychadena* species that we could not assign with certainty to a known species. One is similar to *P. aequiplicata* but differs in having unbroken dorsal ridges. We have genetic evidence that a cryptic *Ptychadena* species, close to *P. aequiplicata*, lives in the West African transition zone between rainforest and savanna (Vences, Kosuch and Rödel, unpubl. data). Our record may belong to this undescribed species.

All other specimens belonged to one *Ptychadena* species that showed characters somehow intermediate between *P. mascareniensis* and *P. oxyrhynchus*. They were characterized by extremely long legs like *P. oxyrhynchus*, but webbing was

much more reduced than in this species. Arrangement of dorsal ridges was similar to *P. mascareniensis*, they had an unbroken whitish dorsalateral ridge, and clear orange or green vertebral bands. Of adults, the largest specimen was a 62 mm female with a deep yellow venter. We found this species frequently in montane grassland, close to small rock pools in the grassland, in secondary forest along small creeks and in a swampy area close to the village of Banko. We heard them calling beside rock pools in the montane grasslands during night. Unfortunately the call could not be recorded. The acoustic impression of the call did not fit *P. mascareniensis* from Côte d'Ivoire (Rödel unpubl. data). We also recorded very small *Ptychadena* tadpoles in these rock pools.

Petropedetes natator Boulenger, 1905. This species, endemic to the western part of the Upper Guinea Forest zone, lives exclusively in torrent waters within forest (Lamotte and Zuber-Vogeli 1954, Lamotte 1966, Guibé and Lamotte 1958a, Böhme 1994b, Rödel 2003). In PFCF the species was not very abundant, but nearly omnipresent in all fast flowing creeks and rivers from 600 to 1,400 m asl. We never recorded *P. natator* farther from water than 3-5 m. Measuring 37-42.5 mm, males were smaller than those from Mount Sangbé National Park, Côte d'Ivoire (41-47 mm, Rödel 2003). The smallest juvenile recorded measured 18 mm. Several adults had endoparasitic mites below the skin on the ventral side of their thighs. So far these mites have been known only from other petropedetid frogs of the genus *Phrynobatrachus* (Spieler and Linsenmair 1999, Rödel 2000a).

Phrynobatrachus alticola Guibé and Lamotte, 1961. This small leaf litter species is restricted to the Upper Guinea Forests (Guibé and Lamotte 1961, Lamotte 1966). Due to its direct development, it is able to survive in forest parts without open water (Rödel and Ernst 2002a). It is characterized by its very small size and three pairs of warts on the back. Frogs from PFCF differed from Ivorian specimens in often having additional warts on flanks and back (see Rödel and Ernst 2002a, Rödel 2003). Some specimens had warts on the neck fused to a short ridge. Most specimens were nearly uniform brown, some had a clear spot on neck, others showed orange, yellow or green vertebral bands or had completely reddish backs. The ventral parts of thighs were yellow, the throat of adult males were spotted blackish to nearly black. The iris was always reddish golden. The webbing was completely reduced, toe tips were slightly enlarged. In contrast to specimens from Côte d'Ivoire many had endoparasitic mites on ventral parts of thighs.

Whereas in known sites in Côte d'Ivoire (Rödel and Ernst 2002a, Rödel and Branch 2002, Rödel 2003) *P. alticola* is easy to detect by its call but very difficult to see, in PFCF this species was by far the most abundant frog on the forest floor. While at our Ivorian sites *P. alticola* was always recorded far away from open water, a behavior that makes sense due to their direct development, the PFCF specimens were always caught close to small rivers from lowland forest to small pockets of gallery forest on the mountain ridge. The vast majority of recorded specimens were juveniles. At night individuals aggregated (often > 20 individuals) on small plants that were growing either in or close to the water, possibly to escape predation. Thus PFCF specimens showed not only differences in morphology and coloration to Ivorian records, but also differed in habitat choice. Unfortunately, we never heard any call. Therefore we cannot decide if our records belong to a species different from *P. alticola*. One possibility would be *P. tokba* that was described from Guinea by Chabanaud (1921). The senior author investigated specimens from the type series of *P. tokba* and specimens assigned to *P. alticola* by Lamotte (type species of *P. alticola*, possibly lost, A. Ohler pers. comm.). Morphologically, museum specimens of both species are indistinguishable (for specimens examined see appendix in Rödel and Ernst 2002b).

Phrynobatrachus fraterculus (Chabanaud, 1921). This leaf litter species was originally described from Macenta, Guinea. So far it is only know from south-eastern Guinea, Liberia, Sierra Leone and westernmost Côte d'Ivoire. We are not aware of a record from Guinea-Bissau (compare Frost 2002). This species is characterized by its slender body shape with a relatively pointed snout, a brown back with or without a clear vertebral stripe, a very conspicuous, well delimited deep black lateral band, reduced webbing and enlarged toe and finger tips. Adult females and most males had a yellow venter, covered with either few black points or larger black spots. The clear delimitation of the black lateral band is the best character to distinguish *P. fraterculus* from *P. gutturosus*. In addition *P. fraterculus* always had very smooth skin, whereas *P. gutturosus* skin varies between relatively smooth and very warty, but always shows at least some small warts (compare figures in Rödel 2000a). Males measured between 17 and 19 mm (Rödel and Ernst 2002b). One male (MOR T02.21) measured 18.9 mm, had a white venter without spots and two clearly marked gular folds running parallel to the mandibles. The throat of this male was grey. Nothing has been published about the biology of this species. In Taï National Park, Côte d'Ivoire we found five specimens, in the course of four years of intense field work, in close vicinity to rivers in swampy parts of primary and secondary forests (Rödel and Ernst unpubl. data). There the largest recorded specimen was a female (MOR T02.25) measuring 24.7 mm. In lowland areas of PFCF we often found several adults and juveniles, sitting on leaves or on the ground, nearly always in close vicinity to a creek. The surroundings were comprised of secondary forest or plantations (cacao, banana). The smallest juveniles, measuring 7.5-8 mm, were found at the creek bank and in a dried up pond. Thus it remains uncertain whether this species reproduces in stagnant or flowing water. Adults were all females, measuring 24-26 mm. One female from PFCF (MOR F068) and one from Taï contained many very small eggs. We assume that *P. fraterculus* reproduces like most other congeners in stagnant or slow flowing water, by depositing a floating egg film (Wager 1986, Rödel 2000a).

Phrynobatrachus cf. *maculiventris* Guibé and Lamotte,

1958. *P. maculiventris* was described from Guinea (forest pond near Doromou). Later on it was put into the synonymy of *P. fraterculus* by Guibé and Lamotte (1963). In PFCF we collected three small *Phrynobatrachus* specimens (MOR B49-51) at a very humid patch of montane grassland (1,300 m asl), close to a small gallery forest. The two larger specimens are a male (14 mm) and a female (19.1 mm). The juvenile and the male have a blackish reticulated venter, the female has a densely black spotted venter and throat. The male's throat is black and covered by a distinct gular flap. All three frogs have a warty brownish back. The female has an irregular yellow vertebral line. Webbing was completely reduced, toe and finger tips were not enlarged. The female contains a few very large eggs. These specimens differ from *P. fraterculus* by smaller size, lack of the conspicuous black lateral line, a warty back, lack of enlarged toe and finger tips, and a slightly different ventral coloration. They differ from *P. gutturosus* by their ventral coloration and general body shape. They differ from both species by fewer but much larger eggs, and thus most likely by a different reproductive behavior. In *P. gutturosus* males are also larger than in *P.* cf. *maculiventris*. It remains to be confirmed if they really are *P. maculiventris* or represent an undescribed species.

Arthroleptis/Schoutedenella spp. As mentioned in earlier papers, with the present state of knowledge it is not possible to determine species of this group with certainty (Rödel and Branch 2002, Rödel and Agyei 2003). Intraspecific variation largely overlaps with interspecific variation. Morphological criteria (even finger length in males) or color patterns (with or without hour glass pattern on back, vertebral lines and others) are unsuitable for species definitions. Reliable characters that allow at least for definition of species boundaries in a given area are advertisement calls. However, we never heard any frog of this group calling during the RAP. We found *Arthroleptis/Schoutedenella* frogs in montane grassland (partly in very dry areas), high altitude forests and primary and secondary lowlands forests. Given this wide range of habitats it is reasonable to assume that our records cover two to three different species. At least one species of the montane grasslands may be differentiated from others by a very warty skin (compare Guibé and Lamotte 1958b). Snout-vent length was moderate in all specimens encountered, ranging from 12 to 25 mm.

Hyperolius picturatus Peters, 1875. According to Schiøtz (1967, 1999) this frog, endemic to the Upper Guinea Forest zone, might comprise two species. However, frogs from various West African localities examined by us to date proved to be genetically identical (Kosuch and Rödel unpubl. data). This applies also to species from PFCF, although known populations largely differ in coloration and habitat use (compare Schiøtz 1967, 1999; Rödel and Branch 2002; Rödel 2003; Rödel and Ernst submitted). In PFCF we found *H. picturatus* to be very common in dense vegetation along fast flowing creeks, both at low and high altitude. Small choruses were still calling during the RAP period. All males had a granular belly, yellow venter, orange throat, red thighs with a longitudinal stripe on the dorsal surface, a brown back covered with black spots and broad yellow lateral bands bordered black. The few females very similarly colored but had a uniform olive back. We found tadpoles and metamorphosing juveniles (11-12 mm) just outside a gallery forest in montane grassland in a jammed up small pond with muddy ground and few water plants.

Afrixalus fulvovittatus (Cope, 1861). The taxonomic confusion in using the names *A. fulvovittatus* and *A. vittiger* (Peters, 1876), both described from Liberia, for two probably widespread West African savannah *Afrixalus* was discussed in detail by Perret (1976), Schiøtz (1999) and Rödel (2000a). We found typical *A. fulvovittatus* (sensu Perret 1976; Rödel 2000a) or Type B (sensu Schiøtz 1999) both in high altitude grasslands as well as in swampy areas of degraded forest edges and plantations close to the village of Banko. *A. fulvovittatus* was also recently recorded from Guinea by Böhme (1994b). It seems that *A. fulvovittatus* has a more western and southern distribution, living closer to forests, than *A. vittiger*, a true savannah species.

Kassina cochranae (Loveridge, 1941). Until recently it was believed that only one species of spotted *Kassina* lives in West Africa (e.g. Schiøtz 1999; Frost 2002), comprising a western (*K. c. cochranae*) and an eastern subspecies (*K. c. arboricola* Perret, 1985). However, our investigations have shown that three species were hidden in older records, thus restricting the range of *K. cochranae* to westernmost Côte d'Ivoire, southern Guinea, Liberia and Sierra Leone (Rödel et al. 2002b). The herein presented records are a confirmation of this distributional range. We recorded *K. cochranae* in PFCF in a small pond in montane grassland and in lowland farmbush close to camp 2. In the pond, already mentioned in the *H. picturatus* account, we caught a metamorphosing frog (28.5 mm) already showing the typical color pattern. Typical *Kassina* tadpoles in that pond differed from *K. schioetzi* Rödel, Grafe, Rudolf and Ernst, 2003 tadpoles by two clear stripes stretching from snout tip to fin base (Rödel 2000a, Rödel et al. 2003), thus resembling partly tadpoles of *K. lamottei* Schiøtz, 1967 (Rödel and Ernst 2001). Their tail fin was spotted red and black, the body uniform olive or spotted black. Their belly was white and turned into grayish spots towards flanks and head. Measurements of tadpoles were (body length/tail length; in mm): 13/32, 19/45 (fin height: 10 mm), 16/42, 18/44, 15/33, 17/38.

Species richness and community composition:

We recorded about 32 amphibian species during the entire RAP period (Table 4.1). A still increasing species accumulation curve indicates that additional amphibian species are still to be confirmed. We calculated 35-39 amphibian species for the area under investigation (Figure 4.1). We thus would have recorded about 82-91 % of the regional species. For several reasons we believe that this estimate is far too optimistic (see discussion).

Only 8 (24.8 %) of the recorded amphibian species

have a range that exceeds West Africa (Table 4.1). Most species are restricted to West Africa. More than half of the species (19, 61.3 %) are endemic to the Upper Guinea rainforest zone. Fifteen species were endemic to the western part of the Upper Guinean forest zone (48.4 %) and at least three species (9.7 %) might be new to science and endemic to the Simandou Range. A new toad is most likely an endemic species to the Simandou Range. By far the most species were connected to forest or at least farmbush habitats (Table 4.1).

We only found 12 reptile species or got good evidence for their presence during this RAP (Appendix 5). We therefore can make no judgments concerning this animal group. However, the snake fauna alone may be suspected to surpass 60 species, as estimated for other West African areas (compare Angel et al. 1954a, b; Barbault 1975; Rödel et al. 1995, 1997, 1999; Böhme 1999; Rödel and Mahsberg 2000; Ernst and Rödel 2002; Ineich 2002; Branch and Rödel 2003).

DISCUSSION

Although the time of the RAP, during the beginning of the dry season, was not favorable for assessing amphibians, we recorded 32 amphibian species and calculated that 3-7 amphibian species were not recorded by us. However, we assume that we only recorded a much smaller portion of the regional species than calculated. Due to the fact that only a few species were regularly calling, it is likely that we overlooked species that had already stopped reproducing. In addition, we investigated only two comparatively small areas within the whole mountain range. During our investigations we continuously added new species to our list (Fig. 4.1). In the course of only two weeks of fieldwork we managed to record several rare or new species. Given the huge areas of the Simandou Range not yet investigated, including the driest (savanna areas in the north) and most humid (forests in the south) parts it is reasonable to assume that many more amphibian species will occur in the whole Pic de Fon area. Several species might be restricted to the grasslands and/or endemic to the Simandou Range.

So far as known the most diverse West African sites with respect to amphibians are Mount Nimba with 58 amphibian species (Guibé and Lamotte 1958a, b, c, 1963) and Taï National Park, Côte d'Ivoire with 56 amphibian species (Rödel 2000b). Taï National Park is exceptional in having a vast number of highly specialized and endemic forest amphibians. Mount Nimba likewise gives home to many endemic species and harbors high amphibian diversity due to its huge variety of different habitats. Like Mount Nimba the Simandou Range is special due to its unique faunal composition. The amphibian community of PFCF comprises a wide range: from local endemics to species with a nearly Africa-wide range; from species that are restricted to primary rain forests to savannah specialists. We recorded all species known from the Upper Guinea Forests that are dependant on flowing forest creeks (*Bufo togoensis, Conraua* sp., *Petropedetes natator,*

Table 4.1. Amphibian species recorded in the Pic de Fon classified forest, including numbers of record sites, habitat preference and African distribution of the species. S = savannah, FB = farmbush (degraded forest and farmland), F = forest, A = Africa (occurs also outside West Africa), WA = West Africa (Senegal to eastern Nigeria), UG = Upper Guinea (forest zone West of the Dahomey Gap), WUG = western Upper Guinea (forest zone of western Côte d'Ivoire and west of that area), E = endemic to Pic de Fon, * = record based on descriptions from villagers (CITES species), ** = records possibly comprise several species, cf. & sp. = determination needs confirmation or new species are involved.

Species	Habitat #	S	FB	F	A	WA	UG	WUG	E
PIPIDAE									
Silurana tropicalis	9		x	x		x			
BUFONIDAE									
Bufo maculatus	6, 7, 13	x	x		x				
Bufo regularis	2	x			x				
Bufo superciliaris *	6			x	x				
Bufo togoensis	6, 7, 7-9			x			x		
Bufo (?) sp.	2	x							x
RANIDAE									
Hoplobatrachus occipitalis	2, 5, 6, 8, 9, 11, 13	x	x		x				
Conraua sp.	2, 6, 7, 12			x				x	
Amnirana albolabris	6, 7, 8, 13		x	x	x				
Amnirana sp.	12			x					x
Ptychadena cf. *aequiplicata*	7			x	x				
Ptychadena sp. aff. *aequiplicata*	6-9			x					?
Ptychadena sp.	2, 5, 6, 9, 6-9, 11		x					x	
PETROPEDETIDAE									
Petropedetes natator	1, 2, 6, 7			x				x	
Phrynobatrachus accraensis	2, 5, 7, 9, 7-9, 11, 13	x	x			x			
Phrynobatrachus alleni	7		x			x			
Phrynobatrachus alticola	1, 2, 5, 6, 7, 8, 9, 14		x	x			x		
Phrynobatrachus fraterculus	5, 7, 8, 9, 7-9		x	x				x	
Phrynobatrachus liberiensis	6, 7, 8			x			x		
Phrynobatrachus phyllophilus	6, 8			x				x	
Phrynobatrachus cf. *maculiventris*	2	x							x
ARTHROLEPTIDAE									
Arthroleptis sp. **	1, 2, 6, 6-9	x	x	x				x	
Cardioglossa leucomystax	6, 7, 6-9			x	x				
ASTYLOSTERNIDAE									
Astylosternus occidentalis	13			x				x	
HYPEROLIIDAE									
Leptopelis hyloides	2, 6, 7, 9, 6-9, 7-9, 13		x	x		x			
Hyperolius chlorosteus	6, 7, 7-9, 13			x				x	
Hyperolius fusciventris	5, 9, 6-9	x	x			x			
Hyperolius concolor	9, 6-9, 7-9	x					x		
Hyperolius picturatus	2, 7, 11	x	x			x			
Afrixalus dorsalis	5, 9, 6-9	x	x	x					
Afrixalus fulvovittatus	2, 5	x						x	
Kassina cochranae	2, 7-9	x						x	

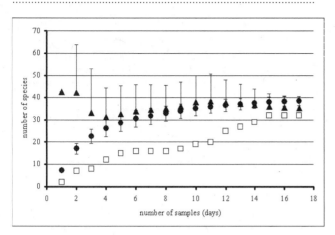

Figure 4.1. Estimated species richness and amphibian species accumulation curve (white squares) of the Simandou Range/ Pic de Fon area, Guinea. Given are mean values and standard deviation (sd) of 500 random runs of the daily species lists. Black triangles: Chao2 estimator; only positive sd indicated (35.3 ± 4.2 species); black circles: Jack-knife 1 estimator (38.5 ± 2 species).

Astylosternus occidentalis, Cardioglossa leucomystax, Hyperolius chlorosteus) and many farmbush species (e.g. *Phrynobatrachus accraensis, Hyperolius fusciventris, H. concolor*). We assume that we especially failed to record true savannah species that reproduce earlier in the season (e.g. several *Ptychadena* and *Phrynobatrachus* species, *Kassina senegalensis, Hyperolius nitidulus, H. lamottei, Leptopelis viridis, Phrynomantis microps*), forest species that reproduce in stagnant waters (e.g. *Phrynobatrachus gutturosus, P. plicatus, Hyperolius zonatus, Chriromantis rufescens*) and fossorial species that are generally hard to assess (*Geotrypetes seraphini, Hemisus* spp.). We therefore estimate that a total of 50-60 amphibian species might be expected for the whole Simandou Range.

The Simandou Range faces many threats possibly affecting the regional herpetofauna. Bushfires and logging may affect savannah and forest species through most savannah species seem to be adapted to fire events or at least have a reproductive potential that is sufficient to replace the part of the population that dies during fires (see Rödel 2000a for a review).

Logging activities, both for commercial purposes as well as for small scale farming, in contrast, have a serious effect on the local and regional species pool. True forest species seem to be unable to prevail in degraded forests, most probably due to an altered microclimate that offers conditions that they cannot cope with (Ernst and Rödel unpubl. data). The surroundings of our second site offered large parts of primary lowland rainforest. The secondary forests and plantations adjacent to these forests gave habitat to an amphibian fauna that naturally occurs in the transition zone from forest to savannah and in natural disturbances within the forest zone, e.g. tree fall gaps (Schiøtz 1967, Rödel and Branch 2003). The whole forest fauna, including the farmbush spe-

cies, speaks in favor of a high potential for conservation, given that the remaining forests are managed in a sustainable way and large parts of still pristine forest remain untouched.

Our first site was mainly made up by a montane grassland type typical for West African higher altitudes (Guibé and Lamotte 1958a, 1963). Amphibians in this type of habitat are normally not very diverse, however comprise very unique and very range-restricted species. For example the highest parts of Mount Nimba are nearly exclusively inhabited by only two amphibian species, the Nimba endemics *Nimbaphrynoides occidentalis* (Angel, 1943) and *Schoutedenella crusculum* (Angel, 1950). A similar situation might exist on the Simandou Range where we found a dwarf toad and an *Arthroleptis/Schoutedenella* species with warty skin, both most likely new to science. This fauna might be highly threatened by any alteration of the higher altitude habitats, including proposed mining activities. The distribution of these species along the mountain ridge and their respective population sizes urgently need more investigation to more accurately predict possible effects of habitat degradation.

Four reptile and one amphibian species are protected by international laws and therefore are of conservation concern. Especially the presence of the African dwarf crocodile (*Osteolaemus tetraspis*) and the largest African toad (*Bufo superciliaris*) need more investigation.

Should current mining exploration activities proceed to more detailed feasibility studies, the hydrological and microclimatological impacts of lowering the mountain ridge, in relation to amphibians and reptiles, should be considered, as part of a detailed Environmental and Social Impact Assessment (ESIA). Potentially at risk from an alteration of the mountain ridge would be all forest species, especially those that are dependant on permanent watercourses. The mountains most likely act as a barrier to south-east winds, resulting in higher precipitation amounts than usual for areas of this latitude. The lowering of the ridge might result in lower precipitation rates. In addition the whole mountain possibly acts as a water reservoir. A reduction of this reservoir in connection with less rainfall would almost certainly result in an altered forest type and a southern move of the savannah. Watercourses, now to 80 % permanent, would likewise become temporary. Beside the effects on the amphibian and other aquatic and forest faunas, reduction in rainfall could also have significant consequences for the local human population.

CONSERVATION RECOMMENDATIONS

For more precise judgments on the potential impact of intensified human activity in the area, more in-depth fieldwork during the rainy season, including other parts of the Pic de Fon classified forest, is highly recommended. Such research would not only provide a more complete picture of the local species pool, but could also provide a better understanding of population sizes and distribution patterns of those species that are of prominent conservation concern, especially local endemics.

REFERENCES

Angel, F., J. Guibe and M. Lamotte. 1954a. La réserve naturelle intégrale du mont Nimba. Fasicule II. XXXI. Lézards. Memoirs de l'Institute fondamental d'Afrique noire, sér. A, 40: 371-379.

Angel, F., J. Guibe, M. Lamotte and R. Roy. 1954b. La réserve naturelle intégrale du mont Nimba. Fasicule II. XXXII. Serpentes. Bulletin de l'Institute fondamental d'Afrique noire, sér. A, 40: 381-402.

Bakarr, M., B. Bailey, D. Byler, R. Ham, S. Olivieri and M. Omland (eds.). 2001. From the forest to the sea: biodiversity connections from Guinea to Togo, Conservation Priority-Setting Workshop, December 1999. Washington, DC: Conservation International. 78 pp.

Barbault, R. 1975. Les peuplements de lézards des savane de Lamto (Côte d'Ivoire). Annales de l'Université d'Abidjan, sér. E, 8: 147-221.

Böhme, W. 1994a. Frösche und Skinke aus dem Regenwaldgebiet Südost–Guineas, Westafrika. I. Einleitung; Pipidae, Arthroleptidae, Bufonidae. Herpetofauna, 16 (92): 11–19.

Böhme, W. 1994b. Frösche und Skinke aus dem Regenwaldgebiet Südost-Guineas, Westafrika. II. Ranidae, Hyperoliidae, Scincidae; faunistisch-ökologische Bewertung. Herpetofauna, 16 (93): 6–16.

Böhme, W. 1999. Diversity of a snake community in a Guinean rain forest (Reptilia, Serpentes). In: Rheinwald, G. (ed.). Isolated Vertebrate Communities in the Tropics. pp. 69-78. Proceedings of the 4th International Symposium, Bonner zoologische Monographien, 46.

Branch, W.R. and M.-O. Rödel. 2003. Herpetological survey of the Haute Dodo and Cavally forests, western Ivory Coast, Part II: Trapping results and reptiles. Salamandra, Rheinbach 39(1): 21-38.

Chabanaud, P. 1919. Énumération des batraciens non encore étudiés de l'Afrique Occidentale Française, appartenant à la collection du muséum. Bull. Mus. Natl. Hist. Nat., 1919: 456–557.

Channing, A. 2001. Amphibians of Central and Southern Africa. New York: Cornell University Press. 470 pp.

Colwell, R.K. 1994-2000. EstimateS, statistical estimation of species richness and shared species from samples. Version 6.0b1. Online. Available: http://viceroy.eeb.uconn.edu/estimates.

Ernst, R. and M.-O. Rödel. 2002. A new Atheris species (Serpentes: Viperidae) from Taï National Park, Ivory Coast. Herpetological Journal, 12: 55-61.

Frost, D.R. 2002. Amphibian Species of the World: an online reference. V2.21. Online. Available: http://research.amnh.org/herpetology/amphibia/index.html. 15 July 2002.

Guibé, J. and M. Lamotte. 1958a. La réserve naturelle intégrale du Mont Nimba. XII. Batraciens (sauf Arthroleptis, Phrynobatrachus et Hyperolius). Memoirs de l'Institute fondamental d'Afrique noire, sér. A., 53: 241–273.

Guibé, J. and M. Lamotte. 1958b. Morphologie et reproduction par développement direct d'un anoure du Mont Nimba, Arthroleptis crusculum ANGEL. Bull. Mus. Natl. Hist. Nat., 2e Sér., 30: 125–133.

Guibé, J. and M. Lamotte. 1958c. Une espèce nouvelle de batracien du Mont Nimba (Guinée Française) appartenant au genre Phrynobatrachus: Ph. maculiventris n. sp. Bull. Mus. Natl. Hist. Nat., 2e Sér., 30: 255–257.

Guibé, J. and M. Lamotte. 1961. Deux espèces nouvelles de batraciens de l'ouest africain appartenant au genre Phrynobatrachus: Ph. guineensis n. sp. et. Ph. alticola n. sp. Bull. Mus. Natl. Hist. Nat., 2e Sér., 33: 571–576.

Guibé, J. and M. Lamotte. 1963. La réserve naturelle intégrale du Mont Nimba. XXVIII. Batraciens du genre Phrynobatrachus. Memoirs de l'Institute fondamental d'Afrique noire, sér. A., 66: 601–627.

Heyer, W.R., M.A. Donnelly, R.W. McDiarmid, L.-A.C. Hayek and M.S. Foster. 1993. Measuring and monitoring biological diversity, standard methods for amphibians. Washington, DC: Smithsonian Institution Press. 364 pp.

Hughes, B. 1988. Herpetology in Ghana (West Africa). British Herpetological Society Bulletin, 25: 29-38.

Ineich, I. 2002. Diversite specifique des reptiles du Mont Nimba. unpublished manuscript.

Knoepffler, L.-P. 1985. Le comportement fouisseur de Conraua grassipes (Amphibien anoure) et son mode de chasse. Biologia Gabonica, 1: 239-245.

Lamotte, M. 1966. Types de répartition géographique de quelques batraciens dans l'Ouest Africain. Bulletin de l'Institute fondamental d'Afrique noire, Sér A, 28: 1140–1148.

Lamotte, M. and J.-L. Perret. 1968. Révision du genre Conraua Nieden. Bulletin de l'Institute fondamental d'Afrique noire, sér. A., 30: 1603–1644.

Lamotte, M. and M. Zuber-Vogeli. 1954. Contribution à l'étude des batraciens de l'Ouest Africain III. Le développement larvaires de deux espèces rhéophiles, Astylosternus diadematus et Petropedetes natator. Bull. Inst. fond. Afr. noire, Sér. A, 16: 1222–1233.

Perret, J.-L. 1976. Identité de quelques Afrixalus (Amphibia, Salientia, Hyperoliidae). Bull. Soc. neuchat. Sci. Nat., Sér. 3, 99: 19–28.

Perret, J.-L. 1977. Les Hylarana (Amphibiens, Ranidés) du Cameroun. Rev. Suisse Zool., 84: 841–868.

Perret, J.-L. 1985. Description of Kassina arboricola n. sp. (Amphibia, Hyperoliidae) from the Ivory Coast and Ghana. S. Afr. J. Sci., 81: 196–199.

Perret, J.-L. and J.-L. Amiet. 1971. Remarques sur les Bufo (Amphibiens Anoures) du Cameroun. Ann. Fac. Sci. Cameroun, 5: 47-55.

Rödel, M.-O. 2000a. Herpetofauna of West Africa, Vol. I: Amphibians of the West African savanna. Frankfurt/M. (Edition Chimaira), 335 pp.

Rödel, M.-O. 2000b. Les communautés d'amphibiens dans le Parc National de Taï, Côte d'Ivoire. Les anoures comme bio-indicateurs de l'état des habitats. *In:* Girardin, O., I. Koné and Y. Tano (eds.). Etat des recherches en cours dans le Parc National de Taï (PNT), Sempervira, Rapport de Centre Suisse de la Recherche Scientifique, Abidjan, 9: 108-113.

Rödel, M.-O. 2003. The amphibians of Mont Sangbé National Park, Ivory Coast. Salamandra, 39: 91-110.

Rödel, M.-O. and A.C. Agyei. 2003. Amphibians of the Togo-Volta highlands, eastern Ghana. Salamandra. Rheinbach, 39(3/4): 207-234.

Rödel, M.-O. and W.R. Branch. 2002. Herpetological survey of the Haute Dodo and Cavally forests, western Ivory Coast, Part I: Amphibians. Salamandra, 38: 245-268.

Rödel, M.-O. and R. Ernst. 2000. *Bufo taiensis* n. sp., eine neue Kröte aus dem Taï-Nationalpark, Elfenbeinküste. Herpetofauna, Weinstadt, 22 (125): 9-16.

Rödel, M.-O. and R. Ernst. 2001. Description of the tadpole of *Kassina lamottei* Schiøtz, 1967. Journal of Herpetology, 35: 678-681.

Rödel, M.-O. and R. Ernst. 2002a. A new reproductive mode for the genus *Phrynobatrachus*: *Phrynobatrachus alticola* has nonfeeding, nonhatching tadpoles. Journal of Herpetology, 36: 121-125.

Rödel, M.-O. and R. Ernst. 2002b. A new *Phrynobatrachus* species from the Upper Guinean rain forest, West Africa, including a description of a new reproductive mode for the genus. Journal of Herpetology, 36 (4): 561-571.

Rödel, M.-O. and D. Mahsberg. 2000. Vorläufige Liste der Schlangen des Tai-Nationalparks/ Elfenbeinküste und angrenzender Gebiete. Salamandra, Rheinbach, 36 (1): 25-38.

Rödel, M.-O., K. Grabow, C. Böckheler and D. Mahsberg. 1995. Die Schlangen des Comoé-Nationalparks, Elfenbeinküste (Reptilia: Squamata: Serpentes). Stuttgarter Beiträge zur Naturkunde, Stuttgart, Serie A, Nr. 528: 1-18.

Rödel, M.-O., K. Grabow, J. Hallermann and C. Böckheler. 1997. Die Echsen des Comoé-Nationalparks, Elfenbeinküste. Salamandra, 33: 225-240.

Rödel, M.-O., K. Kouadio and D. Mahsberg. 1999. Die Schlangenfauna des Comoé-Nationalparks, Elfenbeinküste: Ergänzungen und Ausblick. Salamandra, Rheinbach, 35 (3): 165-180.

Rödel, M.-O., D. Krätz and R. Ernst. 2002a. The tadpole of *Ptychadena aequiplicata* (WERNER, 1898) with the description of a new reproductive mode for the genus (Amphibia, Anura, Ranidae). Alytes, 20: 1-12.

Rödel, M.-O., T.U. Grafe, V.H.W. Rudolf and R. Ernst. 2002b. A review of West African spotted *Kassina*, including a description of *Kassina schioetzi* sp. nov. (Amphibia: Anura: Hyperoliidae). Copeia, Lawrence, 2002 (3): 800-814.

Rödel, M.-O., J. Kosuch, M. Veith and R. Ernst. 2003. First record of the genus *Acanthixalus* Laurent, 1944 from the Upper Guinean rain forest, West Africa, including the description of a new species. Journal of Herpetology, 37: 43-52.

Schiøtz, A. 1967. The treefrogs (Rhacophoridae) of West Africa. Spolia zoologica Musei Haunienses, 25: 1–346.

Schiøtz, A. 1999. Treefrogs of Africa. Frankfurt/M. (Edition Chimaira), 350 pp.

Spieler, M. and K.E. Linsenmair. 1999. The larval mite *Endotrombicula pillersi* (Acarina: Trombiculidae) as a species-specific parasite of a West African savannah frog (*Phrynobatrachus francisci*). Am. Midl. Nat., 142: 152-161.

Tandy, M. and R. Keith. 1972. *Bufo* of Africa. *In:* Blair, W.F. (ed.). Evolution in the genus *Bufo*. Austin & London: University of Texas Press. 119–170.

Wager, V.A. 1986. Frogs of South Africa, their fascinating life stories. Delta, Craighall. 183 pp.

Chapter 5

A rapid survey of the birds of the Forêt Classée du Pic de Fon, Guinea

Ron Demey and Hugo J. Rainey

SUMMARY

During 11 days of field work in the Forêt Classée du Pic, 233 bird species were recorded, 131 in the highland area and 198 in the lowlands. Of these, eight are of conservation concern (two in the highlands and seven in the lowlands), the most important being Sierra Leone Prinia *Schistolais leontica* which has a very limited distribution in the West African highlands. We found six of the 15 restricted-range species that make up the Upper Guinea Forests Endemic Bird Area. A substantial component of the forest-restricted species in the country was found, as 104 of the 153 species of the Guinea-Congo Forests biome now known to occur in Guinea were recorded. Seven species were recorded in Guinea for the first time. Considering the high conservation value of this forest reserve, it is recommended that further surveys be conducted in order to produce a more complete species list. We also make recommendations for managing the forest for the conservation of its biodiversity.

INTRODUCTION

Birds have been proven to be useful as indicators of the biological diversity of a site. Their taxonomy and global geographical distribution are relatively well documented in comparison to other taxa (ICBP 1992), which facilitates their identification and permits rapid analysis of the results of an ornithological study. The conservation status of most species having been reasonably well assessed (BirdLife International 2000), the results and conclusions of such a study can be reviewed and implemented productively. Birds are also among the most charismatic species, which can aid the presentation of conservation recommendations to policy makers and stakeholders.

Previous studies of some of the remaining forests in West Africa have shown that they are of considerable importance for the survival of the birds of the Upper Guinea Forests (e.g. Allport et al. 1989; Gartshore et al. 1995; Demey and Rainey *in press*). However, the avifaunas of the majority of the rapidly decreasing forests in West Africa remain inadequately known. The avifauna of Guinea is comparatively poorly known overall and large areas of the country have yet to be surveyed (Robertson 2001a). Only two areas in south-east Guinea have been surveyed to date (Wilson 1990; Halleux 1994). The Guinean side of Mount Nimba, a UNESCO World Heritage Site, has barely been studied (Brosset 1984; Robertson 2001a), although the Liberian side of the mountain has been the subject of a long term avifaunal survey (Colston and Curry-Lindahl 1986).

We carried out 11 days of field work, four days in the highlands at altitudes of 900 to 1,550 m (08°31.52'N, 08°54.21'W) (27 - 30 November) and seven days in the lowlands at 550-800 m (08°31.29'N, 08°56.12'W) (1 - 7 December).

For the purposes of standardization, we have followed the nomenclature, taxonomy and sequence of Borrow and Demey (2001).

METHODS

The principal method used during this study consisted of observing birds by walking slowly along mining tracks (primitive roads created for mineral exploration activities) and forest trails. Notes were taken on both visual observations and bird vocalizations. Tape-recordings of unknown vocalizations and those of rare species were made for later analysis and deposition in sound archives. Attempts were made to visit as many habitats as possible, particularly those which appeared likely to hold threatened or poorly known species. However, the difficulty of access to most parts of the forest due to the steep and rocky slopes, the dense vegetation and the scarcity or absence of paths meant that we were not able to cover large areas. The main habitats at the highland site consisted of grassland on steep, rugged hills and, in ravines, along small streams and in valleys and depressions, forest bordered by bushes and scrub. At the lowland site the majority of the work was carried out in forest on steep hillsides and, lower down, in forest on level ground. A number of streams ran through the site, bordered by both forest and lower vegetation. Derived savannah covered the area between the lower limit of the forest and the boundary of the reserve. Some areas at the forest edge had been cleared for coffee and cassava and within the forest on level ground there were banana plantations and some cocoa. Field work was carried out from just before dawn (around 06:30) until 13:00, and from 15:00 until sunset (around 18:30). Some further work was carried out at night to obtain records of owls and nightjars and recordings of their vocalizations.

Mist-netting was carried out on six days. The nets used had four shelves and were up to 12 m long and 3 m high when set up. The primary aim was to obtain records of secretive and silent species which can pass unnoticed during general observations. In the highlands nets were set on two days for a total of 13.1 100-meter net hours. The nets were set in grassland, forest edge, forest and across a small forest stream. In the lowlands, nets were set over four days for a total of 19.9 100-meter net hours. They were set in primary forest (canopy 30-40 m), in low forest (canopy 15-20 m), across two small forest streams and at forest edge. One 12 m canopy net was set at a height of 20 m at the forest edge for 17.5 hours.

For each field day a list was compiled of all species recorded. Numbers of individuals or flocks were noted, as were any evidence of breeding (e.g. the presence of juveniles) and basic information on the habitat in which the birds were observed. This enabled us to produce indices of abundance for each species based on encounter rate (numbers of days on which a species was encountered and number of individuals and flocks involved). Comparisons can thus be made between the two sites and other sites in the region. The definitions of the abundance ratings are given in Appendix 6. The conservation threat status of each species was taken into account in our analysis, as was the presence of biome-restricted species and species confined to the Upper Guinea Forests

Endemic Bird Area. Threat status is defined in Table 5.1. A number of species have restricted ranges and are found only within the forests of Upper Guinea. Similarly, many species are only found within forests of the Guinea-Congo biome and the presence of a high proportion of these provides an indication of the quality of the site. For full definitions see Stattersfield et al. (1998), BirdLife International (2000) and Fishpool and Evans (2001).

RESULTS

We recorded a total of 233 bird species during our survey; these are listed in Appendix 6, along with an estimate of relative abundance at each of the two sites and the indication of observed breeding evidence. Also indicated are threat status, endemism to the Upper Guinea Forest block, membership of biome-restricted assemblages and habitat. 104 species restricted to the Guinea-Congo Forest biome were recorded in total.

Table 5.1. Species of global conservation concern recorded during the survey.

Species		Threat Status
Ceratogymna elata	Yellow-casqued Hornbill	nt
Lobotos lobatus	Western Wattled Cuckoo-shrike	VU
Phyllastrephus baumanni	Baumann's Greenbul	DD
Criniger olivaceus	Yellow-bearded Greenbul	VU
Bathmocercus cerviniventris	Black-headed Rufous Warbler	nt
Schistolais leontica	Sierra Leone Prinia	VU
Illadopsis rufescens	Rufous-winged Illadopsis	nt
Lamprotornis iris	Emerald Starling	DD

Threat status (BirdLife International 2000):
VU = Vulnerable: species facing a high risk of extinction in the medium-term future.
DD = Data Deficient: species for which there is inadequate information to make an assessment of its risk of extinction.
nt = Near Threatened: species coming very close to qualifying as Vulnerable.

Highlands

In total, 131 species were recorded at this site (see Appendix 6), of which two are of global conservation concern (Bird-Life International 2000; Table 5.1): Sierra Leone Prinia *Schistolais leontica* is classified as Vulnerable and Emerald Starling *Lamprotornis iris* as Data Deficient. Of the 153 species of the Guinea-Congo Forests biome occurring in the country (Robertson 2001a, this study), 49 (32 %) were recorded in the highlands.

In addition, a number of species were observed which are rare and poorly known in either Guinea or West Africa. These include Long-billed Pipit *Anthus similis*, Grey-winged Robin

Chat *Cossypha polioptera*, Dusky Crested Flycatcher *Elminia nigromitrata*, Preuss's Golden-backed Weaver *Ploceus preussi*, Green Twinspot *Mandingoa nitidula* and Dybowski's Twinspot *Euschistospiza dybowskii*. The distinctive subspecies *henrici* of Rufous-naped Lark *Mirafra africana*, which is restricted to a few highland areas in Sierra Leone and Mount Nimba, was fairly common in the grasslands. Palearctic migrants from Europe were common in both forest and grassland.

Lowlands

At this site 198 species were recorded (see Appendix 6), seven of which are of global conservation concern (BirdLife International 2000; Table 5.1). Two of these are classified as Vulnerable (Western Wattled Cuckoo-shrike *Lobotos lobatus* and Yellow-bearded Greenbul *Criniger olivaceus*), three are Near Threatened (Yellow-casqued Hornbill *Ceratogymna elata*, Black-headed Rufous Warbler *Bathmocercus cerviniventris* and Rufous-winged Illadopsis *Illadopsis rufescens*), while two are considered Data Deficient (Baumann's Greenbul *Phyllastrephus baumanni* and Emerald Starling *Lamprotornis iris*). Of the 153 species of the Guinea-Congo Forests biome occurring in the country (Robertson 2001a; this study), 98 (64 %) were found in the lowlands.

In addition, a number of rare and poorly known species were observed. These include a forest ibis *Bostrychia rara/olivacea*, Blue-headed Bee-eater *Merops muelleri*, Lyre-tailed Honeyguide *Melichneutes robustus*, Grey-winged Robin Chat *Cossypha polioptera*, Yellow-bellied Wattle-eye *Dyaphorophyia concreta*, Dusky Tit *Parus funereus* and Green Twinspot *Mandingoa nitidula*.

At both sites, mist-netting was successful in its aims of finding inconspicuous species that would not otherwise have been observed (see Appendix 7). In total, 181 individuals of 56 species were caught, amongst which were two species of conservation concern: Black-headed Rufous Warbler *Bathmocercus cerviniventris* and Sierra Leone Prinia *Schistolais leontica*. Other significant species included Lemon Dove *Aplopelia larvata* and Forest Scrub Robin *Cercotrichas leucosticta* (see below). Our capture rate of 5.5 birds per 100-meter net hours is much higher than previously reported rates from other West African forests (e.g. Allport et al. 1989; Gartshore et al. 1995).

Six of the 15 restricted-range species, i.e. species which have a global breeding range of less than 50,000 km², that make up the Upper Guinea Forests Endemic Bird Area (Fishpool and Evans 2001, Stattersfield et al. 1998) were found in the reserve. These are Western Wattled Cuckoo-shrike *Lobotos lobatus*, Yellow-bearded Greenbul *Criniger olivaceus*, Black-headed Rufous Warbler *Bathmocercus cerviniventris*, Sierra Leone Prinia *Schistolais leontica*, Sharpe's Apalis *Apalis sharpii* and Rufous-winged Illadopsis *Illadopsis rufescens*. Seven species were recorded for the first time in Guinea (see Table 5.2). Sixteen raptor species were found and these include some of the largest vultures and eagles in Africa including African White-backed Vulture *Gyps africanus*, Crowned Eagle *Stephanoaetus coronatus* and Martial Eagle *Polemaetus bellicosus*.

Table 5.2. Species recorded in Guinea for the first time

Bubo poensis	Fraser's Eagle Owl
Neafrapus cassini	Cassin's Spinetail
Indicator willcocksi	Willcocks's Honeyguide
Smithornis capensis	African Broadbill
Phyllastrephus baumanni	Baumann's Greenbul
Cercotrichas leucosticta	Forest Scrub Robin
Vidua camerunensis	Cameroon Indigobird

NOTES ON SPECIFIC SPECIES

See Table 5.1 for explanation of threat status.

Species of conservation concern

Ceratogymna elata Yellow-casqued Hornbill (nt). This species was seen on four days in the lowlands. One flock numbered 14 birds; the other observations were of two individuals on each occasion. This species was previously known from three other sites in Guinea (Robertson 2001a). Surprisingly few hornbills of any species were observed throughout the whole survey.

Lobotos lobatus Western Wattled Cuckoo-shrike (VU). One male was seen at c. 750 m altitude east of the lowland camp. It was feeding in the canopy and sub-canopy at a height of 15-25 m in primary forest near a clearing. This species was previously only known in Guinea from Ziama Forest Reserve (Halleux 1994, Robertson 2001a).

Phyllastrephus baumanni Baumann's Greenbul (DD). We found one pair at the forest edge at c. 570 m, where the forest graded into derived savannah. What was presumed to be the same pair was found the next day in a mixed species flock at c. 100 m from this site. This is the first observation of this species in Guinea. Until recently there were very few reliable records of it anywhere within its range (Fishpool 2000).

Criniger olivaceus Yellow-bearded Greenbul (VU). A pair was seen in a mixed species flock in primary forest at c. 570 m. It was feeding in the mid-storey at a height of 10-15 m. This species was previously known only from Ziama and Diécké Forest Reserves (Robertson 2001a).

Bathmocercus cerviniventris Black-headed Rufous Warbler (nt). We found four singing males and a duetting pair at 550-580 m in dense vegetation near small streams. We also trapped a female close to the lowland camp, in the territory of one of the four males. Previously this extremely local species was known only from Ziama (Halleux 1994).

Schistolais leontica Sierra Leone Prinia (VU). At least two and probably three pairs were found at 1,300-1,350 m. One pair was found on large bushes inside gallery forest and a second was seen regularly in low dense bushes at the edge of another patch of gallery forest. Two individuals were trapped on the other side of this forest patch. The face and underparts of one individual (a juvenile?) were slightly paler than those of the other bird. This species is now known from four sites throughout its restricted range including only one other site in Guinea: Mount Nimba (Fishpool 2001; Okoni-Williams et al. 2001; Robertson 2001a,b; L.D.C. Fishpool pers. comm.).

Illadopsis rufescens Rufous-winged Illadopsis (nt). This species was singing in primary forest at c. 570, 650 and 1,200 m.

Lamprotornis iris Emerald Starling (DD). We observed a flock of ten in wooded savannah at the Rio Tinto camp and one individual in similar habitat at the lowland site.

Other rare or poorly known species
(Status in West Africa from Borrow and Demey 2001)

Bostrychia rara/olivacea Ibis sp. One individual was seen by I. Herbinger (pers. comm.) at a forest stream at 900 m near the lowland camp. It was not specifically identified but either of the two possible species, which are both rare and local in West Africa, would be new for Guinea.

Aplopelia larvata Lemon Dove. One was trapped at 570 m in forest with a low canopy (c. 15 m). This species is known from few sites in West Africa, from Sierra Leone to western Côte d'Ivoire, where it is rare to uncommon (Demey and Rainey in prep.).

Merops muelleri Blue-headed Bee-eater. A pair was seen at the forest edge at 570 m and one individual of this pair was subsequently trapped at the same site. This is a scarce and local species in West Africa.

Melichneutes robustus Lyre-tailed Honeyguide. One was heard displaying daily above gallery forest at 560 m. Previously reported only from Ziama Forest Reserve (Halleux 1994).

Mirafra africana Rufous-naped Lark. Seen frequently in the grassland and on the mining tracks above 1,300 m. The subspecies in question, *henrici*, is known from only a few highland areas in Upper Guinea. A displaying male was repeatedly seen jumping vertically c. 80 cm off the ground whilst rattling its wings. This behaviour does not appear to have been described before (Colston and Curry-Lindahl 1986, Keith at al. 1992, R. Safford pers. comm.).

Sheppardia cyornithopsis Lowland Akalat. Three individuals were trapped in forest at 570 and 1,350 m and one was seen at the former site in a mixed species flock. This is a relatively high encounter rate for a species that is rarely recorded in West Africa.

Elminia nigromitrata Dusky Crested Flycatcher. Recorded almost daily in both highland and lowland areas. Three were also trapped in the lowlands. This species is generally uncommon in West Africa, but appears to be quite common in south-east Guinea (Halleux 1994, this study).

Dyaphorophyia concreta Yellow-bellied Wattle-eye. Encountered in both highlands and lowlands. Two individuals were caught in the lowlands. This species is generally rare to scarce in West Africa.

Parus funereus Dusky Tit. Three together seen foraging in the canopy of large trees at the edge of a clearing at c. 570 m. Generally rare to scarce in West Africa and previously only reported in Guinea from Ziama Forest Reserve (Halleux 1994).

Ploceus preussi Preuss's Golden-backed Weaver. One seen in forest at 1,350 m. Previously reported from Ziama and Diécké Forest Reserves (Wilson 1990, Halleux 1994). Generally scarce and local in West Africa.

Mandingoa nitidula Green Twinspot. One trapped at 1,350 m and another at 580 m at the forest edge. Uncommon to rare in West Africa.

Euschistospiza dybowskii Dybowski's Twinspot. Three males trapped at 1,350 m at the forest edge. Uncommon to scarce and local in West Africa.

Species new for Guinea
Bubo poensis Fraser's Eagle Owl. One adult was identified from a rattling call recorded on 6 December in high primary forest at the site where *Criniger olivaceus* was observed. Its identity was confirmed from a comparison of the recording with those of Chappuis (2000). This species is uncommon to fairly common throughout the Lower Guinea forest block.

Neafrapus cassini Cassin's Spinetail. One to three individuals seen on almost every day at the lowland site. This is a locally not uncommon resident with irregular distribution in the rainforest zone in West Africa.

Indicator willcocksi Willcocks's Honeyguide. One heard singing in gallery forest at 560 m and another seen at the edge of a forest clearing at 570 m. This is a rare to uncommon forest resident in West Africa.

Smithornis capensis African Broadbill. Five individuals seen and heard displaying in forest at the lowland site. This is a generally scarce to rare resident with a patchy distribution in West Africa.

Phyllastrephus baumanni Baumann's Greenbul. See above.

Cercotrichas leucosticta Forest Scrub Robin. A pair trapped in low forest (canopy c. 15 m) at 570 m. This very shy species is a scarce forest resident, occurring in West Africa from Sierra Leone to Ghana.

Vidua camerunensis Cameroon Indigobird. Four male indigobirds seen at 1,500 m at forest edge and two other males in savannah at 560 m were identified as this species on the basis of the white bill, pale purple legs and brown flight feathers. Two of its potential host species were found in the reserve: Dybowski's Twinspot *Euschistospiza dybowskii* and Blue-billed Firefinch *Lagonosticta rubricata*. The status and distribution of this species are imperfectly known due to its similarity with other indigobirds.

DISCUSSION

The total number of 233 species recorded at both sites is high in view of the short study period and in comparison with the total number of c. 625 species recorded for the whole of Guinea (Robertson 2001a). This gives an indication of the high quality of the reserve. By comparison, 287 and 141 species have been recorded in Ziama and Diécké Forest Reserves respectively, the two other sites in south-east Guinea that have also been studied (Robertson 2001a). After many years of intensive study, 383 species have been found on the Liberian side of Mount Nimba and its surrounding forests (Colston and Curry-Lindahl 1986). Mount Nimba is similar in many respects to Pic de Fon because of its altitudinal range and varied habitats and gives an indication of the potential of the Pic de Fon. The 104 species restricted to the Guinea-Congo Forests biome that we recorded in the reserve constitute 68 % of the species of this biome known from Guinea – a high proportion.

The presence of Sierra Leone Prinia in the highlands was the most important finding of the study. This species is currently only known from three other sites in the world and one of these, Mount Nimba, is also being prospected for mineral deposits on the Guinean side, while habitat on the Liberian side of the mountain has already been partially destroyed by irresponsible mining practices. Sierra Leone Prinia seems to be only found in dense vegetation at forest edge and along streams above 700 m (Borrow and Demey 2001, this study). It could be particularly vulnerable to alteration of the higher altitude habitats in Pic de Fon. Although the threat status of this species is currently given as Vulnerable because it has an inferred adult population of less than 10,000 individuals which is declining and fragmented, it may perhaps be reclassified as "Endangered" (a more severe threat status) as it is likely to have an area of occupancy of less than 500 km^2 and is known from fewer than six locations (BirdLife International 2000). Even though some mountains in the Upper Guinea region where Sierra Leone Prinia might occur have not yet been surveyed for birds, a review of the conservation status of this species appears desirable, given the threat to its known sites.

Very few hornbills, in terms of numbers and species, were encountered in the reserve. Most forest hornbill species have been recorded at the other forest reserves in south-east Guinea (Robertson 2001a). As hornbills are known to be capable of long-distance movements to obtain food (Kemp 1995, HJR pers. obs.) this absence may be a function of the local phenology of the fruiting trees. At this time of year in Côte d'Ivoire, most hornbills are in the southern forests; hornbills may be absent from the Pic de Fon as they are further south in this season, perhaps in the Liberian forests. Hunting of large mammal species was found to be quite intensive in the forest reserve (See Chapter 8, Barrie and Kante 2004). Discussions with our guide A. Camara indicated that birds were also targeted by hunters and this may partially explain the absence or low density of larger species such as guineafowl and hornbills. Yellow-headed Picathartes *Picathartes gymnocephalus* and Nimba Flycatcher *Melaenornis annamarulae*, two species of conservation concern, were not recorded during the survey. However, given the habitats found in the reserve and the presence of these species at similar sites nearby (Robertson 2001a), they may reasonably be expected to occur.

The large number of species of conservation concern recorded during such a short survey indicates the quality and potential of the forest (see Table 5.1). This site qualifies as an Important Bird Area on the basis of the number of threatened species (category A1) and presence of large numbers of both restricted-range (A2) and biome-restricted (A3) species (Fishpool and Evans 2001).

CONSERVATION RECOMMENDATIONS

Considering the high conservation value of the forest, the following recommendations are made:

1. Consideration should be given to keeping a substantial portion of the forest reserve intact over the full altitudinal range, from the peaks to the lowlands, as some species may require forests over a range of altitudes during different seasons. There are few sites in West Africa that have intact forest stretching over such a wide range of altitudes. Hornbills are known to require large areas of forest to obtain sufficient food and as the phenologies of different trees are likely to vary over altitude they are likely to require forest over the full altitudinal range to ensure their survival in the reserve.

2. The potential effects of habitat alteration should be considered prior to initiating any program of mineral exploitation. If mining exploration proceeds to feasibility studies, more detailed studies of the impacts on birds should be undertaken as part of a detailed Environmental And Social Impact Assessment (ESIA). Of particular concern are potential impacts on the Sierra Leone Prinia population, and on species that are restricted to montane grassland such as the Martial Eagle and Rufous-naped Lark.

3. Clearance of the forest for agriculture and setting of fires is widespread. Attempts should be made to rehabilitate those areas that have already been cleared. Almost all forest species are adversely affected by clearance or alteration of forest and this includes most of the threatened species that we found during this survey.

4. Hunting in the reserve is rife and should be reduced. We found that guineafowl and hornbills, amongst the largest birds in West Africa, were scarce and suggest that

these populations will decline further if hunting pressure is not reduced.

5. Further surveys should be conducted at different sites throughout the reserve. These surveys should be carried out in different seasons (e.g. at the start of the rainy season, in April-May, when more species, particularly cuckoos, owls, and certain warblers, may be breeding and are vocally most active) and in habitats that we were unable to cover, such as the lowland rivers. The latter habitat could be of importance to two species of conservation concern, Rufous Fishing Owl *Scotopelia ussheri* and White-crested Tiger Heron *Tigriornis leucolopha*. The former species is very poorly known and should be a target for future surveys. Different species of migratory birds from Europe may be encountered and altitudinal migration of birds within the forest may also be observed.

6. Monitoring programs should be put in place to assess the value of particular management regimes to wildlife. If funding is available, local hunters who know best the forests and their wildlife should be employed through wildlife monitoring programs. Their employment may also reduce hunting rates.

REFERENCES

Allport, G.A., M. Ausden, P.V. Hayman, P. Robertson and P. Wood. 1989. The conservation of the birds of Gola Forest, Sierra Leone. Study Report No. 38 International Council for Bird Preservation. Cambridge, UK.

Barrie, A. and S. Kanté. 2004. A rapid survey of the large mammals of the Forêt Clasée du Pic de Fon, Guinea. *In*: McCullough, J. (ed.) 2004. A Rapid Biological Assessment of the Forêt Classée du Pic de Fon, Simandou Range, South-eastern Republic of Guinea. RAP Bulletin of Biological Assessment 35. Conservation International, Washington, DC.

BirdLife International. 2000. Threatened Birds of the World. Lynx Edicions and BirdLife International. Barcelona, Spain and Cambridge, UK.

Borrow, N. and R. Demey. 2001. Birds of Western Africa. Christopher Helm. London.

Brosset, A. 1984. Oiseaux migrateurs européens hivernant dans la partie guinéenne du Mont Nimba. Alauda 52: 81–101.

Chappuis, C. 2000. Oiseaux d'Afrique - African Bird Sounds, Vol. 2. 11 CDs with booklet. Société d'Etudes Ornithologiques de France, Paris, France and The British Library National Sound Archive, London, UK.

Colston, P.R. and K. Curry-Lindahl. 1986. The birds of Mount Nimba, Liberia. British Museum (Natural History). London.

Demey, R. and H.J. Rainey. *In press*. The birds of Haute Dodo and Cavally Forest Reserves. RAP report. Washington, DC: Conservation International.

Demey, R. and H.J. Rainey. In prep. An annotated checklist of the birds of Mont Sangbé National Park, Ivory Coast.

Fishpool, L.D.C. 2000. A review of the status, distribution and habitat of Baumann's Greenbul *Phyllastrephus baumanni*. Bulletin of the British Ornithologists' Club 120: 213-229.

Fishpool, L.D.C. 2001. Côte d'Ivoire. *In*: Fishpool, L.D.C. and M.I. Evans (eds.). Important Bird Areas in Africa and Associated Islands: Priority sites for conservation. Newbury and Cambridge, UK: Pisces Publications and BirdLife International. Pp. 219-232.

Fishpool, L.D.C. and M.I. Evans (eds.). 2001. Important Bird Areas in Africa and Associated Islands: Priority sites for conservation. BirdLife Conservation Series No. 11. Pisces Publications and BirdLife International. Newbury and Cambridge, UK.

Gartshore, M.E., P.D. Taylor and I.S. Francis. 1995. Forest Birds in Côte d'Ivoire. BirdLife International Study Report No. 58. BirdLife International. Cambridge, UK.

Halleux, D. 1994. Annotated bird list of Macenta Prefecture, Guinea. Malimbus 17: 85-90.

ICBP. 1992. Putting biodiversity on the map: priority areas for global conservation. International Council for Bird Preservation. Cambridge, UK.

Keith, S., E.K. Urban and C.H. Fry (eds.). 1992. The Birds of Africa, Vol. 4. London, UK: Academic Press.

Kemp, A. 1995. The Hornbills. Oxford University Press. Oxford, UK.

Okoni-Williams, A.D., H.S. Thompson, P. Wood, A.P. Koroma and P. Robertson. 2001. Sierra Leone. *In*: Fishpool, L.D.C. and M.I. Evans (eds.). Important Bird Areas in Africa and Associated Islands: Priority sites for conservation. Newbury and Cambridge, UK: Pisces Publications and BirdLife International. Pp. 769-778.

Robertson, P. 2001a. Guinea. *In*: Fishpool, L.D.C. and M.I. Evans (eds.). Important Bird Areas in Africa and Associated Islands: Priority sites for conservation. Newbury and Cambridge, UK: Pisces Publications and BirdLife International. Pp. 391-402.

Robertson, P. 2001b. Liberia. *In*: Fishpool, L.D.C. and M.I. Evans (eds.). Important Bird Areas in Africa and Associated Islands: Priority sites for conservation. Newbury and Cambridge, UK: Pisces Publications and BirdLife International. Pp. 473-480.

Stattersfield, A.J., M.J. Crosby, A.J. Long and D.C. Wege. 1998. Endemic Bird Areas of the World: Priorities for Biodiversity Conservation. BirdLife International. Cambridge, UK.

Wilson, R. 1990. Annotated bird lists for the Forêts Classées de Diécké and Ziama and their immediate environs. Unpublished report commissioned by IUCN.

Chapter 6

Rapid survey of bats (Chiroptera) in the Forêt Classée du Pic de Fon, Guinea

Jakob Fahr and Njikoha Ebigbo

SUMMARY

We report on the results of a bat survey of the Pic de Fon, Simandou Range, southeastern Guinea. We document a speciose bat assemblage characterized by forest species, including bats such as *Epomops buettikoferi*, *Rhinolophus guineensis* and *Hipposideros jonesi* that are endemic to Upper Guinea or West Africa. The sympatric occurrence of three species of *Kerivoula* is noteworthy, with both *K. cuprosa* and *K. phalaena* representing first records for Guinea. Moreover, three individuals of Welwitsch's Mouse-eared bat, *Myotis welwitschii*, were captured during the survey. This is the first record for West Africa and represents a range extension of 4,400 km to the northwest from the nearest known localities. We review the distribution of this species in Africa and conclude that the species shows a paramontane distribution pattern. Seven species or 33.3 % out of the total of 21 species are registered in the latest IUCN Red List: one species as "Vulnerable" (*Epomops buettikoferi*) and six species as "Near Threatened" (*Rhinolophus alcyone*, *R. guineensis*, *Hipposideros jonesi*, *H. fuliginosus*, *Kerivoula cuprosa*, *Miniopterus schreibersii*). Our results of the RAP survey as well as the occurrence of bat species that are endemic to the Upper Guinea Highlands highlight the outstanding regional importance of the montane habitats of West Africa in general, and of the Simandou Range in particular, for the conservation of bats in Africa.

INTRODUCTION

In West Africa, the forest cover has been reduced to 14.4 % of the original forest extent and the remaining forests continue to be degraded or lost. The forests in the Upper Guinea biogeographical region are ranked as biodiversity hotspots of continental and global importance (Myers et al. 2000, Brooks et al. 2001). Consequently, Conservation International held a Priority-Setting Workshop in 1999 to select key areas for conservation in this region based on the consensus of participating experts (Bakarr et al. 2001). The workshop resulted in the delimitation of priority areas for biodiversity protection (ranked from high to exceptionally high), including regions that are currently unprotected. Moreover, the workshop highlighted the need for imminent biological surveys to provide baseline scientific information that is lacking for many areas. It was decided that poorly known priority areas should be surveyed as soon as possible through Rapid Assessment Programs (RAP) to gather the necessary scientific basis for their protection and to build capacity in the region. The distribution of most bat species in Africa is still insufficiently known. This holds especially true for West and Central Africa where few surveys have been carried out to date. In addition, many habitats are vanishing at alarming rates and with them basic but invaluable knowledge of species and their distributions. These base-line data are essential for sound conservation and management plans of target areas and in consequence the effective protection of bats.

Bats (Chiroptera) are one of the ecologically most diverse groups of vertebrates and regarded as keystone species for the maintenance of ecosystem functions. This is due to their high species richness, large biomass and trophic diversity, especially in the tropics where they play important roles as seed dispersers and pollinators as well as predators of insects. They are the only mammalian group that has evolved active flight, opening a wide range of ecological niches. Bats form the second largest mammalian order numbering more than 1,100 species worldwide (N. Simmons *in press*), only exceeded in species richness by rodents. However, on the local scale they are usually the most species-rich group of mammals in tropical communities. Although the body size of bats is comparatively small, they are characterized by remarkably long life spans and extremely low reproductive rates. These life history traits, together with narrow and specialized species-specific habitat requirements, predispose bats to a multitude of actual and potential threats such as: degradation or loss of suitable habitats, alteration or destruction of day roosts or direct exploitation therein, and secondary poisoning through accumulation of toxic substances in the food chain. Globally, almost a quarter (23.8 %) of all bats are threatened (either "Critically Endangered", "Endangered" or "Vulnerable") and 12 species have gone extinct (Hutson et al. 2001, IUCN 2002). Their sensitive reaction to environmental disturbances renders bats an ideal indicator group to assess and monitor the scale and magnitude of habitat alterations.

The rapidly vanishing Upper Guinean forests support a unique and highly diverse bat fauna and the inclusion of bats in the present RAP of the Pic de Fon, Simandou Range, is warranted because of their outstanding ecological importance as keystone species and their high value as very sensitive indicator species.

MATERIAL AND METHODS

Study Site

The Forêt Classée du Pic de Fon, covering a surface of 25,600 ha, was created in 1953 and surrounds the Pic de Fon. The highlands are dominated by montane grasslands. The western slopes show larger tracts of primary forests along ravines, which are connected to the forests found at the base of the mountain range. On the eastern side, only small pockets of gallery forest exist along the streams exiting the Simandou Range. The adjacent plains are dominated by bush-tree savannah with few scattered patches of gallery forest. Annual precipitation varies between 1,700 and 2,000 mm, with a long dry season lasting five to six months (November to April). For a detailed description of the study sites' vegetation, climate and geology, see Chapter 2 (botany) and Chapter 1 (introduction) of this report. Two localities were sampled: the first close to the summit of the Pic de Fon (highland; 8°32'N, 8°54'W, 1350 m), the second on its western slope near the village of Banko (lowland; 8°31'N, 8°56'W, 600 m). The two sites are about 3.6 km apart. The geographical positions of the study sites were determined with a hand-held GPS receiver (Garmin GPS 12).

Sampling and Data Analysis

The fieldwork was accomplished by NE. Mist nets were set for seven nights from 28 November to 6 December 2002 (no nets set on 1-2 December). Two mist net sizes were employed: 12 x 2.8 m and 6 x 2.8 m (16 mm mesh; 2 x 70 d netting) with 5 and 4 shelves, respectively. Most nets were set on poles near ground level or slightly elevated above the surrounding vegetation (herb layer) in forest, savannah and edge habitat. At the lowland site, nets were set both in fairly undisturbed forest and within stands of banana plants. In addition to mist nets near ground level, an elevated net system was set up at the lowland site. A pulley & rope system was used to raise three stacked 12 m nets with the top of the nets reaching a height of 12 m. This net system was installed at the edge of undisturbed *Parinari excelsa* forest. Additionally, a three-bank harp trap (3.4 m^2 capture area) was set at both sites on 10 nights from 27 November to 6 December. The mist nets were usually open from 18:00 to 0:00 hours, while the harp trap was set from 18:00 to 06:00 hours. The total capture effort was 194.8 mist net hours or 16.2 mist net nights (calculated as 12 m-net equivalents) and 111.5 harp trap hours or 9.3 harp trap nights (Tab. 6.1).

A randomized species accumulation curve was calculated with EstimateS 6.0b1 (Colwell 2000). Although there are several statistical methods for estimating the total species number from samples (e.g. Colwell 2000), we were unable to use these methods because sampling effort and methods varied greatly between nights and sites.

Echolocation calls of hand-held bats were recorded with a Pettersson D 240x bat detector (both in 10x and 20x time expansion mode) and transferred to a Sony Walkman Professional WM-D6C. These calls were later analyzed on a standard PC with Avisoft-SASLab Pro 4.2 and used in identifying problematic taxa. In the families Rhinolophidae and Hipposideridae, the constant frequency (CF) component of the echolocation calls is highly species-specific. We measured the CF-frequency (maximum amplitude, second harmonic) with spectrograms (Hanning window, FFT length 512). Twenty-four voucher specimens were sacrificed to check identifications and to document the bat fauna of the area. These specimens are currently deposited in the collection of JF (Dept. of Experimental Ecology, Univ. of Ulm). Taxo-

Table 6.1. Capture effort for each method employed.

Method	Capture Effort	
	Net Hours[1]	Net Nights[2]
12 m mist nets, ground level	144.1	12.0
12 m mist nets, canopy level	28.4	2.4
6 m mist nets, ground level	44.7	3.7
Harp trap	111.5	9.3

[1]one net hour = one mist net/harp trap set for 1 h
[2]one net night = one mist net/harp trap set for 12 h

nomic notes on selected species captured during the present study have been reported elsewhere (Fahr and Ebigbo 2003). We did not compare our data with those of Konstantinov et al. (2000) because their results indicate substantial misidentifications.

RESULTS

The bat fauna of the Simandou Range

In total, we captured 276 bats comprising 21 species, 14 genera, and 6 families (Tables 6.2 and 6.3). For the nets close to ground level, overall capture success was 0.38 bats per 12 m mist net hour (b/nh), for Megachiroptera 0.30 b/nh and for Microchiroptera 0.08 b/nh. For the elevated nets, overall capture success was 2.57 b/nh, for Megachiroptera 2.50 b/nh and for Microchiroptera 0.07 b/nh. The overall capture success for the harp trap was 1.26 bats per trap hour (b/th), for Megachiroptera 0.03 b/th and for Microchiroptera 1.23 b/th. The respective sampling methods yielded capture success with 63 individuals and 12 species in mist nets near ground level, 73 individuals and 7 species in canopy nets, and 140 individuals and 12 species in the harp trap. In terms of sampling effectiveness, we caught seven species or 33.3 % of the total of 21 species exclusively with the harp trap. We sampled five species or 23.8 % of the species total only with the mist nets near ground level and with the canopy nets one species or 4.8 % of the species total. Overall, 13 species or 61.9 % of the 21 species were sampled with one method only (Table 6.2). A single bat cave was found during the survey, with *Rousettus aegyptiacus* (Egyptian fruit bat) as the only species inhabiting the cave (8°32'N, 8°53'W). We estimated the colony to comprise at least a few thousand individuals at the time of the visit.

At the lowland site, we recorded a total of 17 species, including 11 species that were found only at this site. The highland site had a lower total of 10 species, with four species exclusively recorded at this site. Out of the 21 species that were sampled at the Simandou Range, only six species (28.6 %) were recorded both at the highland and lowland sites, demonstrating a comparatively high species turnover between the two sampling areas. It should be noted that the capture effort was very similar for the harp trap at the highland and lowland site with, respectively, 59.2 and 52.3 trap hours. The capture effort for mist nets near ground level was also fairly similar with 89.4 net hours at the highland and 77.0 net hours at the lowland site, respectively (calculated as 12 m-net equivalents). In contrast, we did not employ the canopy system at the highland site, which had a high capture success at the lowland site.

The bat assemblage is comprised of both forest and savannah species although species primarily associated with forest habitat ("F" or "F(S)" category in Table 6.3) dominate the sample (13 species or 61.9 %). Species that are either predominantly found in savanna habitats ("S" or "(F)S") or both in forest and savanna ("FS") constitute smaller frac-

Table 6.2. Capture effort (nh: net/trap hour), capture success (number of individuals; bats per net/trap hour), species coverage (Total: all species; Excl.: exclusively captured with one method) broken down by sampling method.

	Effort [nh]	N° of Indiv.	Bats / nh	Species Total	Species Excl.
Ground nets	166.4	63	0.38	12	5
Canopy nets	28.4	73	2.57	7	1
Harp trap	111.5	140	1.26	12	7
Total	—	276	—	21	13

tions (4 species or 19.0 % for each category, Table 6.3). The assemblage is also characterized by a high number of species that are (partially) dependent on caves as day roosts (*Rousettus aegyptiacus*, *Rhinolophus* spp., *Hipposideros* spp., *Miniopterus schreibersii*). The species accumulation curve rises steeply and no asymptotic plateau is discernible (Figure 6.1), illustrating that the inventory of the bat fauna of the Simandou Range is far from being complete. *Myotis welwitschii*, *Kerivoula phalaena* and *K. cuprosa* represent first records for Guinea.

Myotis welwitschii

Surprisingly, we caught three individuals of *M. welwitschii*, a bat previously known only from East and Southern Africa. This bat is distinguished from all other *Myotis* species occurring in Africa by the distinct coloration of its membranes and pelage. All three individuals (2 males, 1 female) were captured in the upper shelves of 12 m nets (2-3 m above ground level) in edge habitat during the night of 29/30 November at 19:09, 21:27, and 00:10 hours, respectively. The netting sites (highland) were located in a patch of montane grassland

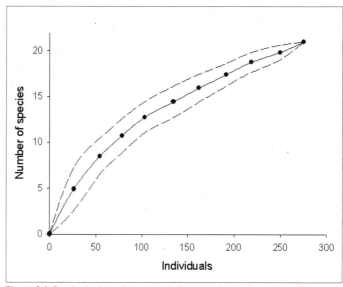

Figure 6.1. Randomized species accumulation curve for the bats sampled during the RAP survey (solid line and dots: randomization, 100 runs; dashed lines: ± 1 sd.).

situated between two stretches of forest running down the western slope. Both forest patches converge below into a single large forest. The males were caught at the edge of a line of trees in the grassland, running parallel to the wider forest. The female was captured just 150 m downhill from where the males were caught, directly at the edge of the wider forest about 200 m above the point of forest convergence.

This exciting record prompted us to review the documented distribution of *M. welwitschii* based on literature, museum specimens, and personal communications (Fahr and Ebigbo 2003). *Myotis welwitschii* is now documented for the following 15 countries: Angola, Burundi, Congo (K.), Ethiopia, Guinea, Kenya, Malawi, Mozambique, Rwanda, South Africa, Sudan, Tanzania, Uganda, Zambia and Zimbabwe. Most of the localities are situated in highlands or mountains (see distribution map, p. 20). The altitudinal range spans from 55 to 2,311 m a.s.l., with a median of 1,209 m (n=61). The analysis of occupied ecoregions revealed a pronounced preference for woodlands and forest-savanna mosaic (Fahr and Ebigbo 2003). To date, no record is known from the closed forest zone of the central Congo Basin.

DISCUSSION

Myotis welwitschii

The record from Guinea corresponds to a range extension of 4,400 km from the nearest known localities along the Albertine Rift (Fahr and Ebigbo 2003). This exceptional range extension emphasizes our very fragmentary knowledge of bat distributions in Africa and highlights the need for additional surveys, especially in regions of high conservation relevance as outlined by the Priority-Setting Workshop (Bakarr et al. 2001).

The distribution pattern of *M. welwitschii* is conspicuously tied to highlands or mountains (see Map p. 20; $\leq$ 500 m alt.: n=9; $\leq$ 1,000 m alt.: n=12; $\leq$ 1,500 m alt.: n=22; > 2,000 m alt.: n=11; > 2,000 m alt.: n=7). This corresponds to a "paramontane" distribution pattern, i.e. species that are tied to mountainous regions although their altitudinal range covers both lower and higher elevations. When viewed in detail, many of the records are situated in areas of topographical complexity, i.e. on the slopes of mountains or highland plateaus. The reason for the paramontane distribution of *M. welwitschii* is difficult to explain. In other bat species, this pattern is probably linked to a high dependency on rock caves as day roosts (Fahr et al. 2002b). The density of caves is usually much higher in mountainous regions, possibly leading in some species to the observed paramontane pattern. However, *M. welwitschii* has been reported to roost unconcealed in vegetation, e.g. on bushes and trees, and has sometimes been found in furled banana leaves (see Fahr and Ebigbo 2003).

Although this species is widely distributed in Africa, it is rarely recorded in bat inventories. The number of known localities almost equals the number of specimens recorded. The reason for this may be either that the abundance of *M. welwitschii* is generally very low or that it is difficult to capture with current sampling methods. Our record from Guinea is remarkable in this respect as we caught three individuals during the same night.

THE BAT FAUNA OF THE SIMANDOU RANGE

Sampling efficiency

Including the results of our study, 63 species are documented for Guinea (Table 6.4; Barnett and Prangley 1997, Ziegler et al. 2002, Fahr et al. 2002b, Fahr and Ebigbo 2003). We recorded 21 species for the Simandou Range, i.e. one-third of the known bat fauna of the country, during the very brief study period of the RAP survey. However, the real species number for both the country and the Simandou Range is certainly much higher than currently known. On the local level, i.e. the Simandou Range, this is indicated by the steeply rising species accumulation curve, which shows that many more species can be expected through continued surveys. Secondly, a comparison of species lists from areas nearby - such as Mount Nimba, Ziama Range and Wonegizi Range - allows us to identify many missing species from those that have only been found in one or several of these areas although the habitat types are very similar (see below, Table 6.5). In this context, it needs to be stressed that species lists for even the best sampled sites such as Mount Nimba are incomplete because a) none of these sites have been surveyed for an extended period of time and spanning all seasons, b) neither harp traps nor canopy nets have been used in previous surveys although these methods are essential to achieve complete inventories (see Fahr 2001 and results of this study), and c) previous sampling has been largely opportunistic, i.e. major habitat types have not been sampled in an exhaustive and consistent way.

Seven species of fruit bats (Pteropodidae) were caught, making up the largest proportion of the species total at the family level (33.3 %, Table 6.4). Vespertilionidae, with 23.8 % of the species total, were the second most speciose family in our survey. No members of the families Rhinopomatidae (Mouse-tailed Bats), Emballonuridae (Sheath-tailed Bats) and Megadermatidae (False Vampire Bats) were recorded during this survey. In West Africa, 120 species have been recorded to date between Mauritania and Nigeria (Fahr et al. 2001), with the Vespertilionidae and Molossidae constituting the main proportion (38.3 % and 17.5 %, respectively; Table 6.4). While most fruit bat species (Pteropodidae) are easily sampled with mist nets near ground level, many species of the families Nycteridae, Rhinolophidae and Hipposideridae detect and avoid mist nets but are captured with harp traps or in their day roosts. A disproportionately high number of additional species are to be expected in the families Vespertilionidae and Molossidae, which are mostly high-flying bats and notoriously difficult to sample. Our sampling effort with the canopy system was too low and too limited in

Table 6.3. Species recorded during the RAP survey of the Simandou Range. GN: ground net; CN: canopy net; HT: harp trap (bold: caught with one method only). Lowland and Highland refers to records at the two different sampling localities (see Material & Methods). Habitat: coarse assignment to preferred habitat types (F: forest; S: savannas & woodlands; in brackets: marginally including the respective habitat type).

Species	N° Ind.	GN	CN	HT	Lowland	Highland	Habitat	
Pteropodidae								
Epomophorus g. gambianus (OGILBY, 1835)	1	x			+			S
Micropteropus pusillus (PETERS, 1868)	19	x	x	x	+	+		S
Epomops buettikoferi (MATSCHIE, 1899)	3	x	x		+		F	(S)
Megaloglossus woermanni PAGENSTECHER, 1885	8	x	x		+		F	
Myonycteris torquata (DOBSON, 1878)	3		x	x	+		F	(S)
Rousettus aegyptiacus unicolor (GRAY, 1870)	87	x	x		+	+	F	S
Eidolon h. helvum (KERR, 1792)	2	x				+	F	S
Nycteridae								
Nycteris grandis PETERS, 1865	1	x			+		F	
Rhinolophidae								
Rhinolophus alcyone TEMMINCK, 1853	1		x		+		F	(S)
Rhinolophus guineensis EISENTRAUT, 1960	7	x		x	+	+	F	
Rhinolophus l. landeri MARTIN, 1838	1		x		+			S
Hipposideridae								
Hipposideros jonesi HAYMAN, 1947	1	x				+	F	S
Hipposideros cf. caffer (SUNDEVALL, 1846)	5		x		+		F	S
Hipposideros cf. ruber (NOACK, 1893)	5		x		+	+	F	(S)
Hipposideros fuliginosus (TEMMINCK, 1853)	116	x	x	x	+	+	F	
Vespertilionidae								
Myotis welwitschii (GRAY, 1866)	3	x				+	(F)	S
Kerivoula lanosa muscilla THOMAS, 1906	1		x		+		F	(S)
Kerivoula phalaena THOMAS, 1912	1		x			+	F	
Kerivoula cuprosa THOMAS, 1912	5		x		+		F	
Miniopterus schreibersii villiersi AELLEN, 1956	4	x		x	+		F	(S)
Molossidae								
Mops spurrelli (DOLLMAN, 1911)	1		x		+		F	

time to capture a sufficient proportion of the potential bat fauna. Consequently, further surveys should include canopy nets as an important sampling component.

Based on the distribution of bats in West Africa and Upper Guinea, and taking into account their respective habitat preferences, we infer that the species total to be expected is minimally 50 species and probably close to 65 species. In particular, we expect additional species in the following families: Pteropodidae – 4 species, Emballonuridae – 2 species, Nycteridae – 6 species, Rhinolophidae – 3 species, Hipposideridae – 3 species, Vespertilionidae – 17 species, and Molossidae – 9 species.

Guinea is potentially a "megadiverse" country within West Africa in general and within Upper Guinea in particular. However, the bat fauna is one of the least known and many additional species are likely to occur in this country. This is exemplified by our captures of *Myotis welwitschii*, *Kerivoula phalaena* and *K. cuprosa*, all of which represent first records for Guinea. Better-explored countries such as Côte d'Ivoire and Ghana are currently known to harbor, respectively, 87 and 82 species although even these totals are by no means complete (J. Fahr in prep.). The special biogeographic position of Guinea includes most of the major biomes of West Africa, e.g. mangrove, coastal savanna, lowland forest, Guinea savanna and Sudan savanna. Additionally, and as detailed below, the montane forests and grasslands of Upper Guinea support species that are restricted or endemic to these habitats. We therefore expect the total number of species for Guinea to be close to 100 species.

On the other hand, our combination of sampling methods (mist nets both near ground and canopy level, harp traps, roost searches) shows that a substantial portion of the local assemblage can be surveyed within a rather short study period. Although the current total of 21 species is certainly much lower than the real species number, it should be kept in mind that we managed to record several rare species although the sampling period was a mere 10 days. Numerous studies emphasize that much longer sampling is necessary to achieve a (near) complete inventory. This survey was conducted at the onset of the dry season. Results from long-term studies in the Comoé and Taï National Parks, Côte d'Ivoire, show that the highest number of species is caught at the beginning of the wet season while capture success during the dry season is much reduced (J. Fahr in prep.). Moreover, the present study surveyed only two comparatively small areas and focused on the forest habitats and montane grasslands in close vicinity while the extensive bush-tree savannas were not sampled. It is recommended that follow-up studies should be conducted at the onset of the wet season (April-June), including habitat types and areas that were not covered by the present study. Additionally, more time and effort needs to be invested in searching for and sampling the numerous caves that are likely to occur in the study site.

We conclude, for the above reasons, that the bat fauna of the Simandou Range is under-sampled and recommend additional surveys using a combination of methods including mist nets near ground and canopy level, harp traps, roost searches (especially cave surveys) and acoustic monitoring (recording of echolocation calls).

Species composition and habitat preferences
The bat fauna of the Simandou Range is largely dominated by forest species, many of which reach their northern distribution limit in this region, e.g. *Megaloglossus woermanni*, *Nycteris grandis*, *Hipposideros fuliginosus*, *Kerivoula* spp. and *Mops spurrelli*. The sympatric occurrence of three species of *Kerivoula*, a genus of very specialized forest bats, is noteworthy since these bats have been rarely recorded in previous surveys. *K. cuprosa* is known from only seven localities world-wide. Several species preferring open habitats were likewise recorded, reflecting the mosaic structure of the study site, which comprises lowland forests, gallery and ravine forests, bush-tree savannahs and montane grasslands. The mixture of habitat types of the area is caused by the elevational gradient and the geographic position of the Simandou Range, being situated in the ecotone of the Guinea Zone where northern savannahs and southern forest formations form a complex mosaic. As many bat species are known to be rather specialized for specific habitat types, we expect that the habitat diversity of the Simandou Range supports an exceptionally species-rich bat assemblage, stressing the uniqueness of this habitat.

The bat fauna is also characterized by the high proportion of species that depend strictly (five species or 23.8 % of the species total: *Rousettus aegyptiacus*, *Rhinolophus landeri*, *R. guineensis*, *Hipposideros jonesi*, *Miniopterus schreibersii*) or partially (likewise five species: *Nycteris grandis*, *Rhinolophus alcyone*, *Hipposideros fuliginosus*, *H.* cf. *caffer*, *H.* cf. *ruber*) on caves. The occurrence of these species at a given site depends to a great extent on the availability of suitable roosting caves, which leads in general to patchy distribution areas.

Endemic species and conservation recommendations
The Simandou Range is bordered by three areas that received an overall "high" priority rank during the Priority-Setting Workshop, i.e. the Fon-Tibé Area (B7), the Wonegizi-Ziama Range (B8) and the Mount Béro-Tetini Area (B9). For mammals only, the Simandou Range is included in an

Table 6.4. Comparison of the number of species in bat families known from Pic de Fon, Guinea, and West Africa (Senegal to Nigeria).

| | No. of Species -- % of Total Species No. | | | | | |
	Pic de Fon		Guinea		West Africa	
Pteropodidae	7	33.3 %	11	17.5 %	14	11.7 %
Rhinopomatidae	0	—	0	—	2	1.7 %
Emballonuridae	0	—	2	3.2 %	5	4.2 %
Nycteridae	1	4.8 %	7	11.1 %	9	7.5 %
Megadermatidae	0	—	1	1.6 %	1	0.8 %
Rhinolophidae	3	14.3 %	9	14.3 %	11	9.2 %
Hipposideridae	4	19.0 %	9	14.3 %	11	9.2 %
Vespertilionidae	5	23.8 %	18	28.6 %	46	38.3 %
Molossidae	1	4.8 %	6	9.5 %	21	17.5 %
Total	21	100.0 %	63	100.0 %	120	100.0 %

Table 6.5. Bat species with isolated occurrences in or endemic to the Upper Guinea (UG) region. #: species dependent on caves.

Species	PdF	Nimba	Ziama	Wonegizi	Distribution
Emballonuridae					
Coleura afra (PETERS, 1852)					isolated occurrence in UG, #
Rhinolophidae					
Rhinolophus guineensis	+	+	+	+	sp. endemic to UG, #
Rhinolophus maclaudi POUSARGUES, 1897					sp. endemic to UG, #
Rhinolophus ziama FAHR et al. 2002			+	+	sp. endemic to UG, #
Rhinolophus denti knorri EISENTRAUT, 1960					ssp. endemic to UG, #
Rhinolophus hillorum KOOPMAN, 1989		+		+	isolated occurrence in UG, #
Rhinolophus simulator alticolus SANBORN, 1936		+		+	isolated occurrence in UG, #
Hipposideridae					
Hipposideros jonesi	+		+	+	sp. endemic to UG, #
Hipposideros marisae AELLEN, 1954		+	+	+	sp. endemic to UG, #
Hipposideros lamottei BROSSET, 1985		+			sp. endemic to UG, #
Vespertilionidae					
Myotis welwitschii	+				isolated occurrence in UG
Myotis tricolor (TEMMINCK, 1832)					isolated occurrence in UG, #
Hypsugo crassulus bellieri DE VREE, 1972				+	ssp. endemic to UG
Miniopterus schreibersii villiersi	+		+		isolated occurrence in UG, #
Miniopterus i. inflatus THOMAS, 1903		+		+	isolated occurrence in UG, #
Molossidae					
Chaerephon b. bemmeleni (JENTINK, 1879)		+			isolated occurrence in UG

Pdf: Pic de Fon, this study; Nimba: Mount Nimba Range, Guinea & Liberia (Aellen 1963, Hill 1982, Brosset 1985, J. Fahr unpubl. data); Ziama: Ziama Range, Guinea (Roche 1972, Fahr et al. 2002b, J. Fahr unpubl. data); Wonegizi: Wonegizi Mountain Range, Liberia (Koopman 1989, Koopman et al. 1995, Fahr et al. 2002b).

area that was ranked to be of "extremely high" conservation priority (Bakarr et al. 2001). These areas are part of the Upper Guinea Highlands, which additionally include Mount Nimba, Diécké Forest, Loma-Tingi Mountains and the Fouta Djallon. A striking feature of the bat fauna in the Upper Guinea Highlands is the disproportionately high number of species that are either endemic to Upper Guinea or represented in this region by isolated populations (Tab. 6.5). Many of these species are also strictly or partially dependent on caves as day roosts (*Coleura afra, Rhinolophus* spp., *Hipposideros* spp., *Myotis tricolor, Miniopterus* spp.), which are offered by the high topodiversity of this region.

Several species are not only endemic to Upper Guinea but also appear to have astonishingly small distribution ranges, i.e. *Rhinolophus maclaudi, R. ziama, Hipposideros marisae* and *H. lamottei*. Of these, *H. marisae* is classified as "Vulnerable" and *H. lamottei* as "Data Deficient" according to the latest IUCN Red List (Hutson et al. 2001, IUCN 2002). Fahr et al. (2002b) proposed to classify *R. maclaudi*

as "Endangered" and the recently described *R. ziama* as "Data Deficient". Both *R. ziama* and *H. lamottei* are known from two localities each, and almost nothing is known about their biology. These species highlight the importance of the mountainous part of the Upper Guinea region for the conservation of bats with isolated or restricted distribution ranges. Furthermore, we point out that this holds for bats with broader distribution ranges as well, as seven species or 33.3 % out of the total of 21 species recorded during this RAP are registered in the latest IUCN Red List of Threatened Species (Hutson et al. 2001, IUCN 2002): one species as "Vulnerable" (*Epomops buettikoferi*) and six species as "Lower Risk: Near Threatened" (*Rhinolophus alcyone, R. guineensis, Hipposideros jonesi, H. fuliginosus, Kerivoula cuprosa, Miniopterus schreibersii*).

It is also remarkable that, out of nine species described from continental Africa within the last 20 years, three (denoted with asterisks) were discovered in the Upper Guinea Highlands: *Epomophorus minimus* Claessen & De

Vree, 1991, *Rhinolophus hillorum**, *R. ziama**, *R. maendeleo* Kock, Csorba & Howell, 2000, *R. sakejiensis* Cotterill, 2002, *Hipposideros lamottei**, *Plecotus balensis* Kruskop & Lavrenchenko, 2000, *Glauconycteris curryae* Eger & Schlitter, 2001, *Scotophilus nucella* Robbins, 1984. The preservation of undisturbed montane habitats in the Upper Guinea region is of outstanding importance, not only on the regional level but also for the conservation of bats on a continental to global scale. Every effort is needed to protect the few and relatively undisturbed areas such as the Simandou Range because of the extremely limited extent of the montane habitats and the large percentage that has already been lost through human land use.

In addition to small- or large-scale habitat degradation, the disturbance or exploitation of bats in their day roosts poses a serious threat (Hutson et al. 2001, Fahr et al. 2002a). While several bat species are known to roost in hollow trees, under bark or between palm fronds that are relatively secure from direct exploitation, it is especially the cave-dwelling species that are endangered by disturbances or direct harvesting during the daytime. The population biology of most bat species does not allow for even low exploitation levels because bats are, for their body size, unusually long-lived and slow reproducing mammals. The vast majority of the African bat species reproduce only once or twice per year and females usually give birth to a single offspring. Exploitation of bats in their day roosts is very easy and effective and leads to a quick depletion of local populations. Any use of cave-dwelling bats as bush meat is therefore highly unsustainable and can lead to the extinction of local populations or even species that have restricted distribution ranges. Frequent and large disturbances in day roosts can also cause the abandonment of roosting sites.

Consequently, an integrated management of the Forêt Classée du Pic de Fon should include as an important component a specific bat cave program, with a special emphasis on the endemic and threatened bat fauna of the Upper Guinea Highlands (Table 6.5). We suggest that this program should focus on the following priorities:

- Continued surveys to assess the entire bat fauna of the Simandou Range.

- Survey of caves: mapping of locations, identification of bat species, and estimation of colony sizes.

- Specific search for threatened bat species endemic to the Upper Guinea Highlands (i.e. *Rhinolophus ziama*, *R. maclaudi*, *Hipposideros lamottei*, *H. marisae*) and in-depth studies of their biology.

- Long-term monitoring program of selected caves with important bat colonies (both in terms of colony size and rare/endemic species).

- Interviews with local communities to evaluate potential exploitation levels.

- Awareness and education programs to counteract unsustainable use of cave bats as bush meat.

CONSERVATION RECOMMENDATIONS

To summarize our data, we have found that 1) the Simandou Range harbors a speciose bat assemblage characterized by forest and cave-dwelling species, 2) of these, one species is considered as "Vulnerable" and six species as "Near Threatened" according to the IUCN Red List, 3) the Upper Guinea Highlands support many species that are endemic to West Africa, in some cases even endemic to these highlands (restricted range size species), 4) several of these species are globally threatened or near threatened, and 5) several new species have been recently discovered in this region, indicating that much remains to be found in this area.

The Simandou Range, together with other larger forest reserves in Guinea (i.e. Ziama, Diécké, Mount Nimba, Mount Béro, Mount Tetini), forms the last stronghold of fairly undisturbed and protected highland habitats in the forest zone of Guinea. Bearing in mind the extremely limited extent of montane habitats in West Africa, we emphasize the uniqueness of these habitats and their conjoined fauna and flora (Lamotte 1998). We expect acute consequences for the bat fauna of the region if these highland habitats should be rendered inhabitable through open cast mining, deforestation or encroaching settlements. This should be explicitly considered as part of a detailed Environmental and Social Impact Assessment (ESIA), if a mine were to proceed to the feasibility stage. The unexpected record of *M. welwitschii*, together with species endemic to the Upper Guinea Highlands or West Africa, essentially highlight the conservation relevance of the region in general and of the Simandou Range in particular. Our findings support the conclusion of the Priority-Setting Workshop which ranks the montane habitats of Guinea as being of extremely high priority for mammals and endorse the Upper Guinea region as a biodiversity hotspot of global importance.

REFERENCES

Aellen, V. 1956. Speologica africana. Chiroptères des grottes de Guinée. Bull. Inst. Fr. Afr. Noire Ser. A Sci. Nat. 18(3): 884-894.

Aellen, V. 1963. La Réserve Naturelle Intégrale du Mont Nimba. XXIX. Chiroptères. Mém. Inst. Fr. Afr. Noire 66: 629-638.

Bakarr, M., B. Bailey, D. Byler, R. Ham, S. Olivieri and M. Omland (eds.). 2001. From the Forest to the Sea: Biodiversity Connections from Guinea to Togo. Conservation International, Washington, DC. 78 pp. www.biodiversity science.org/priority_outcomes/west_africa

Barnett, A.A. and M.L. Prangley. 1997. Mammalogy in the Republic of Guinea: An overview of research from 1946 to 1996, a preliminary check-list and a summary of research recommendations for the future. Mammal Rev. 27(3): 115-164.

Brooks, T., A. Balmford, N. Burgess, J. Fjeldså, L.A. Hansen, J. Moore, C. Rahbek and P.H. Williams. 2001. Toward a blueprint for conservation in Africa. BioScience 51(8): 613-624.

Brosset, A. 1985 [for 1984]. Chiroptères d'altitude du Mont Nimba (Guinée). Description d'une espèce nouvelle, *Hipposideros lamottei*. Mammalia 48(4): 545-555.

Coe, M. 1975. Mammalian ecological studies on Mount Nimba, Liberia. Mammalia 39(4): 523-588.

Colston, P.R. and K. Curry-Lindahl. 1986. The Birds of Mount Nimba, Liberia. British Museum (Natural History). London. 129 pp.

Colwell, R.K. 2000. EstimateS: Statistical estimation of species richness and shared species from samples. Version 6.0b1. Application and user's guide. http://viceroy.eeb.uconn.edu/estimates

Eisentraut, M. 1960. Zwei neue Rhinolophiden aus Guinea. Stuttgarter Beitr. Naturk. (39): 1-7.

Eisentraut, M. H. and Knorr. 1957. Les chauves-souris cavernicoles de la Guinée française. Mammalia 21(4): 321-340.

Fahr, J. 2001. A fresh look at Afrotropical bat assemblages: Combining different sampling techniques and spatial scales. Bat Research News 42(3): 98.

Fahr, J. and N.M.Ebigbo. 2003. A conservation assessment of the bats of the Simandou Range, Guinea, with the first record of *Myotis welwitschii* (GRAY, 1866) from West Africa. Acta Chiropterologica 5(1): 125-141.

Fahr, J., N.M. Ebigbo and P. Formenty. 2002a. Final Report on the Bats (Chiroptera) of Mt. Sangbé-National Park, Côte d'Ivoire. Afrique Nature, Abidjan. 32 pp.

Fahr, J., N.M. Ebigbo and E.K.V. Kalko. 2001. The influence of local and regional factors on the diversity, structure, and function of West African bat communities (Chiroptera). *In:* BIOLOG – German Programme on Biodiversity and Global Change, p. 144-145. Status Report, BMBF & DLR, Bonn. 247 pp.

Fahr, J., H. Vierhaus, R. Hutterer and D. Kock. 2002b. A revision of the *Rhinolophus maclaudi* species group with the description of a new species from West Africa (Chiroptera: Rhinolophidae). Myotis 40: 95-126.

Gatter, W. 1997. Birds of Liberia. Pica Press, The Banks, Mountfield. 320 pp.

Hill, J.E. 1982. Records of bats from Mount Nimba, Liberia. Mammalia 46(1): 116-120.

Hutson, A.M., S.P. Mickleburgh and P.A. Racey (comp.). 2001. Microchiropteran Bats: Global Status Survey and Conservation Action Plan. IUCN/SSC Chiroptera Specialist Group. IUCN, Gland, Switzerland. x + 258 pp.

IUCN. 2002. 2002 IUCN Red List of Threatened Species. www.redlist.org.

Konstantinov, O.K., A.I. Pema, V.V. Labzin and G.V. Farafonova. 2000. Records of bats from Middle Guinea, with remarks on their natural history. Plecotus et al. 3: 129-148.

Koopman, K.F. 1989. Systematic notes on Liberian bats. Am. Mus. Novitates (2946): 1-11.

Koopman, K.F., C.P. Kofron and A. Chapman. 1995. The bats of Liberia: Systematics, ecology, and distribution. Am. Mus. Novitates (3148): 1-24.

Lamotte, M. 1942. La faune mammalogique du Mont Nimba (Haute Guinée). Mammalia 6: 114-119.

Lamotte, M. (ed.). 1998. Le Mont Nimba. Réserve de Biosphère et Site du Patrimoine Mondial (Guinée et Côte d'Ivoire). Initiation à la Géomorphologie et à la Biogéographie. UNESCO Publishing, Paris. 153 pp.

Myers, N., R.A. Mittermeier, C.G. Mittermeier, G.A.B. da Fonseca and J. Kent. 2000. Biodiversity hotspots for conservation priorities. Nature 403: 853-858.

Roche, J. 1972 [for 1971]. Recherches mammalogiques en Guinée forestière. Bull. Mus. natn. Hist. nat. (3)16: 737-781.

Simmons, N.B. In press. Order Chiroptera. *In*: Wilson, D.E. and D.M. Reeder (eds.). Mammal species of the world. A taxonomic and geographic reference. 3rd edition. Smithonian Institution Press, Washington, DC.

Toure, M. and J. Suter. 2001. Workshop report of the 1st trinational meeting (Côte d'Ivoire, Guinea, Liberia), 12-14 September 2001, Man, Côte d'Ivoire. Initiating a Trinational Programme for the Integrated Conservation of the Mount Nimba Massif. Fauna & Flora Int., Conservation International & BirdLife Int., Abidjan. 56 pp. http://www.fauna-flora.org/around_the_world/africa/mount_nimba.htm

Toure, M. and J. Suter. 2002. Workshop report of the 2nd trinational meeting (Côte d'Ivoire, Guinea, Liberia), 12-15 February 2002, N'Zérékoré, Guinea. Initiating a Tri-national Programme for the Integrated Conservation of the Nimba Mountains. Fauna & Flora Int., Conservation International & BirdLife Int., Abidjan. 82 pp.

Verschuren, J. 1977 [for 1976]. Les cheiroptères du Mont Nimba (Liberia). Mammalia 40(4): 615-632.

Wolton, R.J., P.A. Arak, H.C.J. Godfray and R.P. Wilson. 1982. Ecological and behavioural studies of the Megachiroptera at Mount Nimba, Liberia, with notes on Microchiroptera. Mammalia 46(4): 419-448.

Ziegler, S., G. Nikolaus and R. Hutterer. 2002. High mammalian diversity in the newly established National Park of Upper Niger, Republic of Guinea. Oryx 36(1): 73-80.

Chapter 7

A rapid survey of terrestrial small mammals (shrews and rodents) of the Forêt Classée du Pic de Fon, Guinea

Jan Decher

SUMMARY

A survey of the non-flying small mammal fauna during a rapid biological assessment (RAP) of two sites on the west slope of the Pic de Fon (Simandou Range, Guinea) recorded three species of shrews and eight species of rodents. An additional six species of squirrels were observed. No particularly rare or locally endemic small mammals were found. Trapping success was much lower (4-7 %) at the higher elevation site (ca. 1,350 m) than at the lower elevation site (22-32 %; ca. 620 m). Results were characteristic of montane semi-evergreen rain forest with higher levels of small mammal biomass at the lower elevation suggesting increased forest productivity due to an abundance of water, high plant and microhabitat diversity, and in core areas, relatively low levels of disturbance. Local hunters and guides confirmed the presence of the endemic Nimba otter shrew (*Micropotamogale lamottei*) based on pictures and descriptions of the animal, but this species could not be observed or captured during the RAP.

INTRODUCTION

During the 1999 Conservation International Priority Setting Workshop in Ghana (Bakarr et al. 2001) the Guinea Highlands and the Fon-Tibé region were designated as having "extremely high" and "very high" conservation priority on the mammal map and the integrated consensus maps, respectively. The Pic de Fon (Simandou Range) in Eastern Guinea is in close geographic proximity and features some topographic and edaphic similarity to the Nimba Mountain Range just to the south on the border with Côte d'Ivoire and Liberia. Both Mount Nimba and the Ziama Range on the Liberian border near Sérédou are known for a few small mammal endemics and generally high mammalian diversity (Coe 1975, Gautun et al. 1986, Heim de Balsac 1958, Heim de Balsac and Lamotte 1958, Roche 1971, Verschuren and Meester 1977). The Simandou Range fauna also forms an important link to the fauna of the newly established and recently surveyed National Park of Upper Niger, west of Kououssa to the north (Ziegler et al. 2002)

Terrestrial small mammals are an important part of tropical forest ecology. Rodents for example, like bats, primates, and birds, have been shown to be important fruit dispersers of both overstory and understory trees (Gautier-Hion et al. 1985; Longman and Jeník 1987) contributing to the recolonisation of forest gaps.

MATERIALS AND METHODS

Small mammal survey techniques followed those described by Voss and Emmons (1996) and Martin et al. (2001) and complied with recommended guidelines and standard methods for mammalian field work (Animal Care and Use Committee 1998, Wilson et al. 1996).

At both sites on Pic de Fon three and four trap lines were installed, using a total of 56 standard Sherman live traps, 12 Victor rat snap traps and 8 larger Tomahawk live traps. In addition, at 620 m elevation, two pitfall lines were established using nine plastic buckets along plastic driftfences. Trap effort was a total of 606 trap-nights (number of traps times number of nights trapped). At 1,350 m elevation rocky soil conditions did not allow for the digging of pitfall traps. Sherman traps and snap traps were baited with fresh palmnut shavings. Tomahawk traps were baited with smoked fish or pieces of cassava (*Manihot esculenta*).

Traplines were checked every morning and captured animals were measured, identified and released at the capture site, or kept as voucher specimens for further identification. Voucher specimens were deposited at the Museum Alexander Koenig, Bonn, Germany (shrews) and the United States National Museum, Division of Mammals, Washington, D.C. (rodents). For each captured animal some microhabitat data including percentage of canopy cover, distance to nearest tree and nearest fallen log, trap height, slope, and percentage of five ground-cover types were recorded.

The approximate coordinate distribution of the trapping localities, obtained with a Garmin GPS 12 receiving unit, was as follows (end of trap lines only):

Site 1 (1,350 m):   Camp 1:   8° 31' 52.0" N   8° 54' 21.3" W
                    Line A: 8° 31' 52.1" N   8° 54' 23.8" W

Line B: 8° 31' 53.5" N   8° 54' 27.1" W
Line C: 8° 31' 54.6" N   8° 54' 28.2" W

Site 2 (620 m):   Camp 2:   8° 31' 29.2" N   8° 56' 12.2" W
                  Line A: 8° 31' 32.8" N   8° 56' 06.0" W
                  Line B: 8° 31' 35.6" N   8° 56' 04.4" W
                  Line C: 8° 31' 19.6" N   8° 56' 11.0" W
                  Line D: 8° 31' 37.5" N   8° 56' 09.1" W

Scientific names and taxonomy used in this report follow Wilson and Reeder (1993), common names follow Wilson and Cole (2000).

RESULTS

Table 7.1 shows an overview of the distribution of results obtained by standardized trapping over the sampling period. Site 1 was sampled for five nights, Site 2 for four nights. Ninety-nine (99) individual captures were made in 606 trapnights, resulting in 9 species recorded from both sites; 3 species of shrews (Soricidae) and 6 species of rodents (Muridae and Hystricidae). In addition, one individual of Temminck's mouse (*Mus musculoides*) was found dead on a trail near Camp 2 and one individual of Temminck's striped mouse (*Hybomys trivirgatus*) was observed on

Table 7.1. Overview of terrestrial (non-flying) small mammal captures obtained by trapping at two sites on Pic de Fon (Simandou Range), Guinea, from 28 November to 7 December 2002. Two additional species, *Hybomys trivirgatus* and *Mus musculoides*, are not included in this table because they were verified by other methods. For common names see Appendix 8.

Sites:	Site 1 (Elevation ca. 1,350 m)					Site 2 (Elevation ca. 620 m)				
Date:	28 Nov.	29 Nov.	30 Nov.	1 Dec.	2 Dec.	4 Dec.	5 Dec.	6 Dec.	7 Dec.	Totals
ORDER/Genus Species										
INSECTIVORA										
Crocidura foxi	1									1
Crocidura cf. *denti*					1					1
Crocidura grandiceps						1		1		2
RODENTIA										
Hybomys planifrons	1	2		1		2	2	2		10
Hylomyscus alleni			1			2	2	5	2	12
Malacomys edwardsi	1			1						2
Mus setulosus			1							1
Praomys rostratus	1	1	3	3		16	22	10	13	69
Atherurus africanus								1		1
Totals Captured:	4	3	5	5	1	21	26	19	15	99
No. of Species:	4	2	3	3	1	4	4	5	2	9
No. of Traps:	68	72	72	72	14	81	81	77	69	-
Trapnights:	68	72	72	72	14	81	81	77	69	606
Trap Success (%):	5.9	4.2	6.9	6.9	7.1	25.9	32.1	24.7	21.7	16.3

the day of departure along the trail between Site 2 and the village of Banko. One larger rodent, a Brush-tailed porcupine (*Atherurus africanus*) was caught in a Tomahawk trap, with the help of local hunters, near the village of Banko. Squirrel species, observed mainly by RAP participants working on other taxonomic groups, were Green squirrel (*Paraxerus poensis*), Striped ground squirrel (*Euxerus erythropus*), Red-legged sun squirrel (*Heliosciurus rufobrachium*), African giant squirrel (*Protoxerus strangeri*), Slender-tailed squirrel (*Protoxerus aubinii*) and Gambian sun squirrel (*Heliosciurus gambianus*). Only the Slender-tailed squirrel (*Protoxerus aubinii*) is an Upper Guinea endemic (see also large mammal results, Chapter 8).

Comprising 69.7 % of the small mammal capture *Praomys rostratus* was the most frequently captured species followed by *Hylomyscus alleni* (12.1 %) and *Hybomys planifrons* (10.1 %). None of the three shrew species appears to be an Upper Guinea endemic. The shrew identified as *C.* cf. *denti* requires a more detailed examination (R. Hutterer pers. com.). *Crocidura grandiceps* so far was known only from lowland forest in Côte d'Ivoire, Ghana and Nigeria (Hutterer and Happold 1983). Figure 7.1 shows the species accumulation curve over the sampling period for the trapped species including *H. trivirgatus*.

In Figure 7.2 averages of selected microhabitat variables recorded at each trap site are compared for the three shrew and five murid rodent species trapped and averages for all trap sites combined at the higher and at the lower elevation,

respectively. Figure 7.2a shows the relatively high canopy cover at all trap sites with the exception of the single capture of the shrew *Crocidura* cf. *denti* and two captures of the rodent *Praomys rostratus* in Line 1C, placed in grassland at 1,350 m. The average canopy cover at 1,350 m was 79.5 % as compared to 93.6 % at 620 m elevation. Figure 7.2b clearly shows the much higher level of exposed rock at 1,350 m (23.4 %) than at 620 m (6.6 %). Conversely, there was almost twice as much leaf litter ground cover at 620 m (65.6 %) than at 1,350 m (35.6 %).

DISCUSSION

Based on the species accumulation curve and on the 45 small mammal species reported from Mount Nimba (Heim de Balsac 1958, Heim de Balsac and Lamotte 1958, Coe 1975, Verschuren and Meester 1977, Gautun et al. 1986)

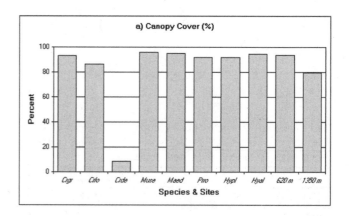

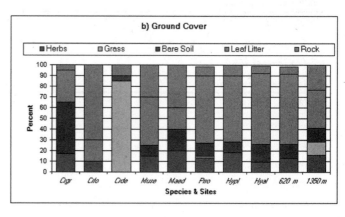

Figure 7.2. Selected microhabitat data for individual small mammal species and for the two elevations (620 m and 1,350 m) sampled on Pic de Fon, Guinea. Abbreviations and sample sizes are: *Crgr = Crocidura grandiceps* (n=2); *Crfo = Crocidura foxi* (n=1); *Crde = Crocidura* cf. *denti* (n=1); *Muse = Mus setulosus* (n=1); *Maed = Malacomys edwardsi* (n=1); *Prro = Praomys rostratus* (n=67); *Hyal = Hylomyscus alleni* (12); *Hypl = Hybomys planifrons* (n=10); 620 m elevation (n=79); and 1,350 m elevation (n=16).

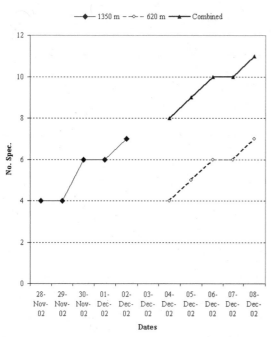

Figure 7.1. Accumulation curve for eleven small mammal species sampled between 28 November and 8 December at 1,350 and 620 m elevation on Pic de Fon, Guinea 2002 (includes *Hybomys trivirgatus* observed on last day).

and the 33 species reported from the Ziama Mountain Range (Roche 1971; see Appendix 8) we can be reasonably certain that the number of small mammal species actually present at Pic de Fon exceeds those recorded during this brief survey. For example, the montane grasslands at 1,350 m elevation were only sampled for two nights during this RAP survey, yielding only one grassland species (*Crocidura* cf. *denti*). The shrews *Crocidura* cf. *denti* and *C. foxi*, collected at 1,350 m, were only recently recorded from one other locality in Guinea, the new National Park of Upper Niger (Ziegler et al. 2002). *Crocidura denti* is also known from Mount Cameroon (Heim de Balsac 1959) and thus may have a relict distribution including certain mountain peaks in West Africa. However, its type specimen is from the Ituri Forest in the DRC (Hutterer 1993). *Crocidura grandiceps*, which was collected at 620 m elevation, is clearly a forest species, originally described from Krokosua Hills in Ghana and also known from Western Côte d'Ivoire and southwestern Nigeria (Hutterer 1983, Hutterer and Happold 1983).

The genera *Malacomys* and *Hylomyscus* are often more common than *Praomys* in larger and more moist tropical forest areas, such as the Haute Dodo and Cavally forests recently surveyed in Côte d'Ivoire (Decher et al. in press). In drier semi-evergreen and more montane forest as on Pic de Fon, *Praomys* is dominant and other genera are less common or missing, which was also shown in forest islands on the Accra Plains of Ghana (Decher and Bahian 1999) and in forest remnants in the Togo Highlands of Ghana (Decher and Abedi-Lartey 2002). *Malacomys*, *Hybomys* and *Lophuromys* seem much more sensitive to forest disturbance or forest insulation than *Praomys* and the arboreal *Hylomyscus*. The occurrence of both *Mus setulosus* and *Mus musculoides* has also been reported from Mount Nimba (Heim de Balsac and Lamotte 1958, Coe 1975) and from the Sérédou region (Roche 1971). The capture of the Brush-tailed porcupine (*Atherurus africanus*) was made closer to the village in secondary forest. Coe (1975:563) already pointed out that this species "is in a bad position in the Nimba area because of the high price it brings as food". The African giant rat (*Cricetomys* sp.) was also reported by villagers from Banko as being commonly hunted in the area. In a 1985 survey of species traded for bushmeat on a market in Accra, Ghana, the African giant rat ranked fourth, with the savannah-dwelling grasscutter or cane rat (*Thryonomys swinderianus*) ranking first and the Brush-tailed porcupine ranking eleventh of 18 recorded species (Ntiamoa-Baidu 1987). Both *Atherurus* and *Cricetomys* are known to be important dispersers of seeds by hoarding or hiding large quantities of fruits and nuts (Gautier-Hion et al. 1985). The continued presence of these larger rodent species might only be insured if the Pic de Fon classified forest remains intact and hunting restrictions are enforced both to provide a protected source area of these species for the more heavily hunted surrounding areas and also in order for these species to assist forest regeneration.

Rarer species that should be sought as potential indi- cators of relatively undisturbed high forest are the shrew *Crocidura nimbae*, endemic to Upper Guinea and previously found in Guinea at Iti, Mount Nimba (Verschuren and Meester 1977), and rodents such as the Upper Guinea endemic *Grammomys buntingi*, the rufous-bellied rat *Lophuromys sikapusi*, the long-tailed Defua Rats *Dephomys defua* (incl. *D. eburnea),* three species of the dormice (*Graphiurus,* Myoxidae) and up to three species of scaly-tailed flying squirrels (Anomaluridae) all reported from Mount Nimba or from the Ziama Range (Coe 1975, Gautun et al. 1986, Heim de Balsac and Lamotte 1958, Roche 1971).

There should be a significant difference in species composition between grassland and forest if both ecosystems are sampled more adequately. Grassland species include the grasscutter (*Thryonomys swinderianus*) mentioned above, a number of additional shrew species of the genus *Crocidura* and grassland rodents such as the gray climbing mouse *Dendromys melanotis*, previously found at Mount Nimba (Heim de Balsac and Lamotte 1958), *Arvicanthis* sp., *Lemniscomys* sp., *Tatera kempi* and *Uranomys ruddi*, all of which were not found during this survey.

The endemic aquatic otter shrew (*Micropotamogale lamottei*) was only claimed to be present by local informants when they were shown pictures of the animal. This species very possibly exists at Pic de Fon since Guth et al. (1959) reported specimens from Ziéla and Sérédou in Guinea, less than 100 km from the present study site. Roche (1971) captured one immature specimen in the Sérédou Region during two month of intensive trapping. Vogel (1983) was more successful using Sherman and aquatic fish traps to obtain live specimens in Côte d'Ivoire.

This short survey at Pic de Fon revealed a typical forest small mammal community that appears to be somewhat species-impoverished. It is more characteristic of montane semi-evergreen forest than of large contiguous moist high forest. However, given the data available for Mount Nimba, Ziama, and the National Park of Upper Niger (Appendix 8), which are based on more extensive surveys, we can be reasonably confident that additional species will be found with a more intensive sampling effort at Pic de Fon.

CONSERVATION RECOMMENDATIONS

Based on this rapid survey the small mammal fauna of Pic de Fon is only incompletely sampled and the sparse data obtained should be viewed as a very preliminary picture of the true small mammal diversity of this area when compared to other mountainous areas in Eastern Guinea (Appendix 8). A continued survey effort is highly recommended in particular to verify other shrew species, including the endemic otter shrew which was caught at Mount Nimba and the Forêt Classée de Ziama, by employing a variety of capture methods for extended periods and at additional sites on Pic de Fon.

The potential large scale removal of surface vegetation and soil on the mountain top that could result from proposed mining activities is as much a concern for small mammal populations as are the rapidly advancing plantations and the use of fire to clear forest land by the local population near Site 2. Already the invasive exotic weed *Chromolaena odorata* is growing close to the forest edge providing an additional fire hazard during the dry season (Gautier 1988-89). Some arboreal species such as the scaly-tailed flying squirrels (Anomaluridae) and the dormice (Myoxidae) may have low tolerance for canopy gaps created by encroaching agriculture in forest margins or other excessive canopy removal. Their density at Pic de Fon may already be too low to be detected during a rapid survey.

There is some evidence from temperate zone studies that certain mining activities may negatively affect small mammal diversity and abundance (Kirkland 1976, 1982). In his article "Mammalian ecological studies from Mount Nimba," Coe (1975) reports that on sloping ground the construction of access roads on ridge tops during rains led to extensive land slides. Also, the dumping of capping waste on steep slopes, when exposing iron ore, after initial successful colonization with the plant *Melistometum*, in later stages led to the siltation of rivers that drain the region on the Côte d' Ivoire side. The author suggests that "spoil material should be used to fill existing excavations and to avoid dumping material on steep slopes" (Coe 1975: 570). Water contamination or reductions in water level may also affect aquatic invertebrates and thus indirectly affect small predators such as the endemic Nimba otter shrew or the two otter species, *Aonyx capensis* and *Lutra maculicollis*, which possibly still occur in the region in low numbers. Small mammals have also been used as bioindicators to check for heavy metal contamination in mining areas with previously unexposed soil on the surface (McBee and Bickham 1990). These studies and others should be considered when planning potential mining activities to reduce potential threats or harm to the resident small mammal population.

A recent critique of science/policy practices around biodiversity in Guinea highlighted the *a priori* conservation bias towards "biodiversity wealth and conservation in reserves, and biodiversity destruction in inhabited areas" (Fairhead and Leach 2002:105). Our experience using local experts to capture *Atherurus africanus* near the village of Banko and our reliance on local informants regarding the presence of the endemic otter shrew highlight the need for closer cooperation with local people and their traditional knowledge base and for viewing the Pic de Fon forest reserve and surrounding villages as parts of a whole that need to coexist in mutual dependence.

In conclusion, in order to better assess the true diversity of small mammals on Pic de Fon and the impact of anthropogenic disturbances resulting from mineral exploration activities and deforestation, and in order to monitor effects of future activities in the area, I recommend a more in-depth survey of small mammals involving additional sampling sites in disturbed and undisturbed habitats and at different elevations in both forest and grasslands on Pic de Fon.

REFERENCES

Animal Care and Use Committee. 1998. Guidelines for the capture, handling, and care of mammals as approved by The American Society of Mammalogists. Journal of Mammalogy, 79:1416-1431.

Bakarr, M., B. Bailey, D. Byler, R. Ham, S. Olivieri and M. Omland. 2001. From the forest to the sea: biodiversity connections from Guinea to Togo. Conservation International, Washington, D. C., 78 pp.

Coe, M. 1975. Mammalian ecological studies on Mount Nimba, Liberia. Mammalia, 39(4): 523-581.

Decher, J. and L. K. Bahian. 1999. Diversity and structure of terrestrial small mammal communities in different vegetation types on the Accra Plains of Ghana. Journal of Zoology, London, 247:395-407.

Decher, J. and M. Abedi-Lartey. 2002. Small mammal zoo-geography and diversity in forest remnants in the Togo Highlands of Ghana. Final Report to the National Geographic Society for Research and Exploration, the Ghana Wildlife Division and Conservation International. 32 pp.

Decher, J., B. Kadjo, M. Abedi-Lartey, E. O. Tounkara and S. Kante. *In press.* Small mammals (shrews, rodents, and bats) from the Haute Dodo and Cavally Forests, Côte d'Ivoire. *In:* F. Lauginie, G. Rondeau, and L.E. Alonso (eds.). A biological assessment of two classified forests in South-Western Cote d'Ivoire. RAP Bulletin of Biological Assessment 34. Conservation International. Washington, DC.

Fairhead, J. and M. Leach. 2002. Practising 'Biodiversity': The articulation of international, national and local science/policy in Guinea. IDS Bulletin, 33:102-110.

Gautier, L. 1988-89. Contact forêt-savane en Côte-d'Ivoire centrale: rôle de *Chromolaena odorata* (L.) King & Robinson dans la dynamique de la végétation. Centre Suisse de Recherches Scientifiques en Côte-d'Ivoire (Adiopodoumé), 1988-89:100-108.

Gautier-Hion, A., J.M. Duplantier, R. Quris, F. Feer, C. Sourd, J.-P. Decoux, G. Dubost, L. Emmons, C. Erard, P. Hecketsweiler, A. Moungazi, C. Roussilhon and J.-M. Thiollay. 1985. Fruit characters as a basis of fruit choice and seed dispersal in a tropical forest vertebrate community. Oecologia, 65:324-337.

Gautun, J.C., I. Sankhon and M. Tranier. 1986. Nouvelle contribution à la connaissance des rongeurs du massif guinéen des Monts Nimba (Afrique occidentale). Systématique et aperçu quantitatif. Mammalia, 50:205-217.

Guth, C., H. Heim de Balsac and M. Lamotte. 1959. Recherches sur la morphologie de *Micropotamogale lamottei* et l'évolution des Potamogalinae. I. Écologie, denture, anatomie crâniénne. Mammalia, 23:423-447.

Heim de Balsac, H. 1958. XIV. Mammifères Insectivores. Memoires de l'Institut Français d'Afrique Noire, 53 (Fasc. IV): 301-357.

Heim de Balsac, H. and M. Lamotte. 1958. La réserve
naturelle intégrale du Mont Nimba: 15. Mammifères
Rongeurs. Memoires de l'Institut Fondamental d'Afrique
Noire, 53:339-357.

Hutterer, R. 1983. *Crocidura grandiceps*, eine neue Spitzmaus
aus Westafrika. Revue Suisse de Zoologie, 90: 699-707.

Hutterer, R. 1993. Order Insectivora. *In*: Wilson, D.E. and
D.M. Reeder (eds.). Mammal species of the world: a taxo-
nomic and geographic reference. 2nd edition. Smithonian
Institution Press, Washington, DC. 1,206 pp.

Hutterer, R. and D.C.D. Happold. 1983. The shrews of
Nigeria (Mammalia: Soricidae). Bonner Zoologische
Monographien, 18:1-79.

Kirkland, G.L., Jr. 1976. Small mammals of a mine waste
situation in the Central Adirondacks, New York: A case of
opportunism by Peromyscus maniculatus. The American
Midland Naturalist, 95:103-110.

Kirkland, G.L., Jr. 1982. Ecology of small mammals in iron
and titanium open-pit mine wastes in the central Adirondacks mountains, New York. National Geographic Society
Research Reports, 14:371-380.

Lim, B.K. and P.J. van de Groot Coeverden. 1997. Taxonomic report of small mammals from Côte d'Ivoire.
Journal of African Zoology, 111:261-279.

Longman, K.A. and J. Jeník. 1987. Tropical forest and its
environment. 2nd ed. Longman Singapore Publishers
Ltd., Singapore, 347 pp.

Martin, R.E., R.H. Pine and A.F. DeBlase. 2001. A manual
of mammology with keys to families of the world. 3rd ed.
McGraw Hill, Boston, xi+333 pp.

McBee, K. and J.W. Bickham. 1990. Mammals as bioindicators of environmental toxicity. Pp. 37-88. *In:* Genoways,
H.H. (ed.). Current Mammalogy. Vol. 2. Plenum Publishing Corporation.

Ntiamoa-Baidu, Y. 1987. West African wildlife: a resource in
jeopardy. Unasylva, 39:27-35.

Roche, J. 1971. Recherches mammalogiques en Guinée
forestière. Bulletin du Muséum Nationale d'Histoire
Naturelle 3eme serie(16) :737-781.

Verschuren, J. and J. Meester. 1977. Note sur les Soricidae
(Insectivora) du Nimba libérien. Mammalia. 41(3): 291-
299.

Vogel, P. 1983. Contribution a l'écologie et a la zoogéographie
de *Micropotamogale lamottei* (Mammalia, Tenrecidae).
Revue de Ecologie (La Terre et la Vie), 38:37-49.

Voss, R.S. and L. H. Emmons. 1996. Mammalian diversity
in Neotropical lowland rainforests: a preliminary assessment." Bulletin of the American Museum of Natural
History, 230:1-115.

Wilson, D.E. and F. R. Cole. 2000. Common names of
mammals of the world. Smithsonian Institution Press,
Washington, xiv+204 pp.

Wilson, D.E., F.R. Cole, J.D. Nichols, R. Rudran and M.S.
Foster. 1996. Measuring and monitoring biological
diversity. Standard methods for mammals. Smithsonian
Institution Press, Washington, xxvii + 409 pp.

Ziegler, S., G. Nikolaus and R. Hutterer. 2002. High mammalian diversity in the newly established National Park
of Upper Niger, Republic of Guinea. Oryx 36(1): 73-80.

Chapter 8

A rapid survey of the large mammals of the Forêt Classée du Pic de Fon, Guinea

Abdulai Barrie and Soumaoro Kanté

SUMMARY

This paper presents the results of a Rapid Assessment Program survey conducted in Pic de Fon classified forest from 27 November to 7 December 2002. The purpose of the survey was to assess the biological diversity of large mammals in the region. We used tracks, sound and visual observations, and camera phototraps to survey for the presence of large mammals. We confirmed the presence of 31 and 34 large mammals in Sites 1 and 2 of Pic de Fon classified forest, respectively. In total, we confirmed the existence of 39 mammals in these forests. While Site 1 of the Pic de Fon classified forest is currently under prospecting for iron ore, Site 2 is over-exploited for bushmeat and encroached by farmers. Despite federal laws restricting hunting, we found evidence for active hunting in both locations. Large mammals such as primates and duikers were only rarely directly observed.

INTRODUCTION

The classified forest of Pic de Fon is found in the Simandou Range. Very little information exists for this region. This region is located within the "Guinean Forest Hotspot" as designated by CI, and consists of forests that run perpendicular to the coast of West Africa. Of the Hotspots designated by CI, the Guinean Forest Hotspot has the highest diversity of mammals with an estimated 551 species known to occur (Myers 1998; Bakarr et al. 2001). Although the number of endemic species is relatively low, the forest is still important to the global conservation of mammals (Sayer et al. 1992; Kingdon 1997; Mittermeier et al. 1999) and is one of the two highest priority regions in the world for primate conservation.

Barnett and Prangley (1997) listed 190 species of mammals recorded for Guinea from 26 papers reviewed. None of the papers covered the species of Pic de Fon, as very little ecological information exists for the Simandou region. The majority of mammalogical field studies in Guinea have beeen conducted in Mount Nimba and complemented by studies in adjacent areas of Liberia and nearby Sierra Leone (Barnett and Prangley 1997). IUCN (1988) gives a partial list of mammals recorded in Mount Nimba and lists 10 species as threatened for Guinea. Though large mammals such as Bongo (*Tragelaphus euryceros*) and endemic carnivores such as Johnston's civet (*Genetta johnstoni*) and Liberian mongoose (*Liberiictis kuhnii*) are known from small populations or expected be present in Guinea (Rosevear 1974; Coe 1975), Liberia and Côte d'Ivoire, large mammals have not yet been systematically surveyed. In the order Artiodactyla, three "Vulnerable" duikers in the genus *Cephalophus* (*C. jentinki*, *C. niger* and *C. zebra*) and the "Near threatened" small Royal antelope, *Neotragus pygmaeus,* are endemic (Kingdon 1997) reinforcing the importance of the Upper Guinea Hotspot (Table 8.1).

Other important large mammals include the Leopard (*Panthera pardus*) and Elephant (*Loxodonta africana*). The species composition of Pic de Fon is similar to that of Mount Nimba. However, no systematic surveys of large terrestrial mammals have been undertaken in the Pic de Fon area.

Table 8.1. Endemic and near-endemic large mammalian species of West Africa.

Order	Family	Species	Common name
Primates	Cercopithecinae	*Cercopithecus (mona) campbelli*	Campbell's Monkey
		Cercopithecus diana diana	Diana Monkey
		Cercopithecus (c.) p. buettikoferi	Lesser spot-nosed Monkey
	Colobidae	*Procolobus verus*	Olive Colobus
		Colobus p. polykomos	Western Pied Colobus
		Colobus vellerosus	Geoffroy's Pied Colobus
		Piliocolobus badius badius	Western red Colobus
Carnivora	Herpestidae	*Liberiictis kuhni*	Liberian mongoose
	Viverridae	*Genetta johnstoni*	Johnston's genet
		Genetta bourloni	
		Poiana leightoni	West African linsang
Artiodactyla	Hippopotamidae	*Hexaprotodon liberiensis*	Pygmy hippopotamus
	Notragini	*Neotragus pygmaeus*	Royal antelope
	Cephalophini	*Cephalophus jentinki*	Jentink's duiker
		Cephalophus zebra	Zebra duiker
		Cephalophus niger	Black duiker

MATERIALS AND METHODS

Study Area

We conducted our surveys at the beginning of the dry season in two sites of the Pic de Fon: Site 1 which included the mountain range or highlands (08°31'52.0"N, 08°54'21.3"W) from 27 – 30 November 2002 and Site 2, lowlands (08°31'29.2"N, 08°56'12.2"W) from 2 – 7 December 2002. Site 1 and 2 were approximately 1,350 m and 600 m respectively. Classified forests in Guinea are production forests. Site 1 is currently being prospected for iron ore and the forests around Site 2 are being cleared for farming (coffee, cocoa, plantains and rice). Interior forest streams were flowing normally for this time of year.

Methods

We used active and passive methods to document the presence of large mammals. The active methods included direct observation of species, track and sound identification, nests, dung and other indirect information to determine presence of large non-volant mammalian species in the two study areas. Direct observations and track and sound identification were made during daily excursions from base camp. Surveys were carried out at night using a spotlight. Because our colleagues also collected large mammal records opportunistically and some observations may have been repeated we used this information only to document species presence.

The passive method of observation included the use of nine CamTrakker phototraps (CamTrakker Atlanta, Georgia) operated at each study site. CamTrakker phototraps are triggered by heat-in-motion. Each CamTrakker uses a Samsung Vega 77i 35mm camera set on autofocus and loaded with ASA 200 print film. Time between sensor reception and a photograph was 0.6 secs. Cameras were set to operate continuously (control switch 1 on) and to wait a maximum 20 seconds between photographs (control switches 6 and 8 on). Cameras were placed at sites suspected of being frequented by various mammalian species. Den sites, trails and feeding stations such as fruiting trees were typically chosen for camera placement. Cameras were located approximately 500 m apart and at least 500 m from base camp. We used this method to calculate observation rates for each site just as standard transects are used. Instead of the observer making observations along a route, "observations" moved along routes in front of fixed cameras (observers). For shy mammals under severe hunting pressure camera trapping methods might be more effective than walking transects, especially when observers have different and varied levels of expertise.

RESULTS

We observed, identified by sound or photographed 31 species of large mammals in Site 1 of the Pic de Fon classified forest (Table 8.2) and 34 species in Site 2 (Table 8.3) for a combined total of 39 species of large mammals. The camera phototraps obtained one photograph of two chimpanzees in Site 1 for a photographic rate of one photograph every three days. For Site 2 we obtained a total of three photographs of one species for a photographic rate of one photograph every 1.7 days. The most common large mammals photographed by camera phototraps were *Atherurus africanus,* Brush-tailed porcupine (3 of 4 photographs) and *Pan troglodytes verus,* Western chimpanzee (1 of 4 photographs). No Leopards, Aardvark or Bongo were observed but local hunters reported that these species still occur in Pic de Fon classified forest. Other species reported by the hunters but that were not quite clear from their descriptions and identification were ignored. Using track, dung and hunters we documented the presence of the African buffalo in the Pic de Fon classified forest.

We observed primates on two occasions and our colleagues observed primates on six occasions during our survey. Sightings of seven chimpanzees, calls and nests confirmed the continued presence of chimpanzees in both Sites 1 and 2 of Pic de Fon classified forest.

Table 8.2. Large mammals whose presence was confirmed in Site 1 of the Pic de Fon, 2002 (**H** = heard, **S** = seen, **T** = tracks, **P** = photographed, **M** = many, or **O** = other evidence). Species in bold were documented only at this site. Scientific names are based on Kingdon (1997).

Order	Family	Species	Common name	H	S	T	P	O
Primates	Hominidae	*Pan troglodytes verus*	Chimpanzee					nests
	Cercopithecidae	*Cercocebus atys atys*	Sooty Mangabey	1				
		Cercopithecus diana diana	Diana Monkey	1				
		Cercopithecus campbelli campbelli	Campbell's monkey	2				
		Cercopithecus petaurista buettikoferi	Lesser spot-nosed monkey					
	Galagonidae	*Galagoides demidoff*	Demidoff's Galago					
Rodentia	Sciuridae	**Euxerus erythropus**	Striped ground squirrel		5			
		Paraxerus poensis	Green squirrel		3			
		Heliosciurus rufobrachium	Red-legged sun squirrel		2			
	Hystricidae	*Hystrix cristata*	Crested porcupine					spines
		Atherurus africanus	Brush-tailed porcupine			4		
	Muridae	*Cricetomys emini*	Giant-pouched rat					dung
	Thryonomyidae	*Thryonomys swinderianus*	Marsh cane rat					trapped
Carnivora	Herpestidae	*Herpestes ichneumon*	Ichneumon mongoose		1			
		Herpestes sanguinea	Slender mongoose		2			
		Crossarchus obscurus	Cusimanse		1			
	Viverridae	*Civettictis civetta*	African civet			2		
		Genetta geneta	Common genet					Rio Tinto staff
		Nandinia binotata	African palm civet	2				
Pholidota	Manidae	*Uromanis tetradactyla*	Long-Tailed pangolin			2		
		Phataginus tricuspis	Tree pangolin					feeding
Hyracoidea	Procaviidae	**Procavia capensis**	Rock hyrax	1				
Artiodactyla	Suidae	*Potamochoerus porcus*	Red river hog		M			
	Bovidae	*Syncerus caffer*	African buffalo		2			
		Tragelaphus scriptus	Bushbuck		3			
	Antelopinae	*Sylvicapra grimmia*	Bush duiker		3			
		Cephalophus maxwelli	Maxwell's duiker		M			dung
		Cephalophus niger	Black duiker		4			
		Cephalophus dorsalis	Bay duiker		3			dung
		Cephalophus silvicultor	Yellow-backed duiker		5			
	Neotragini	*Neotragus pygmaeus*	Royal antelope					dung

Mammals documented to occur at only one forest were Striped Ground Squirrel, Common Genet, African Palm Civet, Cusimanse and Rock hyrax from Site 1, and from Site 2, Putty-nosed Monkey, Senegal Galago, Potto, Gambian Sun Squirrel, African Giant Squirrel, Slender-tailed Squirrel,

Giant Pangolin and the Common Warthog. We believe these differences were due to the short duration of our survey and not to fundamental differences in mammalian faunas between the study areas.

Table 8.3. Large mammals whose presence was confirmed in Site 2 (**H** = heard, **S** = seen, **T** = tracks, **P** = photographed, **M** = many, or **O** = other evidence). Species in bold were documented only at this site. Scientific names are based on Kingdon (1997).

Order	Family	Species	Common name	H	S	T	P	O
Primates	Hominidae	*Pan troglodytes verus*	Chimpanzee	2	7		2	
	Cercopithecidae	*Cercocebus atys atys*	Sooty Mangabey		6			
		Cercopithecus diana diana	Diana Monkey	2				
		Cercopithecus campbelli campbelli	Campbell's monkey		3			
		Cercopithecus nictitans	Putty-nosed monkey	2				
		Cercopithecus petaurista buettikoferi	Lesser spot-nosed monkey	2				
	Galagonidae	*Galagoides demidoff*	Demidoff's Galago	4	7			
		Galagoides senegalensis	Senegal galago		1			
	Loridae	***Perodicticus potto***	Potto		1			
Rodentia	Sciuridae	*Paraxerus poensis*	Green squirrel		4			
		Protoxerus aubinnii	Slender-tailed squirrel		1			
		Heliosciurus rufobrachium	Red-legged sun squirrel		3			
		Protoxerus stangeri	African giant squirrel	1				
		Heliosciurus gambianus	Gambian sun squirrel		1	2		
	Hystricidae	*Hystrix cristata*	Crested porcupine			2		spines
		Atherurus africanus	Brush-tailed porcupine		2		3	decomposed
	Muridae	*Cricetomys emini*	Giant-pouched rat					dung
	Thryonomyidae	*Thryonomys swinderianus*	Marsh cane rat					trapped
Carnivora	Herpestidae	*Herpestes ichneumon*	Ichneumon mongoose		1			
		Herpestes sanguinea	Slender mongoose		2			
	Viverridae	*Civettictis civetta*	African civet		1			
Pholidota	Manidae	*Uromanis tetradactyla*	Long-Tailed pangolin			1		
		Phataginus tricuspis	Tree pangolin					feeding
		Smutsia gigantica	Giant pangolin			3		
Artiodactyla	Suidae	*Potamochoerus porcus*	Red river hog		7	M		
		Phacochoerus africanus	Common warthog					rooting
	Bovidae	*Syncerus caffer*	African buffalo			2		
		Tragelaphus scriptus	Bushbuck			5		
	Antelopinae	*Sylvicapra grimmia*	Bush duiker			3		
		Cephalophus silvicultor	Yellow-backed duiker			7		
		Cephalophus maxwelli	Maxwell's duiker			M		dung
		Cephalophus niger	Black duiker		1			
		Cephalophus dorsalis	Bay duiker			2		
	Neotragini	*Neotragus pygmaeus*	Royal antelope					dung

DISCUSSION

Large mammals in Guinea and throughout much of West Africa are extremely rare as a result of unregulated exploitation, habitat loss and the increasing demand for bushmeat (Lowes 1970, Davies 1987, Starin 1989, Martin 1991, McGraw 1998). Much of the forest in this region has undergone vast changes in area and composition as a result of habitat fragmentation. What remains of the high forest is a mix of evergreen and semi-evergreen species, mostly in secondary forests. The increase in human population is accelerating the conversion of the remaining forest habitats into human-dominated settlements and agricultural landscapes. Local pressure for bushmeat, farm- and cropland is reducing the size and future potential of remaining forests throughout West Africa and farming (especially plantation agriculture) has accelerated forest fragmentation and the loss of large mammals. Primary and secondary forests outside of protected areas are targeted for resource extraction (Lebbie 2002) and global resource demands can lead to impacts equally grave to those of hunting and agriculture.

The Pic de Fon classified forest is also situated near to the borders of Liberia and Côte d'Ivoire and the population has increased as refugees and returnees escape from civil conflicts in these countries. Even though the forest is classified and a number of hunting restrictions apply, the pressure to hunt is very great. Based on discussions with local hunters and guides, primates and duikers are the most frequently hunted and trapped species, a phenomenon also reported to be common in other West African countries (Ausden and Wood 1990, Lebbie 1998, Eves and Bakarr 2001). Most hunters and trappers exploit these species for the commercial market rather than for subsistence use, and the large body size of these species make them ready targets. Bushmeat is an important protein source in West Africa and the demand for it is also high (Asibey 1976, Jeffrey 1977, Ajayi 1979, Martin 1983, Falconer and Koppell 1990, Njiforti 1996, Bowen-Jones and Pendry 1999). The demand for bushmeat has been fueled by urban populations in regions where alternative protein costs are high (Wilkie et al. 1992). Most households in rural and urban areas in some West African countries like Ghana, Liberia, Senegal, Equatorial Guinea and Sierra Leone consume bushmeat on a regular basis, with hunting and poaching reported to be a lucrative business (Cremoux 1963, Ajayi 1979, Martin 1983, Addo et al. 1994, Barrie 2002)

Bushmeat hunting parallels habitat loss as a major threat to the survival of mammals in West Africa (Bakarr et al. 2001). Recently, the apparent extinction of *Procolobus badius waldroni* has been attributed to hunting and the demand for bushmeat in this region (Oates et al. 2000). Bushmeat is a critical protein source, as well as a source of livelihood, for many people in the region and a large variety of species are hunted. Antelopes, forest pigs and primates dominate the bushmeat trade, while Marsh cane rat (*Thryonomys swinderianus*) and Giant rat (*Cricetomys* spp.) are preferred by rural people. The extent of such hunting has prompted governments to enact hunting bans, though the legislation is often impractical and cannot be enforced (Sayer et al. 1992).

To undertake our survey of Site 1 we utilized a road system created for mineral exploration activities by Rio Tinto. Roads cut perpendicular to the slope of the hill have caused erosion and in some cases have exposed bedrock. Within the forests we also made use of an extensive network of trails that were familiar to our guides in both Sites 1 and 2.

Although hunting during the closed season and at night is federally prohibited in Guinea, we found 11 shotgun shells, 2 snares and a number of hunting camps in Site 1, and in Site 2 we found 36 shotgun shells, 38 snares or traplines, and heard gunshots during our daily routine and at least four times at night. Control of hunting activities is beyond the jurisdiction of the government agency responsible for logging in Guinea, Direction Nationale des Eaux et Forêts. Several villages were located within or near these forests. The combined forces of illegal hunting and human encroachment act strongly against the perpetuation of large mammals in Pic de Fon classified forest. Our results are suggestive of and consistent with the empty forest syndrome whereby large mammal populations are, one by one, reduced in density and finally exterminated from large areas (Sanderson et al. in prep).

Our results suggest the full biologically rich assortment of large mammals present in Mount Nimba and Taï and Sapo National Parks are also found in the Pic de Fon classified forest. During our brief visit we documented the presence of several species listed as threatened by IUCN: *Syncerus caffer* (African Buffalo) listed as "Conservation Dependent" and *Cephalophus maxwelli* (Maxwell's Duiker), *Cephalophus niger* (Black Duiker), *Cephalophus dorsalis* (Bay Duiker), *Cephalophus silvicultor* (Yellow-backed Duiker), and *Neotragus pygmaeus* (Royal Antelope) listed by IUCN as "Near Threatened" (Tables 8.2 and 8.3). These records suggest that the inclusion of Pic de Fon forest into a protected area system can act to increase the regional populations of these species.

CONSERVATION RECOMMENDATIONS

Lack of enforcement in classified forests regarding illegal activities such as encroachment and bushmeat hunting is no more unusual than that found within national park systems in West Africa, where many parks lack adequate protection and are so-called "paper parks." The policy framework for forest conservation in Guinea has been established. The issues now revolve around the translation of policy into practical action. Mining activities, bushmeat hunting and encroachment activities require careful evaluation and monitoring so that the Pic de Fon classified forest does not quickly become open woodlands that support none of the rich biodiversity found in primary forests.

REFERENCES

Addo, F., E. O. A. Asibey, K. B. Quist and M. B. Dyson. 1994. The economic contribution of women and protected areas: Ghana and the bushmeat trade. Pages 99-115 *In:* M. Munasinghe and J. McNeely, (eds.). Protected area economics and policy: linking conservation and sustainable development. World Bank and World Conservation Union, Washington, D.C.

Ajayi, S. S. 1979. Food and animal production from tropical forest: utilization of wildlife and by-products in West Africa. FAO, Rome, Italy.

Asibey, E. O. A. 1976. The effects of land use patterns on future supply of bushmeat in Africa south of the Sahara. Working Paper on Wildlife Management and National Parks, 5th Session.

Ausden, M. and P. Wood. 1990. The wildlife of the Western Area Forest Reserve, Sierra Leone. February 22nd - April 23rd 1990. RSPB.

Bakarr, M.I. 1992. Sierra Leone: Conservation of Biological Diversity. An assessment report prepared for the Biodiversity Support Program, Washington, DC.

Bakarr, M.I., G.A.B. da Fonseca, R. Mittermeier, A.B. Rylands and K.W. Painemilla (eds.). 2001. Hunting and Bushmeat Utilization in the African Rain Forest. Advances in Applied Biodiversity Science Number 2. Conservation International. Washington, DC.

Barnett, A. A. and Prangley, M. L. 1997. Mammalogy in the Republic of Guinea: An overview of research from 1946 to 1996, a preliminary check-list and a summary of research recommendations for the future. Mammal Rev. 1997; 27(3): 115-164.

Barrie, A. 2002. Post conflict conservation status of large mammals in the Western Area Forest Reserve (WAFR), Sierra Leone. M. Sc Dissertation, Njala University College, Freetown.

Bowen-Jones, E. and S. Pendry. 1999. The threat to primates and other mammals from the bushmeat trade in Africa, and how this threat could be diminished. Oryx 33(3):233-246.

Coe, M.J. 1975. Mammalian ecological studies on Mount Nimba, Liberia. Mammalia, 39: 523-587.

Cremoux, P. 1963. The importance of game meat consumption in the diet of sedentary and nomadic peoples of the Senegal River Valley. Pp. 127-129 in Conservation of Nature and Natural Resources in Modern African States (G. G. Watterson, editor). IUCN publications new series No. 1.

Davies, A.G. 1987. The Gola Forest Reserves, Sierra Leone: Wildlife Conservation and Forest Management. IUCN, Gland, Switzerland. 126p.

Eves, H.E. and M I. Bakarr. 2001. Impacts of bushmeat hunting on wildlife populations in West Africa's Upper Guinea Forest Ecosystem. *In:* Bakarr, M.I., G.A.B. da Fonseca, R. Mittermeier, A.B. Rylands and K.W. Pain-

emilla (eds.). Hunting and Bushmeat Utilization in the African Rain Forest: Perspectives Toward a Blueprint for Conservation Action. Advances in Applied Biodiversity Science Number 2:39-57. Conservation International, Washington, DC.

Falconer, J. and C. Koppell. 1990. The Major Significance of Minor Forest Products: The Local Use and Value of Forests in the West African Humid Forest Zone. FAO, Community Forests Note 6. Rome.

IUCN. 1988. Guinea: Conservation of Biological Diversity. World Conservation Monitoring Centre, Cambridge.

IUCN. 1990. 1990 IUCN Red List of Threatened Animals. IUCN, Gland, Switzerland and Cambridge, UK. 228 pp.

Jeffrey, S. 1977. How Liberia uses wildlife. Oryx 14:168-173.

Kingdon, J. 1997. The Kingdon Field Guide to African Mammals. Harcourt Brace & Company, New York.

Lebbie, A. 1998. The No. 2 River Forest River, Sierra Leone: Managing for Biodiversity and the Promotion of Ecotourism. Report Prepared for the United Nations (UN); Project No. SIL/93/002.

Lebbie, A. 2001. Distribution, Exploitation and Valuation of Non-Timber Forest Products from a Forest Reserve in Sierra Leone. PhD Dissertation, University of Wisconsin-Madison, USA.

Lebbie, A. 2002. Western Guinea Lowland Forest. New Map of the World. National Geographic Society - World Wildlife Fund Publication. Washington, DC.

Lee, P.J., J. Thornback and E.L. Bennett. 1988. Threatened Primates of Africa. The IUCN Red Data Book. IUCN, Gland, Switzerland and Cambridge, UK.

Lowes, R.H.G. 1970. Destruction in Sierra Leone. Oryx 10(5):309-310.

Martin, C. 1991. The rainforests of West Africa: Ecology, Threats and Conservation. Birkhauser Verlag, Boston.

Martin, L.G.H. 1983. Bushmeat in Nigeria as a natural resource with environmental implications. Environmental Conservation 10(2):125-132.

McGraw, W.S. 1998. Three Monkeys nearing extinction in the forest reserves of eastern Cote d'Ivoire. Oryx 32(3): 233-236.

Mittermeier. R.A., N. Myers, and C.G. Mitteermeier (eds.). 1999. Hotspots: Earth's Biologically Richest and most Endangered Terrestrial Ecoregions. Cemex, Conservation International. 430p.

Myers, N. 1998. Threatened biotas: 'hotspots' in tropical forests. Environmentalist 8:187-208.

Njiforti, H.L. 1996. Preferences and present demand for bushmeat in north Cameroon: some implications for wildlife conservation. Environmental Conservation 23(2):149-155.

Oates, J.F. 1986. Action Plan for African Primate Conservation 1986-1990. IUCN/SSC Primate Specialist Group. Stony Brook, New York, USA.

Oates, J.F., M. Abedi-Lartey, S. McGraw, T.T. Struhsacker and G.H. Whitesides. 2000. Extinction of a Western African Red Colobus Monkey. Conservation Biology. 14(5):1526-1533.

Rosevear, D.R. 1974. The carnivores of West Africa: British Museum (Natural History), London.

Sayer, J.A., C.S. Harcourt and N.M. Collins. 1992. The Conservation Atlas of Tropical Forests: Africa. IUCN and Simon & Schuster, Cambridge.

Starin, E.D. 1989. Threats to the monkeys of The Gambia. Oryx 23(4): 208-214 24.

Wilkie, D.S., J.G. Sidle and G.C. Boundzanga. 1992. Mechanised logging, market hunting, and a bank loan in Congo. Conservation Biology 6(4):570-580.

Chapter 9

A rapid survey of Primates in the Forêt Classée du Pic de Fon, Guinea

Ilka Herbinger and Elhadj Ousmane Tounkara

SUMMARY

A rapid assessment of the primate fauna was conducted between 27 November and 7 December 2002 in the semi-evergreen classified forest of Pic de Fon in the Simandou Range, south-eastern Guinea. At two sites, presence and abundance of primate species were estimated using a line transect method. A total of eight primate species were recorded, including two prosimians *(Perodicticus potto* and *Galagoides demidoff)*, five anthropoid monkeys *(Cercocebus atys atys, Cercopithecus campbelli campbelli, Cercopithecus petaurista buettikoferi, Cercopithecus nictitans* and *Cercopithecus diana diana)* and one hominoid ape, the West African Chimpanzee *(Pan troglodytes verus)*. The presence of an additional five species has been described by local villagers and hunters and is very likely *(Papio anubis, Erythrocebus patas, Cercopithecus aethiops sabaeus, Colobus polykomos polykomos* and possibly *Procolobus verus)*. Four out of the 13 taxa *(Cercocebus atys atys, Cercopithecus diana diana, Pan troglodytes verus* and *Procolobus verus)* are listed as Near Threatened or Endangered primate species. Although densities for most monkey species seemed rather low (encounter rates of < 0.25 per hour), density estimates for the chimpanzee population (0.64 chimpanzees/ km²) were above those observed for degraded forests (0.4 chimpanzees/ km²) and indicate that the Pic de Fon classified forest is presumably still holding an important number of individuals of several primate species. Thus, the Simandou primate population, with its many species, holds an important representation of the regional primate diversity for the Upper Guinea Region and their conservation should be of high priority.

INTRODUCTION

Primates are a major component of tropical ecosystems and play an important role, e.g. as seed dispersal agents, in structuring forested habitats (Chapman 1995, Chapman and Onderdonk 1998, Lambert and Garber 1998, Chatelain et al. 2001). They prey on other mammals and are prey for large carnivores, raptors, and snakes. Their presence or absence has implications for the continuance of a variety of plant, invertebrate and vertebrate species. In some forests in West and East Africa, primates form so-called poly-specific associations where as many as nine primate species co-exist, interact, and live at high densities (Galat-Luong and Galat 1978, Whitesides et al. 1988, Bshary 1995). Primates are the primary contributor to the mammalian diversity of such forests.

In the classified forest of Pic de Fon in the Simandou Range, Guinea, a total of 15 primate species may be expected to contribute to the mammalian diversity. The Simandou Range is part of the ecosystem of the Upper Guinea Region, which includes forests from Eastern Sierra Leone to Eastern Togo and is considered one of the world's 25 priority conservation areas because of its high degree of biodiversity and endemism (Mittermeier et al. 1999). Unfortunately, the Upper Guinea Region is also highly threatened and has suffered from a dramatic rate of deforestation in the recent past, with an estimated 80 % of the original forest cover gone by the 1980s (Martin 1989). Likewise, many primate populations have been declining drastically and some species have vanished completely from certain areas in some countries (e.g. several primate species in Ghana (e.g. Miss Waldron Red Colobus), Sierra Leone and Liberia; Lee et al. 1988). The combined effects of habitat destruction and high hunting pressure have resulted in the fact that today the Upper Guinea Region is amongst those regional communities where most of the threatened African primates are located (Lee et al. 1988). The West African Chimpanzee already faces extinction in four West African countries (Togo, Benin, Gambia and Burkina Faso) and Guinea is one of the few countries, along with Côte d'Ivoire, Liberia, Mali and Sierra Leone, that still has populations that might be viable in the long term (Kormos and Boesch 2003).

The aim of this study was to gain information regarding what species of primates occur in the Pic de Fon classified forest, Guinea, and to provide a preliminary estimation of their relative abundance. Moreover, we wanted to assess the current threats to the primate population and propose necessary measures for their protection.

METHODS

The census was conducted at two sites in the southwestern part of the Pic de Fon classified forest between 27 November and 7 December 2002. The team, consisting of the two authors and local guides, visited different forest valleys around the two sites for three and five days, respectively.
Site 1: Pic de Fon (08°31'52"N, 08°54'21"W; 1,350 m; 28-30 November) included three higher elevation forest valleys. Site 2: 'Banko' (08°31'29"N, 08°56'12"W; 600 m; 2-7 December) included six relatively lower elevation forest valleys.

All surveys were conducted on foot by walking slowly (approximately 0.5 km/h) along line transects of varying length (500-4,000 m) and pausing regularly to look and listen for primates. We recorded direct signs of primates, such as sightings, and indirect signs, such as vocalizations and in the case of chimpanzees, nests. Each transect was surveyed only once and we either utilized pre-existing trails (mainly from hunters) or chose a given compass bearing direction. Due to the steep and sometimes very dense terrain we could not always follow a straight line but we tried to keep the transect as straight as possible. We measured the distance covered for each transect with the help of a topofil (hip chain) and recorded the habitat type. We concentrated all surveys in the forest zones and neglected the savannah, because the grass vegetation was up to 2 m high and prevented any sightings of primates. We also watched for primates in the savannah while driving between sites but never sighted any. All forest habitats could be characterized as gallery and ravine forests along streams that were patchily distributed in different valleys.

When we observed or heard monkeys we attempted to determine the species, the number of groups or individuals, and their sex. We also noted time and the position on the transect and estimated the perpendicular distance to the individual seen or heard. In the majority of cases we identified monkeys by their specific long-distance alarm calls given by male individuals. When we detected chimpanzee nests, the following measurements were taken: position on the transect, perpendicular distance to the nest, estimated height of the nest, dbh (diameter at breast height) of the nest tree, and age of the nests (Fresh: only green leaves in an intact nest, sometimes urine, feces; Recent: nest intact, but starting to dry, presence of yellow leaves; Old: nest fairly intact, only yellow leaves; Very old: gaps in the cup of the nests or leaves gone). We also determined nest groups (defined as a cluster of nests that are not further than 50 m apart and are of the same age) and their size.

Diurnal censuses took place between 6:30 a.m. and 17:00 p.m. On several occasions we listened for vocalizations of primates on elevated hills in the early mornings and evenings. We also walked along trails at night between 20:00 p.m. – 23:00 p.m. to census nocturnal prosimian species by picking up the eye shine with the help of a headlamp. At the first site we spent a total of 26 hours and at the second site a total of 45 hours to survey primates (this also includes time spent in the forest but not on a transect, e.g. whilst walking back to camp). Out of the 71 hours of total survey time, 7 hours (2 h at the first and 5 h at the second site) were used for nocturnal surveys.

The low number of direct and indirect signs of monkeys prevents us from calculating density estimates for the different species. We did, however, determine the density of chimpanzees by applying the 'standing crop nest count' method (Plumptre and Reynolds 1996). This method requires only one census for each area and allows density estimate calculations by taking into account the perpendicular distance to nests along transects and the rate of nest decay so that the counts can be corrected to the number of nests produced daily. We used a nest decay rate of 221 ± 22 days (validated for 21 nests in the Fouta Djallon region, Guinea by R. Kormos, personal communication) and a daily nest production rate of 1.15 ± 0.047 for corrections (chimpanzees build more than one nest per day because of day nest constructions and rare reuse of old nests). We applied the software program DISTANCE to analyze the data according to standard line transect analyses, in which the drop in the number of sightings with increasing perpendicular distance is modeled to obtain a probability estimate of sighting an object (Buckland et al. 1993).

We also questioned local people and hunters about the presence of primate species. People were first asked to describe the primates (color, shape, arboreal, terrestrial), imitate their vocalizations and name them in their local language before they were given pictures from which to identify them. Only if description, the local name (verified via a list of all primate species in several local languages), and the picture chosen agreed did we consider that primate to occur in the Simandou Range.

RESULTS

In the Pic de Fon classified forest, we were able to confirm the presence of two prosimian species (*Perodicticus potto*, Potto and *Galagoides demidoff*, Demidoff's Galago), five anthropoid monkey species (*Cercocebus atys atys*, Sooty Mangabey, *Cercopithecus campbelli campbelli*, Campbell's Monkey, *Cercopithecus petaurista buettikoferi* and *Cercopithecus nictitans*, Lesser Spot-nosed and Greater Spot-nosed Guenon, and *Cercopithecus diana diana*, Diana Monkey) and one hominoid ape species, *Pan troglodytes verus*, the West African Chimpanzee (Table 9.1). Moreover, by questioning local villagers, we assume the likely occurrence of four to five additional primate species (*Papio anubis*, Olive Baboon, *Erythrocebus patas*, Patas Monkey, *Cercopithecus aethiops sabaeus*, Green Monkey, *Colobus polykomos polykomos*, Western Black-and-white Colobus and possibly *Procolobus verus*, Olive

Colobus, Table 9.2). We were unable to record these species during our survey, but several local people and hunters provided precise descriptions. One more prosimian species, the Northern Lesser Bush Baby, *Galago senegalensis*, was not clearly described and not recorded but is likely to occur in the savannah habitat. The Pic de Fon classified forest therefore hosts 8 or possibly up to 14 different primate species out of 15 that potentially occur, an important representation of the regional primate diversity. The only primate species that might occur within this range and was neither observed nor known by the local population is the Western Red Colobus (*Procolobus badius*). Moreover, of the 14 primate species that (potentially) occur in Pic de Fon, four are Near Threatened or Endangered (*Cercocebus atys atys, Cercopithecus diana diana, Pan troglodytes verus,* and *Procolobus verus*).

Despite a high number of species, the abundance of the different primates seemed low. During 64 hours of daytime

Table 9.1. The primate species of the Pic de Fon classified forest in the Simandou Range, Guinea, recorded during survey walks and outside census work listed by site.

Species	Vernacular name	Pic de Fon	Banko Forest	Confirmation (N)
Perodicticus potto	Potto	-	+	S(1)
Galagoides demidoff	Demidoff's Galago	+	+	S(17), H(37)
Cercocebus atys atys	Sooty Mangabey	+	(+)	S(3), H(2)
Cercopithecus campbelli campbelli	Campbell's Guenon	+	(+)	H(6)
Cercopithecus petaurista buettikoferi	Lesser spot-nosed Guenon	+	(+)	S(1), H(2)
Cercopithecus nictitans	Putty-nosed or Greater Spot-nosed Guenon	-	(+)	S(1)
Cercopithecus diana diana	Diana Monkey	-	+	H(1)
Pan troglodytes verus	West African Chimpanzee	+	+	S(2), H(28), N(117)
Total number		5	8	S(25), H(76), N(117)

+   species present   S   sighted
-    species absent   H   heard
(+)   species recorded outside census work by other members of the RAP team   N   nest

Table 9.2. Likely occurrence of primate species in the Pic de Fon classified forest, not recorded during survey but identified by local villagers and hunters

Species	Vernacular name	Pic de Fon	Banko forest
Papio anubis	Olive Baboon	+	long ago
Erythrocebus patas	Patas Monkey	+	+
Cercopithecus aethiops sabaeus	Vervet, Grivet, or Green Monkey	+	+
Colobus polykomos polykomos	Western Black-and-White Colobus	+	+
(?) *Procolobus verus*	Olive Colobus	-	+
(?) *Galago senegalensis*	Northern Lesser Bush Baby	-	+

+   species present
-    species absent
(?)   awaiting confirmation

surveys in the forest we encountered (heard or saw) all the diurnal monkey species between one to a maximum of six times (encounter rates per hour ≤ 0.25; Tables 9.3, 9.4). Chimpanzees on the contrary were encountered more often (encounter rates per hour ≤ 0.63; Tables 9.3, 9.4). Nocturnal prosimians seem to occur in higher densities than the diurnal primate species (encounter rates per hour ≤ 1.43; Tables 9.4, 9.5). Whereas we were able to encounter eight different species at the second site 'Banko' and recorded only

five species at the first site Pic de Fon, we confirmed most of the species at the Pic de Fon by hearing or seeing them several times in different valleys but only by one single hearing or sighting at Banko (Tables 9.3, 9.5). Only Demidoff's Galago and the chimpanzee were heard or seen many times at the second site (Tables 9.3, 9.4, 9.5).

Concluding which primate species are most abundant based on the number of observations (direct and indirect) should be considered with care. Species such as the Potto or

Table 9.3. Diurnal censuses: number of sightings (S), hearings (H), or nest counts (N) per km transect in the Pic de Fon classified forest. All transects have been conducted in gallery forest habitat.

Trail	Transect length (m)	Cercocebus atys atys	Cercopithecus campbelli campbelli	Cercopithecus petaurista buettikoferi	Cercopithecus nicitans	Cercopithecus diana diana	Pan troglodytes verus
PIC DE FON							
I a (valley SW of camp)	2,100	2 H, (1 S, 3 ind.)	(I H)	-	-	-	-
I b (4 valleys S of camp)	1,500	-	unidentified group*		-	-	-
I b (adjacent valley S to Ib)	Listening point	1 H	2 H	1 H	-	-	-
I c (Pic Dalbatini valley NW)	550	-	2 H	1 H	-	-	25 N
BANKO							
II a (valley NE of camp)	4,000	-	-	-	-	1 H	5 H (1 party)
II b (valley NW of camp, West of river)	2,650	-	-	-	-	-	2 H (1 party), 35 N
II c (valley W of camp)	2,500	-	-	-	-	-	9 H, (3 parties), 10 N
II d (valley NW of camp, East of river)	1,400	-	-	-	-	-	1 H, 37 N
II e (2 valleys NW of camp)	2,500	-	-	-	-	-	4 H & 1 S (1 male), 1 S (7 ind., females & young), 10 N
Total (during transect)	17,200	3 H, (1 S)	5-6 H	2-3 H	-	1 H	21 H, 2 S, 117 N
Total (outside transect)**		2 S	1 H	-	1 S	-	4 H (1 party), 1 N
Total	17,200	3 H, 3 S	6-7 H	2-3 H	1 S	1 H	25 H, 2 S, 118 N

ind.	individuals
*	guide gave loud call upon sighting monkeys and monkeys fled (presumably *campbelli* and *petaurista*)
(S), (H)	sightings and hearings by other RAP members along transect (sighted same *Cercocebus* group that we heard)
**	Additional sightings and hearings by other RAP members
	one group of *Cercocebus* (4 ind.) sighted next to Pic de Fon camp (NW)
	one mixed species group of *Cercocebus* (10 ind.), *Cercopithecus petaurista* and *nictitans* sighted NE of Banko camp (area II b)
	one campbell's male monkey heard next to Banko camp (NW)
	one group of chimpanzee's (*Pan troglodytes verus*) heard next to Banko camp (NW)
	one nest sighted (valley NE of camp, area II b)

Table 9.4. Number of direct observations and encounter rates per hour for all the primate species surveyed.

	Whole area	Pic de Fon	Banko	Whole Area
	heard/seen	encounter rate per hour		
Perodicticus potto	1	0	0.2	0.14
Galagoides demidoff	10	0.5	1.8	1.43
Cercocebus atys atys	3	0.13	0	0.05
Cercopithecus campbelli campbelli	5-6	0.21-0.25	0	0.08-0.09
Cercopithecus petaurista buettikoferi	2-3	0.08-0.13	0	0.03-0.05
Cercopithecus diana diana	1	0	0.03	0.02
Pan troglodytes verus	8 parties	0	0.2	0.13
Pan troglodytes verus	25 individuals	0	0.63	0.39

Calculations based on total of 64 hours daytime and 7 hours nighttime surveys.
Cercopithecus nictitans once sighted by RAP mammal team.

the Lesser Spot-nosed Guenon are more quiet and/or cryptic compared to Demidoff's Galago or the Sooty Mangabey. However, we were able to confirm the presence of Demidoff's Galago, the Sooty Mangabey, the Campbell Monkey and the chimpanzee more than three times by either direct or indirect signs and this might indicate that these species are more abundant than others.

Besides hearing chimpanzees sometimes several times a day at the second site 'Banko', we were also able to observe them directly twice: on the first occasion a single male passed by closely, and on the second occasion we found a party of seven chimpanzees (an estrous (receptive) female and females with offspring) feeding on *Nauclea diderrichii*. Besides *Nauclea* we also observed other fruits during our survey that are known to be consumed by chimpanzees, such as *Parinari excelsa*, *Vitex doniana*, *Detarium microcarpum*, *Cola cordifolia*, *Irvingia gabonensis* and *Pseudospongias microcarpa*. In total, during six days of surveying the 'Banko' area, we heard chimpanzees calling 25 times, encountered presumably 9 different subgroups or parties of chimpanzees and heard 3 different males drumming (Table 9.3).

On 17.2 km of transect we observed a total of 117 nests (10 additional nests were only sighted after we left the transect to conduct measurements). Despite hearing chimpanzees in lower altitudes of 600 m, nests were found only above 900 m altitude but up to over 1,400 m. We identified 33 different nest groups, with a mean nest group size of 3 and a maximum nest group size of 12 nests (Table 9.6). We might however have overestimated the number of nest groups and underestimated mean nest group size, because nests that grouped together closely often showed different degrees of decay and were therefore counted as separate nest groups, although they might have been from the same group and only the decay of the tree species differed. The large majority of nests (90 %) were old or very old and we observed only nine recent nests and two that were fresh (probably day nests). Despite the relatively low number of fresh nests, chimpanzees clearly ranged within the area surveyed during our study period, indicated by sightings and daily hearings in the second study site. On average, chimpanzees constructed their nests at 18 m height on trees that were middle sized with a diameter of 37 cm. Generally they preferred slopes as nesting sites.

Table 9.5 Nocturnal observations of primates in the Pic de Fon classified forest, including observations from all RAP members.

Species	Pic de Fon	Banko forest	Habitat
Perodicticus potto	-	1 S	Gallery Forest
Galagoides demidoff	2 H*	17 S, 35 H**	Gallery & Secondary Forest
S	sighted		
H	heard		

*1 H by other RAP member
**11 S, 32 H from different members of RAP team

By applying DISTANCE and choosing the best model ('Negative exponential curve'), the density of chimpanzees for the forested areas (ca. 10,000 ha out of the total 27,000 ha, estimated from satellite images) of Pic de Fon is estimated as 0.64 chimpanzees/km², with a total population mean of 64 chimpanzees (18-226, 95 % confidence limits). Because only adult and weaned chimpanzees build nests, we need to correct for the proportion of the population that did not build nests (ca. 17.5 %, Ghiglieri 1984, Plumptre and Reynolds 1996), which leads to a total population mean of 75 chimpanzees (21-246, 95 % confidence limits). Community sizes in chimpanzees range from around 10 to over 100 individuals (Goodall 1986, Nishida et al. 1990, Boesch and Boesch-Achermann 2000, Herbinger et al. 2001). It is therefore likely that the Pic de Fon classified forest holds between one and possibly up to three or four different communities of chimpanzees.

DISCUSSION

Primates are behaviorally complex animals. Many live in structured social groups where they recognize kin over several generations, form long-lasting relationships with other group members, and in many cases require the exchange of individuals between groups before reproducing. Environmental changes, such as habitat destruction or hunting pressures, decrease the reproductive potential of primates, and because primates are long-lived, large-bodied, and slowly reproducing animals this can lead to rapid local extinction. Populations that are small and genetically and socially isolated are highly threatened in their survival and are unlikely to withstand problems of disease, inbreeding or human pressures. Since most primates tend to live in or near the area of their birth, they rarely have the possibility of migration to escape the effects of habitat changes.

In the Pic de Fon classified forest, the primate population is currently threatened by hunting, habitat destruction due to agricultural activity from the local population, and possibly also by noise pollution and increased access via mining roads to previously little-disturbed gallery forests as a result of mining exploration activities. The effects of agricultural activity are seen only in the lower altitudes while mineral exploration activities affect only the higher altitudes. The very elevated hunting pressure in the area is most likely affecting the primate population, despite the fact that the large majority of the population are Moslem and do not consume primates. Hunting practices (wire trapping) are unselective and primates that use the ground for locomotion (especially chimpanzees and mangabeys, but also other cercopithecine species) presumably fall victim often. Adult chimpanzees are known to be able to free themselves from wires but often either lose the concerned limb or die from related bacterial infections (Goodall 1986, Boesch and Boesch-Achermann 2000). Infant or juvenile chimpanzees are not able to free themselves from wires and are known to refuse the help of their mother and most often die from the

Table 9.6 Chimpanzee nest parameter for the different transects.

Trail	Altitude (m)	# of Nests	Nest Groups	Mean Nest Group Size	Max. Nest Group Size	Nest Age Classes Fresh	Recent	Old	Very old	Mean Height (m)	Mean dbh (cm)
PIC DE FON											
BANKO											
I c (550 m)	1,100-1,417	25	9	3	6	-	2 (8%)	5 (20%)	18 (72%)	12	26
II b (2,650 m)	900-1,077	35	9	4	8	1 (3%)	-	17 (48.5%)	17 (48.5%)	20	41
II c (2,500 m)	900-1,050	10	3	3	4	-	-	5 (50%)	5 (50%)	17	36
II d (1,400 m)	900-1,100	37	9	4	12	-	-	20 (54%)	17 (46%)	20	43
II e (2,500 m)	900-1,100	10	3	3	7	1 (10%)	7 (70%)	2 (20%)	-	20	38
TOTAL	900-1,417	117*	33	3	12	2 (1.7%)	9 (7.7%)	49 (41.9%)	57 (48.7%)	17.8	36.8

* 10 nests were sighted when we left the transect for measurements; they have not been included in the density estimation calculation.

resulting infections (Boesch and Boesch-Achermann 2000). Moreover, a minority of people whose religious beliefs do not prohibit them from consuming primates target all the primate species while hunting (personal communication of the local population).

We observed a much more elevated hunting pressure in the lower altitudes of the Pic de Fon classified forest. Whereas the primate and the mammal RAP teams found two wire traps, several hunter camp sites and trails, and a total of 11 cartridges around the higher elevation site Pic de Fon, the two groups collected a total of 35 wire traps and 32 cartridges and observed many fire and camp sites and trails all over the lower elevation site near Banko. We also heard several gunshots during the day and night despite the fact that the villages had pledged to avoid all hunting activities during our visit. The lower observation rate of all the anthropoid primate species at the second site Banko could be a reflection of the much more pronounced hunting pressure in this area. Either monkeys remained more quiet and cryptic or they in fact occur in lower densities at lower elevations.

The fact that we did not find any nests lower than 900 m altitude indicates that chimpanzees presumably have their range center or core area, where they spend most of their time, in the higher elevations. Due the high hunting pressure, they might consider the lower elevations to be too insecure (chimpanzees have been observed in low and high altitude habitats of around 150 m to more than 1,500 m). As a result, any future activities, potentially including mining, conducted in the higher altitude elevations of the forest could have a significant impact on the chimpanzee population, especially if hunting and agricultural pressure from the lower elevations continue unabated and chimpanzees are forced to continue moving to higher elevations to escape such pressures. If mining exploration proceeds to feasibility studies, these potential impacts should be considered in depth as part of a detailed Environmental and Social Impact Assessment (ESIA).

In the higher elevation forest patches, the presence of chimpanzees was confirmed only in valleys that had no immediate road access and were therefore less disturbed, suggesting the possibility that habitat alteration or noise disturbance from ongoing mineral exploration activities may have already had some effect on the chimpanzee population.

Despite the various threats, the Pic de Fon classified forest in the Simandou Mountain Range is still holding an important representation of the biodiversity of the primate order for the Upper Guinea Region. However, although the number of species is relatively high (8 and possibly 14), their low abundance (encounter rates of < 0.25 per hour for all diurnal monkey species) indicates that they are already highly threatened in their survival. Similar or slightly higher encounter rates for two of the species surveyed here are known from a rapid assessment in Marahoué National Park in Côte d'Ivoire (*Cercopithecus campbelli* and *C. petaurista*: 0.05-0.42 per hour, Struhsaker and Bakarr 1999). In this study, as in our survey, hunting pressure was negatively correlated

with encounter rates. The abundance of monkey species in other protected West African sites, for example in the Taï National Park in Côte d'Ivoire, seem much more elevated (one sees and hears a given species several times a day, due to high densities of 2 to over 100 ind./km², dependant on the species; Zuberbühler and Jenny 2002). Primate densities in the very simliar habitat of the nearby Mount Nimba region have not been studied extensively, but a preliminary report suggests densities of 1 to up to 30 ind./km², dependant on the monkey species (Galat-Luong and Galat unpublished report). As a result of more effective protection, population densities of the various species might therefore increase again in the Pic de Fon.

From other surveys, mostly conducted in lowland rainforests (Marchesi et al. 1995, Plumptre and Reynolds 1996), densities counting 1-2 chimpanzees/km² are known from intact primary forest, whereas densities are estimated lower for degraded forests (0.4 chimpanzees/km²) or human encroached forests and mosaic habitats (0.09 chimpanzees/km²). With a density (0.64 chimpanzees/km²) in between primary and degraded forest, the chimpanzee population in the Pic de Fon classified forest is most likely under threat and declining but is still holding an important number of individuals. A similar average population density (0.45 chimpanzees/km², varying from 3.0 ind./km² for gallery forest to 0.01 ind./km² for savannah) has also been found for the chimpanzees in the Mafou forest within the Haut Niger National Park, further north in a drier savannah habitat in Guinea (Fleury-Brugiere and Brugiere 2002). From a nationwide survey Ham (1998) estimated an average population density of 0.16-0.34 chimpanzees/km² for potential chimpanzee habitats within Guinea, indicating that the chimpanzee population of Pic de Fon classified forest with an above average density is an important one within the country.

Chimpanzees are listed as "Endangered" on IUCN's Red List and under Appendix I of CITES (Most Critically Endangered species), and out of the three subspecies the West African Chimpanzee is the most threatened by habitat destruction, hunting pressure, bushmeat and pet trade, as well as disease transmission. Currently there are presumably between 25,000 to 58,000 chimpanzees left in all of West Africa, with the majority living in unprotected areas (Kormos and Boesch 2003). It is therefore of high importance to protect the remaining chimpanzee populations and to establish additional protected areas. Moreover, even in areas where chimpanzees are more densely populated, like for example in western equatorial Africa, the recent spread of Ebola haemorrhagic fever rivals hunting as a threat to apes and has led to the near extinction of the local ape population (e.g. Minkébé forest in northern Gabon where the ape density dropped by about 99 % over the past decade, Walsh et al. 2003). Therefore, conserving smaller populations of chimpanzees becomes more and more important when large populations are equally threatened to become extinct.

Different chimpanzee populations have been shown

to be culturally distinct (they have a unique set of behavioral patterns that is passed on to the next generation), a trait that they share only with humans in its complexity (Whiten et al. 1999, Whiten and Boesch 2001). Protecting the chimpanzee population of the Simandou Mountain Range might therefore enable us to save a number of unique behavioral patterns that are either still undiscovered or only known for chimpanzees from this region. During long-term observations of a chimpanzee community in Bossou, the Mount Nimba region around 100 km further south, and a similar habitat to the Simandou region, a number of unique behaviors have been observed that are unknown for other regions (e.g. pestle pound (mash palm crown with petiole), insect pound (probe used to mash insect), resin pound (extract resin by pounding), branch hook (branch used to hook branch), dig (stick used as spade to dig termite nests), termite fish using leaf midrib, and algae scoop (scoop algae using wand); Whiten et al. 1999).

CONSERVATION RECOMMENDATIONS

To guarantee the survival of the primates in the Pic de Fon classified forest, we recommend that:

- Hunting should be totally banned for all Near Threatened and Endangered species and severely restricted for other species at lower risk, not only legally but also practically.

- Agricultural activities within the borders of the classified forest need to be identified and halted and programs developed enabling the local human population to maintain their standard of living without degrading the forest.

- A sufficiently high number of guards are urgently needed to effectively enforce current legislation protecting this classified forest, with equipment enabling them to survey the entire area.

- Conservation education is essential to alter hunting practices as well as attitudes so that primates are not perceived as abundant meat species or crop-pests. Due to the recognized closeness of chimpanzees and humans, chimpanzees could play an important role as a flagship species in educational campaigns.

- Hydrological studies are essential prior to any activities that might result in changes to the local watershed or regional climate change, including lowering the crest of the Simandou Range by several hundred meters. Climate changes and the disturbance of the ground water resources in the mountain range are likely to result in a reduction of forest cover, and as primates depend heavily on forested habitats this would have major impacts on the whole population.

- The effects of noise and habitat disturbance on primate populations require further analysis, as do the potential consequences to primate populations of lowering the crest of the Simandou Range.

- Immediate and full protection are required, especially for the four Near Threatened or Endangered primate species of Pic de Fon, the Sooty Mangabey, the Diana Monkey, the Olive Colobus and the West African Chimpanzee, to guarantee their survival, given the many current threats they face. If no immediate conservation measures are supported and implemented, we fear that most of the primates will disappear from this classified forest in the near future.

- Further surveys are strongly recommended during different seasons and covering longer time spans to confirm the distribution, ranging patterns, abundance, and status of the primate population along the Simandou Mountain Range, in order to provide adequate protection of the primates of Pic de Fon classified forest.

REFERENCES

Boesch, C. and H. Boesch-Achermann. 2000. The Chimpanzees of the Taï Forest: Behavioural Ecology and Evolution. Oxford University Press, Oxford.

Bshary, R. 1995. Rote Stummelaffen, Colobus Badius und Diana Meerkatzen, Cercopithecus Diana, im Taï-Nationalpark, Elfenbeinküste: wozu assoziieren sie? Dissertation, Universität München.

Buckland, S.T., D.R. Anderson, K.P. Burnham and J.L. Laake. 1993. Distance Sampling: Estimating Abundance of Biological Populations. Chapman & Hall, London.

Chapman, C.A. 1995. Primate seed dispersal: Coevolution and conservation implications. Evol. Anthrop. 4 (3): 74-82.

Chapman, C.A. and D.A. Onderdonk. 1998. Forests without primates: Primate/plant dependancy. Am. J. Primatol 45 (1): 127-141.

Chatelain, C., B. Kadjo, I. Kone and J. Refisch. 2001. Relations Faune-Flore dans le Parc National de Taï: une étude bibliographique. Tropenbos-Côte d'Ivoire Série 3.

Fleury-Brugiere, M.-C. and D. Brugiere. 2002. Estimation de la population et analyse du comportement nidificateur des chimpanzés dans la zone intégralement protégée Mafou du Parc national du Haut-Niger. Report to the Parc National du Haut-Niger/AGIR project, Faranah.

Galat-Luong, A. and G. Galat. 1978. Abondance relative et associations plurispécifiques des primates diurnes du parc national de Taï (Côte d'Ivoire). ORSTOM.

Galat-Luong, A. and G. Galat. unpublished report. Les Primates des Monts Nimba. Operation Pertubations et grande faune sauvage, Institut de Recherche pour le Developpement (IRD), Senegal, 1999.

Ghiglieri, M.P. 1984. The Chimpanzees of Kibale Forest: A Field Study of Ecology and Social Structure. Columbia University Press, New York.

Goodall, J. 1986. The Chimpanzees of Gombe. Belknap Press, Harvard University, Cambridge, MA.

Ham, R. 1998. Nationwide chimpanzee survey and large mammal survey, Republic of Guinea. Unpublished report for the European Communion, Guinea-Conakry.

Herbinger, I., C. Boesch and H. Rothe. 2001. Territory characteristics among three neighboring chimpanzee communities in the Taï National Park, Côte d'Ivoire. Int. J. Primatol. 22: 143-167.

Kormos, R. and C. Boesch. 2003. Regional Action Plan for the Conservation of Chimpanzees in West Africa. IUCN/SSC Action Plan. Washington, DC: Conservation International.

Lambert, J.E. and P.A. Garber. 1998. Evolutionary and ecological implications of primate seed dispersal. Am. J. Primatol. 45 (1): 9-28.

Lee, P.C., J. Thornback and E.L. Bennett. 1988. Threatened Primates of Africa, The IUCN Red Data Book. IUCN Gland, Switzerland and Cambridge, U.K.

Marchesi, P., M. Marchesi, B. Fruth and C. Boesch. 1995. Research Report: Census and distribution of chimpanzees in Côte d'Ivoire. Primates 36: 591-607.

Martin, C. 1989. Die Regenwälder Westafrikas: Ökologie – Bedrohung - Schutz. Birkhäuser Verlag, Basel.

Mittermeier, R.A., N. Myers, C.G. Mittermeier and P.R. Gil. 1999. Hotspots: Earth's Biologically Richest and Most Endangered Terrestrial Ecoregions. CEMEX.

Nishida, T., H. Takasaki and Y. Takahata. 1990. Demography and reproductive profiles. In: Nishida, T. (ed.). The Chimpanzees of the Mahale Mountains. Tokyo: Tokyo Univ. Press, Pp. 63-97.

Plumptre, A. J. and V. Reynolds. 1996. Censusing Chimpanzees in the Budongo Forest, Uganda. Int. J. Primatol. 17: 85-99.

Struhsaker, T.T. and M.I. Bakarr. 1999. A Rapid Survey of Primates and Other Large Mammals in Parc National de la Marahoue, Cote d'Ivoire. RAP Working Papers 10. Conservation International, Washington, DC.

Walsh, P.D., K.A. Abernethy, M. Bermejo, R. Beyers, P. De Wachter, M.E. Akou, B. Huijbregts, D.I. Mambounga, A.K. Toham, A.M. Kilbourn, S.A. Lahm, S. Latour, F. Maisels, C. Mbina, Y. Mihindou, S.N. Obiang, E.N. Effa, M.P. Starkey, P. Telfer, M. Thibault, C.E.G. Tutin, L.J.T. White and D.S. Wilkie. 2003. Catastrophic ape decline in western equatorial Africa. Nature advance online publication, 6 April 2003 (doi:10.1038/nature01566).

Whiten, A., J. Goodall, W.C. McGrew, T. Nishida, V. Reynolds, Y. Sugiyama, C.E.G. Tutin, R.W. Wrangham and C. Boesch. 1999. Cultures in chimpanzees. Nature 399: 682-685.

Whiten, A. and C. Boesch. 2001. The cultures of chimpanzees. Sci. Am. 284 (1): 60-67.

Whitesides, G.H., J.F. Oates, S.M. Green and R.P. Kluberdanz. 1988. Estimating Primate Densities from Transects in a West African Rain Forest: A Comparison of Techniques. J. Anim. Ecol. 57 (2): 345-367.

Zuberbühler, K. and D. Jenny. 2002. Leopard predation and primate evolution. J. Hum. Evol. 43: 873-886.

Chapter 10

A socio-economic threats and opportunities assessment for the Forêt Classée du Pic de Fon, Guinea

Eduard Niesten and Léonie Bonnehin

SUMMARY

This chapter presents an analysis of socio-economic factors contributing to biodiversity threats in Guinea's Pic de Fon classified forest. The following section describes the methodologies employed in this analysis, namely a survey of available literature and an information exchange workshop convened by CI in Conakry. The chapter then reports the findings derived from the literature survey and the workshop. Following a brief discussion of analysis and conclusions based on these findings, the chapter closes with broad recommendations that set the stage for the Initial Biodiversity Action Plan.

INTRODUCTION AND METHODOLOGY

A thorough search of the published and gray literature generated very little socio-economic data specific to the Pic de Fon classified forest. Even more discouraging was the dearth of historic records and time-series studies, making an analysis of socio-economic trends almost impossible. Substantial further fieldwork is required to develop a robust, quantitative understanding of the threats and opportunities in the Pic de Fon classified forest, beyond the scope of the current effort with respect to both budget and timeline. This is particularly true of socio-economic data, which will be critical to design of appropriate conservation strategy.

Given the lack of documentation on relevant socio-economic trends in the area, Conservation International convened a workshop in Conakry on Dec. 12-13, 2002 to elicit information on threats and opportunities for biodiversity conservation in the Pic de Fon region from a broad spectrum of stakeholders and experts. This workshop sought to complement literature-based research with information from parties directly engaged in conservation in the immediate project area, Guinea, and the broader region. The exercise also afforded an opportunity to solicit additional documentation not obtained through the initial literature searches. Thus, the overarching purpose of the workshop was to facilitate information exchange.

The workshop was facilitated by CI (Léonie Bonnehin) using the ZOPP approach (Zielorientierte Projektplanung, or OOPP- Objective Oriented Project Planning - in English).[1] This approach relies on a systematic structure for stakeholder identification and issues analysis in a workshop setting. The analysis typically proceeds in four principal steps:

1. Stakeholders: an overview of persons, groups and organizations connected to a project and their interests, attitudes, and implications for project planning.
2. Threats: enumeration of major threats, including cause and effect and prioritization.

---

[1] For a more complete presentation of the methodology, see http://web.mit.edu/urbanupgrading/upgrading/issues-tools/tools/ZOPP.html.

3. Goals: a reformulation of the problems into realistically achievable goals.
4. Opportunities: identification of opportunities and feasibility assessment of alternatives.

The defining characteristic of this methodology is that it takes place without pre-formulated programs of action or conclusions. The results of the Conakry workshop thus emerged from a participatory process with contributions from all participants. Since the workshop outputs benefitted from input from all participants, they are likely to enjoy broad-based support.

Participants were selected to represent a broad range of perspectives and expertise regarding conservation efforts throughout Guinea, the Greater Nimba Highlands, and the Pic de Fon classified forest. In total, 33 participants attended from 17 organizations including national government, multilateral donors, bilateral donors, NGOs and scientific research centers, in addition to RTM&E (Table 10.1 lists the parties in attendance at the workshop). The Conakry workshop would have benefited from the contribution of perspectives from representatives of communities located surrounding the Pic de Fon classified forest. PGRR is the institution that has made the most progress thus far in establishing links with these communities, but this engagement is only in its early stages and has not yet had sufficient time to lay the foundation for this type of exchange.

RESULTS

Literature Survey
Threats to Biodiversity in the Upper Guinea Forest
Extreme poverty, rapid increase in human population density and under-funded environmental governance characterize the Upper Guinea Forest region. Remaining habitat continues to deteriorate due to commercial logging and the spread of agriculture and agroforestry, while fauna is threatened by intense bushmeat hunting. Meanwhile, civil unrest limits development of human capacity and weakens environmental enforcement in a setting that already features a low level of institutional capacity. Moreover, refugee flows intensify pressure on forest resources, with over 600,000 refugees located in Guinea alone (Bakarr et al. 2001). Thus, daunting challenges for conservation throughout the Hotspot include reliance on slash-and-burn cultivation practices, heavy commercial mining pressure, a deeply entrenched bushmeat hunting sector and persistent civil conflict (Bakarr et al. 2001).

Threats to Biodiversity in Guinea
The National Biodiversity Action Plan of Guinea emphasizes the following drivers of biodiversity loss in the country: human population growth, a forced focus on short-term economic development needs, reliance on inappropriate technologies, limited means for actualizing biodiversity values, a lack of community resource governance, human migration, political instability and civil war (MMGE 2002).

Table 10.1. Conakry Workshop Participants

Organization	# of participants
Ministry of Mines, Geology, and the Environment, Guinea	2
National Directorate of Water and Forests, Guinea	4
Global Environment Facility (GEF) Guinea	1
USAID	2
PGRR/N'Zérékoré Forestry Center	2
CEGEN (Nimba Environmental Management Center)	4
UNDP (United Nations Development Program) Guinea	1
UNESCO MAB (Man and Biosphere Reserve Program)	1
Rio Tinto Mining and Exploration Limited	1
AGIR (European Union Integrated Resource Management Program)	1
Conservation International	3
University of Würzburg (Germany)	1
Guinée Ecologie	3
PEGRN/USAID (Expanded Natural Resource Management Program)	1
CERE (Ecology Research Center, University of Conakry)	1
Winrock International	1

Perhaps the most daunting fact confronting conservation planning is that some of the poorest people in the world live in Guinea, and particularly in Guinée Forestière. Moreover, their means for sustenance and livelihood are limited almost entirely to forest-based income. At least 90 % of all energy consumption in Guinea is in the form of wood and charcoal (MMGE 2002). Agriculture, which accounts for around 85 % of employment in the region, depends on conversion of forest areas to cultivation (MMGE 2002). Nearly 140,000 ha of forest are destroyed each year in Guinea (MMGE 2002). Just two years earlier, Konomou and Zoumanigui (2000) estimated annual deforestation at 120,000 ha. Bushfires damage vast areas of the country every year. At least 17 of 190 mammal species in Guinea are threatened with extinction, and at least 16 of 625 bird species are threatened. At least 36 plant species are Endangered. Bushmeat hunting, urbanization, refugee flows and illiteracy rates approaching 70 % also pose obstacles to conservation efforts.

The four pillars of the national strategy to conserve biodiversity in the face of these threats are: the creation of a representative protected area system, inclusion of local communities through participatory management arrangements, the development of human capacity to fulfill a wide range of conservation roles and the reinforcement of local, regional and international cooperation in conservation efforts (MMGE 2002).

The Pic de Fon classified forest

The Pic de Fon classified forest is surrounded by 24 villages whose inhabitants directly impact the forest through converting forested land to agricultural land, setting bushfires to create pasture and artisanal mining (PGRR 2002b). Camara and Guilavogui (2001) suggest that villagers are well aware of the Pic de Fon classified forest boundaries, but this does not deter them. Uncontrolled use of bushfires (for hunting or the preparation of land for cultivation or pasture) and artisanal mining have been identified as the principal threats to the remainder of the forest block (PGRR 2001, Camara and Guilavogui 2001). Thus, biodiversity in the Simandou Range and the Pic de Fon classified forest in particular faces a multitude of threats of anthropogenic origin. However, of the many variables and trends that contribute to the complex dynamics behind these pressures, demographic change may be the most fundamental.

Population Profile

The Pic de Fon classified forest overlaps with the prefectures of Beyla and Macenta. At present, population density in these prefectures remains relatively low, documented at 9 to 16 inhabitants per km^2 (Hatch 1998). However, population density may be higher in the southern foothills of the Pic de Fon classified forest (Konomou and Zoumanigui 2000), in the area that PGRR has identified as of principal interest for conservation and natural resource management efforts (PGRR 2001). Hatch (1998) reported population growth

rates as high as 2.9 % per year for the prefectures of Beyla and Macenta, suggesting that these population densities are rapidly increasing, doubling in fewer than 25 years. However, it remains to be verified whether demographic trends among populations living in the immediate vicinity of the Pic de Fon classified forest are comparable to those of the larger prefectures.

Several ethnic groups are represented in the three prefectures, including Konianké, Peul, Kouronko, Toronka, Kissi, Malinké, Guerzé, Toma and Manian (Hatch 1998, Konomou and Zoumanigui 2000). The indigenous populations of the 24 villages immediately surrounding the Pic de Fon classified forest consist principally of Toma, Manian, Konianké and Guerzé ethnic groups. Peul and Malinké people also operate in the area in the trade, pastoralism and artisanal mining sectors (Camara and Guilavogui 2001).

Konomou and Zoumanigui (2000) report a refugee population of 629,275, or about 40 % of the population in Guinée Forestière (Konomou and Zoumanigui 2000). Indeed, dramatic influxes of refugees from Liberia and Sierra Leone may account for the high population growth rates recorded in the Kérouané, Beyla and Macenta prefectures.

In sum, impending problems related to population include: increasing pressure on the resource base due to population growth and continued refugee flows; and eroding community-level resource management institutions due to increasing ethnic diversity as a consequence of migration.

Economic Activity

The principal economic activities in the Pic de Fon classified forest area are agriculture and pastoralism, accompanied by some involvement in small-scale artisanal mining for gold as well as iron and diamonds (Camara and Guilavogui 2001). Several sources including Conakry workshop participants mention artisanal mining as a biodiversity threat in the Pic de Fon classified forest. However, the RAP team reported only one observation of artisanal mining activity (see Appendix 3, description for site coded FONBA3), so the extent to which it occurs remains to be verified. Bushmeat hunting is also a prominent activity.

Agriculture in Guinée Forestière revolves around rice production. In the region as a whole, approximately 120,000 ha are sown annually for rainfed rice; rice occupies more than three-quarters of the total cultivated area in the region (Konomou and Zoumanigui 2000). Farmers rely predominantly on shifting cultivation methods that include burning in preparation for planting. In relatively fertile locations such as valleys and depressions in the areas surrounding the Pic de Fon classified forest, the spread of permanent rice cultivation is a leading cause of deforestation (Konomou and Zoumanigui 2000).

Agriculture accounts for about 90 % of employment in the area surrounding Pic de Fon classified forest. Cultivation concentrates on rice, the principal staple for local consumption, in combination with a wide range of other crops (fonio, manioc, potatoes, yam, maize, peanuts, taro, etc.).

Forested areas also feature agroforestry production of cash crops such as coffee, palm oil, cocoa, banana and kola nuts (Camara and Guilavogui 2001).

Cultivation occurs within the boundaries of the Pic de Fon classified forest. Among a subset of 10 villages in Beyla prefecture alone, PGRR has documented at least 561 ha of cultivation within the Pic de Fon classified forest (PGRR 2002b). Rice accounts for the greatest proportion (45 %) followed by manioc (17 %), bananas (11 %) and coffee (9 %). Thus, agricultural encroachment into the Pic de Fon classified forest includes both subsistence and cash crops. PGRR (2001) estimates that as much as 35-40 % of the original forest block has been affected by agricultural encroachment.

Extensive livestock activities in the area include the husbandry of cattle, sheep and goats. A minority of the communities also raise pigs. Livestock is raised on natural pastures, and bushfires are used to stimulate pasture regeneration (Camara and Guilavogui 2001).

Hunting also ranks among the threats to biodiversity in the area (Hatch 1998, Camara and Guilavogui 2001). Although Guinean law accommodates traditional hunting rights for subsistence, there are indications that hunting also takes place on a commercial basis to service the bushmeat trade (PGRR 2001). Such hunting is permitted under existing law for species that are not under special protection, but subject to well-defined restrictions. Chimpanzees and other threatened primates and large mammals are 'integrally protected', meaning that their killing or capture is prohibited by law. However, the absence of an enforcement presence leaves a distinct possibility that the various regulations are ignored. The prevailing practice of burning bush to drive game is particularly destructive to habitat, although no specific evidence was found during the RAP to indicate that at present this practice is widely used in the Pic de Fon classified forest.

The literature survey did not yield direct information regarding the question of whether bushmeat is an important component of diets among the villages surrounding the Pic de Fon classified forest. However, we do know that pigs are raised in some of these villages and religious or cultural dietary restrictions against bushmeat typically accompany similar restrictions against pork. Therefore, it is unlikely that restrictions against bushmeat operate in all villages around the Pic de Fon classified forest. Moreover, of the four main regions in Guinea, bushmeat consumption is most prevalent in Guinée Forestière. The RAP team found snares and traps, heard gunshots, and found shell casings, leaving no doubt that bushmeat hunting takes place in the Pic de Fon classified forest. To what extent this hunting serves subsistence needs or is driven by commercial bushmeat markets remains to be determined.

In summary, impending problems related to economic activity include: continued conversion of forest to agriculture in the Pic de Fon classified forest; continued burning for forest clearing, pasture regeneration, and hunting; and an increasing scale of hunting to service a growing population.

Logging

Commercial logging is accelerating throughout Guinea and indeed the entirety of the Upper Guinea Forest ecosystem (see Figure 10.1). This reflects both efforts to generate foreign exchange through timber exports, and the pressure of a large and growing fuelwood demand. Fuelwood supplies about 77 % of household energy needs, and charcoal another 3 % (Diawara 2001). Logging is identified as one of the principal threats to biodiversity in the National Biodiversity Action Plan (MMGE 2002). In the two sites surveyed, the RAP expedition found little indication of the extent to which logging has impacted the Pic de Fon classified forest thus far. However, analysis by the FAO suggests that by 2020 in Guinea demand will outstrip supply of wood for both lumber and fuel, and also for non-timber forest products (NTFPs) (Diawara 2001). Therefore, within the next 15 years, classified forests without formal protected status and adequate enforcement capacity are likely to be under threat throughout Guinea.

Infrastructure

The prefectures that overlap with the Simandou Range are not well served with respect to infrastructure. Road access to the region generally is poor, and electricity and telecommunications infrastructure are limited and unreliable. High transportation costs and weak communications linkages result in a limited degree of integration of the area into the national economy. However, pending infrastructure investments will raise the returns to cash-generating activity (cash crops, logging, bushmeat) and thus strengthen incentives for forest conversion in the Pic de Fon classified forest and surrounding areas.

Institutional Context

The State is virtually a non-presence in the area. Ostensibly, the Centre Forestier de N'Zérékoré and CEGEN operate under government mandates, but rely heavily on German assistance in the former case and UNESCO and the UNDP in the latter. Therefore, if the government is to take an active role, a substantial investment in capacity building will be required. Institutions at the local community level do not exert

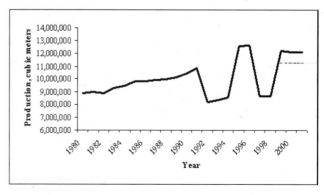

Figure 10.1. Annual Roundwood Production, Guinea (1980-2001)

noticeable influence on resource management within the Pic de Fon classified forest. This is not surprising, as activities within the Forêt Classée are extralegal and insecurity of tenure is not conducive to investment in management regimes or cooperation. However, substantial field research is required to develop a more comprehensive assessment of institutional presence, capacity and need with regard to natural resource management at the community level.

Regional Political Context

Over the past three years, relations between Lansana Conté and other leaders in the region have improved considerably, perhaps indicating a reduced likelihood of armed conflict between Guinea and any of its neighbors. However Guinea and Liberia continue to exchange accusations of supporting rebel movements against the other's ruling regime. Political stability within Guinea appears to have solidified over the past few years as well, but is by no means certain. Thus, refugee influxes and internal displacement of Guineans remain distinct possibilities in the foreseeable future. Events in Côte d'Ivoire starkly illustrate how apparent stability and prosperity can be illusory. Liberia remains a cause for concern, a low-intensity insurrection in Senegal continues to simmer, and it remains to be seen if Sierra Leone can sustain progress toward peace as the UN withdraws its presence.

As noted above, Konomou and Zoumanigui (2000) report a refugee population of nearly 630,000 in Guinée Forestière. Camps operate near N'Zérékoré, Seredou and Kouankan, housing refugees from both Liberia and Sierra Leone. UNEP (2000) documents the environmental challenges that have accompanied the presence of a large refugee population in south-eastern Guinea, and the potential for still further increases in the refugee population must be accommodated in any comprehensive Biodiversity Action Plan for the area. For instance, if the situation in Côte d'Ivoire further deteriorates, streams of new refugees may flow into Beyla prefecture (Tahirou 2002).

Workshop Results

The workshop opened with a presentation of the initial results obtained from the RAP exercise in the Pic de Fon classified forest, including an introduction to the overall context for biodiversity conservation in the Upper Guinea Forest Hotspot. The participants then agreed on the objectives of the workshop, as stated below:

1. To identify and confirm the principal stakeholders in the region
2. To identify and characterize threats to biodiversity and obstacles confronting conservation in the region
3. To develop an initial prioritization of these biodiversity threats
4. To identify opportunities and possible solutions to these biodiversity threats

Workshop Objective 1: To identify and confirm the principal stakeholders in the region

The participants at the Conakry workshop identified 14 categories of stakeholders relevant to the Pic de Fon region, and listed concerns and expectations on the part of each of these stakeholder groups with respect to biodiversity conservation. Stakeholders were defined as any actor that could influence, in either a positive or negative manner, the pursuit of biodiversity conservation in the Pic de Fon classified forest. Stakeholders and their concerns and expectations as identified at the workshop are listed in Table 10.2.

Workshop Objective 2: To identify and characterize threats to biodiversity and obstacles confronting conservation in the region

The participants were asked to identify and give evidence for the principal threats to biodiversity in the Pic de Fon classified forest. This exercise did not explicitly differentiate between proximate causes and ultimate trends. Instead, participants listed broad-scale threats to biodiversity and obstacles to conservation, and identified factors contributing to these threats and obstacles. Table 10.3 lists the principal existing threats to biodiversity in the Pic de Fon classified forest, concluded at the Conakry workshop. The list reflects the informed opinion of carefully selected experts, but does not imply that firm evidence or documentation is available for each threat. Many of the threats listed in Table 10.3 therefore remain to be ground-truthed and quantified.

Workshop Objective 3: To develop an initial prioritization of biodiversity threats in the Pic de Fon

After identifying the principal threats to biodiversity, workshop participants sought to reach consensus on an initial ranking of these threats. However, they concluded that insufficient data is available to develop a robust ranking. Instead, the workshop elected to prioritize elements of a possible conservation strategy for the Pic de Fon classified forest, discussed in the following section, and strongly emphasized that further research and data collection is required as an immediate first step.

Workshop Objective 4: To identify opportunities and possible solutions to biodiversity threats in the Pic de Fon

The dearth of firm data on biological and socio-economic trends in the Pic de Fon classified forest and surrounding area severely constrains our ability to devise appropriate conservation strategies for the Pic de Fon classified forest. Nevertheless, based on their experience and knowledge as experts on the region, the workshop participants developed several recommendations to address the threats identified in Table 10.3. Table 10.4 lists the Conakry workshop recommendations for components of a conservation strategy for the Pic de Fon classified forest. The prioritization of these actions is discussed in the section of this chapter on Recommendations.

Stakeholder	Concerns	Goals/Expectations Pertaining to Conservation
Government of Guinea	Impact of natural resource exploitation on resource base Needs of electorate	Increased biodiversity protection and sustainable management of natural resources Economic development Increased financial resources to support conservation
CERE/University of Conakry		Improved research facilities and infrastructure Involvement in research activities Financial support for research staff Financial support of research center Development of scientific expertise
Donors (e.g., USAID, World Bank)	Poor financial management Misdirection of funds	Conservation investment strategies Biodiversity protection
Communities surrounding the Pic de Fon classified forest	Expropriation of lands Restrictions on resource use rights	Participation in resource management decisions Financial benefits at community levels Equitable sharing of benefits Sustainable resource management Economic alternatives/opportunities to replace destructive activities
Mining sector; Rio Tinto	Constraints on resource extraction Limitation or closure of mining concessions	Defined area for protection based on impact study Coordinate investments with protection plan Reconcile resource extraction with conservation Biodiversity conservation
Timber sector		Improved exploitation methods Sustainable resource management Protection of timber concessions Organized marketing channels for timber Nearby markets
Environmental NGOs	Insufficient participation of local communities Lack of adequate funding Conflict between NGOs and other institutions Failure to observe agreements Diversion of conservation funds Exclusions of NGOS from decisions-making processes	Protection of endangered species Environmental education Strengthen NGO capacity Sustainable resource management Participation in planning, implementation, and tracking of conservation strategy Enhance biodiversity values Restore degraded habitat Reinforce protection of high priorities areas Promote respect for rule of law
CEGEN/N'Zérékoré Forestry Center		Improved knowledge of biodiversity Strengthening institutional capacity Development of partnerships Development of regional management plan Marketing of mining concession boundaries Co-management of resources with local communities
Handicraft manufactures	Restrictive regulation Barred access to resources	Improve access to credit/micro-credit Sound management of raw materials Resource access rights Enhance value of artisanal products Training and assistance to improve marketing Access to improved resource management methods and technology Development of export channels
Apiculturalists		Modernization of methods Marketing of honey

Stakeholder	Concerns	Goals/Expectations Pertaining to Conservation
Farmers from communities neighboring the Pic de Fon classified forest	Limitations on farming area Expropriation of land	Introduction of new methods Improvement of methods Enhance value of produce Increased returns/yields Agricultural credit Financial security Value adding in production
Fishermen from communities neighboring the Pic de Fon classified forest	Restrictive regulation Reduced resource access	Improved fishing techniques Watershed protection Demarcation of fishing zones Collaborative management Enhance value of catch
Birds hunters operating in the Pic de Fon classified forest	Restrictive regulation Prohibition of bird capture	Increased quotas Sustainable use of bird resource
Bushmeat hunters operating in the Pic de Fon classified forest	Hunting ban Halt to wildlife use Restrictive regulation Regulation of hunting zones	Sustainable use of wildlife resources Definition of hunting zones Participation in wildlife management

Table 10.3. Principal Biodiversity Threats in the Pic de Fon

Threat/Problem	Contributing Factors
Deforestation	Timber exploitation, aggressive logging Slash-and-burn shifting cultivation Bushfire Clearing for pasture
Agricultural pressure	Cultivated plots in the Forêt Classée / demand for fertile land Tenure conflicts Immigration to the region, human population growth Uncontrolled spread of cash crops
Declining fauna and flora	Over-exploitation Poaching, unregulated fishing Over-harvesting of forest products Unregulated collection of herbs, medicines
Ecosystem destruction	Noise pollution from heavy machinery associated with Rio Tinto exploration activity Air pollution and pollution of waterways Destruction of river/stream beds for artisanal gold and diamond mining Deforestation near springs Introduction of alien species
Institutional weakness	Lack of respect for laws / contradictory regulations Weakness in human capacity / lack of monitoring
Human population growth	Population movements in response to exploitation Demographic growth Presence of plots in the Forêt Classée Refugee migration
Widespread poverty	Very low standard of living Illiteracy Inadequate health and education infrastructure

Table 10.4. Conservation Actions Needed in the Pic de Fon classified forest

Solutions	Actions
Improve knowledge of the Pic de Fon classified forest	Scientific research into biodiversity Support for environmental evaluation of the Pic de Fon classified forest and a biological inventory Implement a structure for ecological tracking in the Pic de Fon classified forest
Development of a sustainable biodiversity management strategy	Adopt appropriate models for timber exploitation Protect exceptional sites such as hot springs Implement sustainable resource management Include local communities in management Develop a management plan Promote permanent settlement as opposed to shifting modes of production
Strengthen institutional capacity for conservation	Strengthening legal conservation status of the Pic de Fon classified forest Enforce legislation Financial involvement of the State Support harmonization of relevant laws
Public awareness building	Build public awareness of importance of biodiversity conservation Disseminate knowledge of relevant laws Environmental education
Support local community development	Support economic alternatives that are compatible with biodiversity conservation Promote wildlife based tourism Disseminate ecologically compatible agricultural techniques Create community centers and health clinics

DISCUSSION

The preceding sections present the findings from two methods of data collection – the literature survey and the workshop held in Conakry. Qualitatively, the overall picture that emerges from each of these methods is consistent. However, neither exercise yielded adequate information regarding the scale, intensity and time horizon characterizing the various types of threats to biodiversity in the Pic de Fon classified forest. Without reliable quantitative information on these aspects of the trends affecting the Pic de Fon classified forest, any prioritization of threats and the responses to them remains tentative.

Agriculture appears to be one of the most pressing threats to the integrity of the Pic de Fon classified forest. Agricultural encroachment from communities surrounding the Pic de Fon classified forest is already evidenced by the presence of plots for both subsistence and cash crops. The planting of perennial tree crops such as coffee requires a substantial investment on the part of farmers, indicating a perception that long-term cultivation within Pic de Fon classified forest boundaries is unlikely to provoke a reaction from authorities. Moreover, the reliance on shifting cultivation methods necessarily will exert persistent pressure on remaining habitat as areas are cleared for new plots.

The literature survey and the workshop both identified bushmeat hunting as a leading threat to biodiversity in the Pic de Fon classified forest. This conclusion is supported by the RAP team, who observed numerous snares and traps, heard rifle shots and found shell casings left by hunters. The widespread use of fire as a means to drive game in this region also can destroy substantial areas of habitat, although the RAP team did not find indications of the use of this method in the areas surveyed.

The use of fire to prepare pasture for livestock was identified as a threat in the general region in the literature and the workshop, but there are no specific references to livestock encroachment in the Pic de Fon classified forest. However, each of the 24 villages surrounding the Pic de Fon classified forest engages in raising livestock. Therefore, combined with the use of fire to clear forest for agriculture and to drive game, uncontrolled bushfires to stimulate pasture regeneration remain a potent threat to habitat within the Pic de Fon classified forest.

Several sources mention illegal artisanal mining inside the Pic de Fon classified forest boundaries, and Conakry workshop participants also identified such mining as a threat. In particular, the literature and participants note the destructive impact of artisanal or small-scale mining on streambeds and waterways. As the RAP team only reported one observation of artisanal mining activity, it is identified as a possible threat, pending further research and verification.

Although various materials reviewed in the literature survey and the Conakry workshop both mention timber, the RAP team saw no indication that the area has been subjected to large-scale logging in the past. However, commercial logging is expanding in Guinea, and demand for fuelwood rapidly is outpacing the regenerative capacity of the resource base. Collection of wood for fuel, construction and the like by members of communities neighboring the Pic de Fon classified forest is in all likelihood taking place, but the scale and impact of such activity remains to be determined. The RAP team noted that the area from the entrance of the classified forest up to the foot of the mountain near Camp 2 is stripped of trees (a few skeletons stand out), likely due to local wood use for houses, furniture, firewood, etc. in addition to farmland clearing. At this stage, timber exploitation is categorized as a threat that may be significant now, and, if not, is likely to become increasingly acute over time.

Finally, unregulated extraction of other resources such as fish and non-timber forest products (NTFPs) from the Pic de Fon classified forest was also identified in the literature survey and the workshop. However, neither exercise yielded information regarding the fish species typically caught or the range of NTFPs typically collected, not to mention extraction rates or overall impact. Without further information on the nature or scale of extraction, fishing and NTFP collection can only be considered possible threats to biodiversity.

Thus, agricultural encroachment and bushmeat hunting appear to be the most important threats to biodiversity in the remaining forests of the Pic de Fon. The use of uncontrolled fires in both of these activities as well as livestock pasture preparation can be particularly devastating for natural habitat. Other threats such as illegal artisanal mining, over-fishing, timber exploitation and NTFP collection remain to be verified and quantified. Table 10.5 summarizes conclusions regarding threats to biodiversity in the Pic de Fon classified forest.

RECOMMENDATIONS

At the workshop held in Conakry participants prioritized needed conservation actions. They agreed that in the immediate term (1 to 2 year time horizon) urgently needed actions include strengthening of enforcement capacity, particularly with regard to hunting regulations, further research on several socio-economic and biological aspects of the area,

an awareness-building campaign among local communities, and maintenance of Pic de Fon classified forest boundary markers. It was also emphasized that support for local community development is essential in order to promote sustainable livelihoods, decrease pressure on forest resources, and to achieve long-term conservation objectives.

Efforts in Mount Nimba serve as a suitable model for designing strategy for the Pic de Fon classified forest. The program of CEGEN, the body responsible for coordinating conservation efforts in the Mount Nimba area, includes protection of core areas, complemented by measures to ensure local community benefits. These measures include experimentation with alternatives to bushmeat hunting to supply protein, water sanitation projects, and the provision of health services. Moreover, CEGEN has entered into discussions with the Government of Guinea that has recently resulted in the incorporation of the Simandou Range within its mandate.

The recommended socio-economic data collection necessarily must take place at the community level, through surveys, interviews and direct exchange between researchers and villagers. Conducting such research requires extensive efforts to convey the purpose of the research to local people while also managing expectations that might emerge regarding eventual benefits at the local community level. Doing so requires a considerable investment of time and resources in relationship building, and a fair degree of prior commitment to continued follow-up work once the research is complete. Data collection at the community level that does not scrupulously adhere to such principles is irresponsible on many levels. As an example of the resources and time required for a full consultation process, a proposal to the Global Environment Facility to fund a similar exercise at Mount Nimba budgeted $560,000 to finance activities over a period of 12 months (GEF 1999).

It will be critical to formulate a conservation strategy for the Pic de Fon within the broader context of the Greater Nimba Highlands. The inclusion of the Pic de Fon under the mandate of CEGEN is an important step in this direction. Many of the biodiversity threats noted for the Pic de Fon classified forest are part of dynamics that operate at the regional level, including bushmeat hunting, extensive livestock raising, uncontrolled use of fire and refugee flows.

Table 10.5. Biodiversity Threats in the Pic de Fon classified forest

Principal Threats to Biodiversity		
Spread of agriculture	Unregulated hunting	Forest clearing for pasture
Slash-and-burn cultivation	Uncontrolled bushfires	
Possible Threats to Biodiversity		
Illegal artisanal mining	Timber exploitation	Unregulated NTFP collection
Over-fishing	Commercial mining operations	

Therefore planning of conservation investments must also proceed on a regional basis. Regional perspectives are crucial, as a major threat to larger mammals in particular is habitat fragmentation. The natural corridor being constructed to link the chimpanzee population of Bossou to those of the Nimba Mountains serves as an example of such regional planning and action (Ham et al. in process).

MATERIALS CONSULTED

[Author/institution uncertain]. 1995. Livre Blanc/ Monographie Nationale sur la Diversité Biologique. Guinea/UNBio: Conakry. (494 pp.)

Arid Lands Information Center (University of Arizona). [date uncertain]. Draft Environmental Profile – Guinea. United States National Committee for Man and the Biosphere [UNMAB]. USAID/Office of Forestry, Environment, and Natural Resources: Washington DC. (229 pp.)

Bakarr, M., B. Byler, R. Ham, S. Olivieri and M. Omland. 2001. From the Forest to the Sea: Biodiversity Connections from Guinea to Togo. Conservation Priority-Setting Workshop, December 1999. Conservation International: Washington, DC.

Bonnehin, L. 2002. "Analyse des Menaces et Opportunités de Conservation de la Biodiversité au Pic de Fon, Massif de Simandou, Guinèe". Rapport de l'Atelier ZOPP, Conakry Dec. 12-13 2002. Conservation International.

Camara, L. 2000. "Revue et Amélioration des données relatives aux produits forestiers en République de Guinée". Programme de partenariat CE-FAO (1998-2002). Food and Agriculture Organization of the United Nations: Rome.

Camara, N. 1999. "Revue des données du Bois-Energie en Guinée". Programme de partenariat CE-FAO (1998-2002). Food and Agriculture Organization of the United Nations: Rome.

Camara, W. and K. Guilavogui. 2001. "Identification des villages riverains autour des forets classes du Pic de Fon et Pic de Tibe". PGRR/Centre Forestier N'Zérékoré, Mesures Riveraines.

CEGEN. (no date). "The Nimba Mountains: balancing environment and sustainable development". Produced by Fauna & Flora International.

Diawara, D. 2000. "Les Donnees Statistiques sur les Produits Forestiers Non-Ligneux en Republique de Guinèe". Programme de partenariat CE-FAO (1998-2002). Food and Agriculture Organization of the United Nations: Rome.

Diawara, D. 2001. "Document National de Prospective – Guinée". L'Etude prospective du secteur forestier en Afrique (FOSA). Food and Agriculture Organization of the United Nations: Rome.

Direction Générale des Services Economiques. 1953. Arrete Portant Classement la forêt de "Fon".

[GEF] Global Environment Facility. 1999. "PDF-B Proposal – Conservation of biodiversity through integrated participatory management in the Nimba Mountains."

Government of the Republic of Guinea. 1999. "Global Environment Facility PDF-B Proposal – Conservation of Biodiversity through Integrated Participatory Management in the Nimba Mountains".

Government of the Republic of Guinea. 2000. "Interim Poverty Reduction Strategy Paper". International Monetary Fund: Washington DC.

Hatch & Associates, Inc. 1998. "Preliminary Environmental Characterisation Study: Simandou Iron Ore Project Exploration Programme, South-Eastern Guinea". Montreal, Canada.

Konomou, M. and K. Zoumanigui. 2000. "Zonage Agro-Ecologique de la Guinée Forestière". Centre de Recherche Agronomique de Seredou. Ministere de l'Agriculture, des Eaux et Forets: Guinée.

Kormos, R., T. Humle, D. Brugière, M.-C. Fleury-Brugière, T. Matsuzawa, Y. Sugiyama, J. Carter, M. S. Diallo, C. Sagno and E.O. Tounkara. 2003. Chapter 9 The Republic of Guinea. Pp. 63-76. In: Kormos, R., C. Boesch, M.I. Bakarr and T.M. Butynski (eds.). West African Chimpanzees. Status survey and recommendations for the Conserviation Action Plan.Gland, Switzerland and Cambridge, UK. IUCN/SSC Primate Specialist Group. IUCN,. , ix + 219.

Ministere de l'Agriculture, des Eaux et Forets/Direction Nationale des Forets et de la Faune. 199?. "Expose des Motifs: Projet de Loi portant Code de la Protection de la Faune Sauvage et règlementation de la Chasse. Guinea: Conakry.

Ministere des Traveaux Publics et de l'Environnement/ Direction Nationale de l'Environnement. 1997. Monographie Nationale sur la Diversité Biologique. Guinea/ PNUE: Conakry. (146 pp.)

[MMGE] Ministry of Mines, Geology, and the Environment. 2002 (January). National Strategy and Action Plan for Biological Diversity; Volume 1: National Strategy for Conservation Regarding Biodiversity and the Sustainable Use of these Resources. Guinea/UNDP/GEF: Conakry.

PGRR. 2001. "Pic de Fon – Situation Actuelle, Analyse et Recommandations (Proposition d'axes de collaboration)". Centre Forestier N'Zérékoré.

PGRR. 2002a. "Proposition de stratégie pour la surveillance du Pic de Fon – Attente PGRF". Centre Forestier N'Zérékoré, Section Conservation – Biodiversité.

PGRR. 2002b. "Rapport d'Activites – Delimitation du versant sud-ouest de la Forêt Classée du Pic de Fon". Centre Forestier N'Zérékoré (Antenne Bero).

Rio Tinto. 2002. Reglements Concernant l'Environnement, la Sante, la Securite et les Relations avec les Communautes et Dispositions Concernant l'Environnement et la

Securite Destinees aux Entrepreneurs. Rio Tinto Mining and Exploration Ltd Africa/Europe Region.

Robertson. 2001. Guinea. *In:* Fishpool, L.D.C. and M.I. Evans (eds.). Important Bird Areas in Africa and Associated Islands: Priority sites for conservation. Newbury and Cambridge, UK: Pisces Publications and BirdLife International. Pp 391-402.

Szczesniak, P. 2001. "The Mineral Industries of Côte d'Ivoire, Guinea, Liberia, and Sierra Leone". U.S. Geological Survey Minerals Yearbook.

Tahirou, B. 2002. "Collecte de données bibliographiques sur la Forêt Classée du Pic de Fon et du perimeter de prospection de Rio Tinto". Draft report submitted to Conservation International.

United Nations Environment Programme (UNEP, in collaboration with UNCHS and UNHCR). 2000. Environmental Impact of Refugees in Guinea. UNEP/Regional Office for Africa: Nairobi, Kenya.

USAID/Guinea. 2000. "Briefing Book for Invited Partners". Synergy Workshop.

USAID/Guinea Natural Resources Management project. 1996. "Scoping statement – Nyalama classified forest co-management initiative". (28 pp.)

U.S. Department of State [USDOS]. 2000. FY 2001 Country Commercial Guide: Guinea. U.S. & Foreign Commercial Service, U.S. Department of State: Washington DC.

World Bank. 2001. Country Assistance Strategy Progress Report for the Republic of Guinea. Report No. 22451. World Bank: Washington DC.

Wright, L. and M. Sylla. 2001. "Mining Annual Review 2001 – Guinea". The Mining Journal.

Wright, L. and M. Sylla. 2002. "Mining Annual Review 2002 – Guinea". The Mining Journal.

Ziegler, S., G. Nikolaus and R. Hutterer. 2002. "High Mammalian Diversity in the Newly Established National Park of Upper Niger, Republic of Guinea". *Oryx.* Vol. 36, No. 1: pp. 73-80.

Table des matières

Préface

Dans de nombreuses régions du globe, le développement des ressources naturelles va de pair avec des considérations économiques, sociales et environnementales. Cette compétition se produira de plus en plus fréquemment, à mesure que de nouvelles régions sont touchées par ce développement. Des meilleurs mécanismes de dialogue et de coopération doivent être mis en place au niveau des principales parties prenantes pour empêcher que le développement des ressources ne compromette inutilement les autres facteurs. C'est dans cet esprit que Rio Tinto Mining and Exploration Limited (RTM&E) et Conservation International ont négocié en novembre 2002 un protocole d'accord pour initier des activités à l'intérieur et aux alentours de la Forêt Classée du Pic de Fon dans la chaîne du Simandou.

Il existe très peu de données publiées sur la chaîne du Simandou, et plus particulièrement sur le Pic de Fon. Cependant, le Pic de Fon, avec sa mosaïque unique d'habitats et notamment des habitats de montagnes rares, a été pressenti pour receler une importante biodiversité, et a été désigné comme zone prioritaire pour l'évaluation de la biodiversité au cours de l'Atelier pour la Sélection des Priorités de CI qui s'est tenu à Elmina au Ghana en 1999. Outre sa richesse en biodiversité, la chaîne du Simandou abrite aussi une richesse minéralogique potentielle, et en particulier des gisements importants de minerai de fer. Depuis 1997, RTM&E mène des activités d'exploration minière sur des sites contigüs de la chaîne du Simandou où la compagnie détient une licence d'exploration.

L'Evaluation Biologique Rapide (RAP), financée par Rio Tinto, a été menée en novembre et décembre 2002 par une équipe pluridisciplinaire de 13 scientifiques internationaux et et régionaux, des spécialistes de la biodiversité terrestre en Afrique de l'Ouest. Conjointement, une évaluation des menaces et des opportunités socio-économiques a été réalisée, avec une revue de la littérature grise et l'organisation d'un atelier regroupant toutes les parties prenantes. Cet atelier a réuni 33 participants de 17 organisations, dont des représentants du gouvernement national, des bailleurs de fonds multilatéraux et bilatéraux, des ONG, des centres de recherche scientifique et de la société RTM&E. L'atelier a permis d'identifier les menaces pesant sur la biodiversité de la région, ainsi que les opportunités pour la conservation.

Ce travail initial améliorera la connaissance scientifique dans la région tout en apportant des informations essentielles pour la gestion environnementale de RTM&E sur le site. Les données contribueront à fournir des bases plus détaillées à d'autres études et à l'Evaluation de l'Impact Environnemental et Social (ESIA) qui sera conduit en parallèle aux études de faisabilité globales, si le projet se poursuit au-delà de la phase d'exploration en cours.

Ces activités ont également créé les bases d'un Plan d'Action Initial pour la Biodiversité (IBAP) pour le Pic de Fon et pour la Grande Chaîne du Nimba (qui inclut à la fois la chaîne du Simandou et la chaîne du Nimba). Selon l'IBAP, l'objectif global est l'amélioration de la gestion des ressources naturelles dans la Région du Nimba, par la lutte contre les menaces pesant sur la biodiversité, par la promotion de pratiques agricoles saines et par le développement de sources de revenus alternatives et durables pour les communautés qui dépendent largement des forêts pour vivre.

RTM&E et CI espèrent que les résultats de ce travail permettront d'améliorer la gestion des ressources et les efforts de conservation non seulement au Pic de Fon, mais aussi dans d'autres zones de la Grande Chaîne du Nimba. Ce rapport ne devrait pas être considéré comme une fin en soi, mais plutôt comme un point de départ destiné à susciter l'intérêt, la connaissance et le soutien de la promotion de la conservation régionale et du développement durable, à travers une coopération et un dialogue renforcés avec les principales parties prenantes locales et régionales.

Colin Harris
Rio Tinto Mining and Exploration Limited
Africa/Europe Division

Glenn Prickett
Conservation International
Center for Environmental Leadership in Business

Participants et auteurs

Mohamed Alhassane Bangoura (Reptiles et Amphibiens)
Chercheur Independent affilié au Centre de Gestion de
l'Environnement des Monts Nimba
BP 1869, Conakry
Guinée
Email. Mohamed_alhassane@yahoo.fr

Abdulai Barrie (Grands Mammifères)
Department of Biological Sciences
Faculty of Environmental Sciences
Njala University College
University of Sierra Leone
PMB, Freetown
Sierra Leone
Email. ahbarrie@yahoo.com

Jan Decher (Micromammifères)
Department of Biology
University of Vermont
Burlington, VT 05405
États-Unis
Email. Jan.Decher@uvm.edu

Nicolas Londiah Delamou (Plantes)
Chef d'Antenne Forêt Classée Mont Béro
P.G.R.R./Centre Forestier N'Zérékoré
BP 171, N'Zérékoré
Guinée
Email. cfzpgrr@sotelgui.net.gn

Ron Demey (Oiseaux)
Van der Heimstraat 52
2582 SB Den Haag
Pays-Bas
Email. rondemey@compuserve.com

Mamadou Saliou Diallo (Coordinateur)
Guinée-Ecologie
210, rue DI 501, Dixinn
B.P. 3266, Conakry
Guinée
Email. dmsaliou@mirinet.com

Njikoha Ebigbo (Chiroptères)
Department of Experimental Ecology
University of Ulm
Albert-Einstein-Allee 11
89069 Ulm
Allemagne
Fax. +49731 5022683
Email. njikoha.ebigbo@biologie.uni-ulm.de

Jakob Fahr (Chiroptères)
Department of Experimental Ecology
University of Ulm
Albert-Einstein-Allee 11
89069 Ulm
Allemagne
Fax. +49731 5022683
Email. jakob.fahr@biologie.uni-ulm.de

Ilka Herbinger (Primates)
Max Planck Institute for Evolutionary Anthropology
Department of Primatology
Deutscher Platz 6
04103 Leipzig
Allemagne
Email. herbinger@eva.mpg.de

Jean-Louis Holié (Plantes)
Consultant botaniste
P.G.R.R./Centre Forestier N'Zérékoré
BP 624, Conakry / BP 171, N'Zérékoré
Guinée

Soumaoro Kante (Grands Mammifères)
Division Faune et Protection de la Nature
Direction Nationale des Eaux et Forêts
BP 624, Conakry
Guinée
Email. dfpn@sotelgui.net.gn

Jennifer McCullough (Coordinateur, Editeur)
Conservation International
1919 M Street NW, Suite 600
Washington, DC 20036
États-Unis
Email. j.mccullough@conservation.org

Piotr Naskrecki (Insectes: Tettigoniidae)
Director, Invertebrate Diversity Initiative
Conservation International
Museum of Comparative Zoology
Harvard University
26 Oxford St.
Cambridge, MA 02138
États-Unis
Email. p.naskrecki@conservation.org

Hugo Rainey (Oiseaux)
School of Biology
Bute Medical Building
University of St. Andrews
St. Andrews, Fife KY16 9TS
United Kingdom
Email. hjr3@st-andrews.ac.uk

Mark-Oliver Rödel (Reptiles et Amphibiens)
Department of Animal Ecology and Tropical Biology
Biocenter
Am Hubland, D-97074 Würzburg
Allemagne
Email. roedel@biozentrum.uni-wuerzburg.de

Elhadj Ousmane Tounkara (Primates)
ENRM Project
Winrock International
BP 26, Conakry
Guinée
Email. ourypdiallo@yahoo.com
jfischer@sotelgui.net.gn

Profil des organisations

CONSERVATION INTERNATIONAL

Conservation International (CI) est une organisation non-gouvernementale internationale à but non-lucratif basée à Washington, DC, États-Unis. CI est convaincu que l'héritage naturel de notre planète doit être préservé pour que les générations futures puissent prospérer spirituellement, culturellement et économiquement. CI a pour mission de conserver l'héritage naturel vivant sur terre, notre biodiversité globale et de démontrer que les sociétés humaines peuvent co-exister de manière harmonieuse avec la nature.

Conservation International
1919 M Street NW, Suite 600
Washington, DC 20036
ÉTATS-UNIS
tel. 800 406 2306
fax. 202 912 0772
web. www.conservation.org
www.biodiversityscience.org

RIO TINTO MINING ET EXPLORATION LIMITED

Rio Tinto Mining and Exploration Limited (RTM&E) est la division en charge de l'exploration de Rio Tinto p.l.c, une société internationale d'exploitation minière basée au Royaume-Uni. Les activités de Rio Tinto permettent de fournir les minéraux et les métaux essentiels aux besoins mondiaux et contribuent à améliorer les conditions de vie. Rio Tinto affirme l'obligation d'excellence pour assumer ses responsabilités en matière de santé, de sécurité, d'environnement et auprès des communautés.

Rio Tinto Mining and Exploration
6 St James Square
Londres
SW1Y 4LD
Royaume-Uni
tel. 0207 9302399
fax. 0207 9303249
web. www.riotinto.com

CENTRE FORESTIER DE N'ZÉRÉKORÉ

Le Centre Forestier de N'Zérékoré (CFN) est la structure de la Direction Nationale des Eaux et Forêts (DNEF) en charge du travail sur le terrain dans la région forestière guinéenne. A travers le Projet de Gestion des Ressources Rurales (PGRR), le CFN est responsable actuellement de la gestion forestière dans la région du Pic du Fon, et s'occupe notamment de la délimitation des forêts.

PGRR/ Centre Forestier de N'Zérékoré
BP 171, N'Zérékoré
GUINÉE
tel. (224) 91 15 03
email. cfzpgrr@sotelgui.net.gn

GUINÉE ECOLOGIE

Guinée Ecologie est une organisation non-gouvernementale de droit guinéen enregistrée en 1990, sur base de volontariat et à but non lucratif. La vision de l'organisation se base sur les préoccupations globales relatives à l'état actuel de la terre et à la pression de nombreuses activités humaines non durables. La mission de Guinée Ecologie est de contribuer à travers la recherche, l'éducation, l'information, la communication et la gestion à la protection de l'environnement et à la conservation de la nature et de la biodiversité selon les principes du développement durable.

Guinée Ecologie
210, rue DI 501, Dixinn
BP 3266, Conakry
GUINÉE
tel. (224) 46 24 96
email. dmsaliou@afribone.net.gn

Remerciements

L'assistance de nombreuses personnes a permis le succès de cette mission RAP. Les participants remercient Mamadou Saliou Diallo et les membres de Guinée Ecologie pour la remarquable organisation logistique en Guinée. Sans leur sérieuse assistance, l'ensemble de la mission n'aurait sans doute pas pu se réaliser. Dans la région du Pic de Fon, nous avons reçu tout le support possible de la part du Centre Forestier de N'Zérékoré et de son Conservateur Cécé Papa Condé. Merci aussi à l'équipe de Rio Tinto Mining and Exploration Limited (RTM&E) représentée par John Merry et Luc Stévenin, pour nous avoir fourni le carburant, la nourriture et toute l'aide dont nous avions besoin sur la montagne. Travailler sur le second site n'aurait pas été possible sans l'assistance et le soutien des villageois de Banko. Nos assistants locaux, Bernard Doré et Kaman Camara, parmi tant d'autres, ont été d'une aide inestimable durant les travaux sur le terrain, tout comme l'agent de la Direction Nationale des Eaux et Forêts Ouo Ouo Bamanou. Nous aimerions aussi remercier tous les participants du RAP pour leur agréable et stimulante compagnie. Le Ministère de l'Agriculture et des Eaux et Forêts, République de Guinée, a fourni les autorisations de prélèvement et d'exportation des échantillons (Code 100, Nr. 000741- 742). Nous sommes redevables envers Conservation International (CI) en général, et en particulier envers Jennifer McCullough et Leeanne E. Alonso pour leur invitation à participer à ce Programme d'Evaluation Biologique Rapide ainsi que pour son organisation. Merci aussi à Francis Lauginie et Guy Rondeau pour leur assistance dans la préparation de ce RAP.

Nous souhaitons souligner les contributions de John Merry, Colin Harris et Sally Johnson de RTM&E ainsi que David Richards de Rio Tinto p.l.c. Sans leur aide matérielle et financière, ce projet n'aurait pas été possible et leurs commentaires sur ce rapport ont été précieux. Nous souhaitons remercier le Centre for Environmental Leadership in Business (CELB) de CI, et en particulier Assheton Carter, Marielle Canter, et Greg Love pour avoir facilité la collaboration avec RTM&E et piloté le Plan Initial d'Action pour la Biodiversité. Leonie Bonnéhin de CI – Côte d'Ivoire, a représenté un atout inestimable pour ce projet, à la fois avant et après les travaux sur le terrain, en assurant la gestion logistique, en organisant et pilotant parfaitement l'atelier des menaces et opportunités (en coopération avec Mamadou Saliou Diallo), et en réalisant des travaux complémentaires à cette étude. Nous devons aussi remercier les autres membres du Programme Afrique de l'Ouest de CI, notamment Jessica Donovan, Mohamed Bakarr, Olivier Langrand et Tyler Christie pour leur implication et leurs conseils avant et après l'expédition. De plus, sans l'aide de Rebecca Kormos, cette expédition aurait pu ne jamais avoir lieu. Nous souhaitons la remercier sincèrement pour son enthousiasme et pour nous avoir fait profiter de ses nombreux contacts et de sa connaissance de la Guinée. Nous devons aussi remercier Karl Morrison du Régional Conservation Strategies Group de CI pour sa participation et son assistance durant l'atelier et pour le Plan Initial d'Action pour la Biodiversité, Eduard Niesten pour son aide dans la formulation de la méthodologie d'évaluation des menaces et opportunités, Mark Denil du Conservation Mapping Program de CI et Glenda Fabregas pour la patience dont elle a fait preuve dans la conception des publications du RAP.

L'équipe chiroptère est très reconnaissante envers les conservateurs suivants pour avoir fourni des données sur les spécimens de leur collection: Peter Taylor (Durban), William Stanley (Chicago), David Harrison (Sevenoaks), Georges Lenglet (Bruxelles), Manuel Ruedi (Genève),

Wim van Neer (Tervuren), Nico Avenant (Bloemfontein), Susan Woodward (Toronto), Dieter Kock (Frankfort/M.), Fritz Dieterlen (Stuttgart) et Teresa Kearney (Pretoria). Meredith Happold (Canberra), Ernest Seamark (Pretoria), Paul van Daele (Livingstone, Zambie) et P. Taylor ont gentiment mis à notre disposition des données non publiées sur la distribution de *Myotis welwitschii*. Le présent travail a été partiellement financé par le programme BIOLOG du Ministère Allemand de l'Education et de la Recherche (BMBF; Project W09 BIOTA-West, 01LC0017). L'équipe petits mammifères non volants souhaite remercier le Dr. R. Hutterer du Museum Alexander Koenig de Bonn, Allemagne ; L. Gordon et Dr. M. Carleton du United States National Museum, Washington ; et Dr. C. W. Kilpatrick de l'Université du Vermont pour leur assistance dans l'identification et la conservation des spécimens échantillons.

Rapport succinct

UNE EVALUATION BIOLOGIQUE RAPIDE DE LA FORÊT CLASSÉE DU PIC DE FON, CHAÎNE DU SIMANDOU, GUINÉE

Dates de l'expédition
27 novembre – 7 décembre 2002

Description de la zone

La chaîne du Simandou est située au sud-est de la Guinée et s'étend sur 100 km, de Komodou au nord à Kouankan au sud. A l'extrémité sud de la chaîne du Simandou se trouve la Forêt Classée du Pic de Fon. En 1953, environ 25 600 ha de forêts ont reçu un statut de protection limitée avec comme objectif la conservation de la forêt et des sols. La chaîne du Simandou constitue également une barrière naturelle contre les feux de savane et régule les cours d'eau ayant leur source dans le massif (notamment les rivières Diani, Loffa et Milo). Cette réserve située dans la zone de transition entre la forêt et la savane présente une variété de types d'habitat allant de la forêt pluviale à la savane guinéenne humide, en passant par les forêts galeries de montagne et de ravins. L'altitude de la Forêt Classée du Pic de Fon varie de 600 m à plus de 1 600 m au-dessus du niveau de la mer (le Pic de Fon est le point le plus élevé de la chaîne à 1 656 m d'altitude et le deuxième sommet de la Guinée), ce qui donne lieu à un type d'habitat supplémentaire, la prairie montagneuse, un habitat rare et menacé en Afrique de l'Ouest. La Forêt Classée du Pic de Fon est entourée de 24 villages dont les habitants dépendent de la forêt comme source de nourriture, d'eau, de bois de chauffe et de produits médicinaux. Deux sites ont été étudiés au cours de cette expédition RAP : l'un près du sommet du Pic de Fon et l'autre à plus basse altitude, près du village de Banko.

Justification de l'expédition

Il existe très peu de données publiées sur la chaîne du Simandou, et plus précisément sur le Pic de Fon. Cependant, les experts ont pressenti que le Pic de Fon, avec sa mosaïque unique d'habitats et notamment des habitats de montagne rares, abriterait une importante biodiversité. Le Pic de Fon a été désigné comme zone prioritaire pour l'évaluation de la biodiversité au cours de l'Atelier pour la Sélection des Priorités de Conservation International (CI) qui s'est tenu à Elmina au Ghana en 1999. Outre sa richesse biologique, la chaîne du Simandou abrite aussi une richesse minéralogique potentielle, et en particulier d'importants gisements de minerai de fer. Rio Tinto Mining and Exploration Limited (RTM&E) mène actuellement des activités d'exploration sur quatre sites contigüs de la chaîne du Simandou où ils détiennent une licence d'exploration.

Ce RAP a été réalisé à l'instigation de Rio Tinto pour contribuer à leurs évaluations environnementales et sociales initiales. Les résultats seront incorporés aux études de base et aux évaluations plus détaillées lors de l'Etude d'Impact Social et Environnemental (ESIA) qui sera menée parallèlement à l'évaluation générale de faisabilité, si le projet se poursuit au-delá de l'actuelle phase d'exploration. Compte tenu de l'importance potentielle de la biodiversité de la

Forêt Classée du Pic de Fon, RTM&E a entériné un accord avec CI, accord arrangé par le « Center for Environmental Leadership in Business » (CELB), pour évaluer la biodiversité de la région, ainsi que les menaces potentielles et les opportunités pour la conservation de la Forêt Classée du Pic de Fon. Ce partenariat vise à réaliser d'importantes avancées pour la conservation de la biodiversité, l'industrie, les communautés qui dépendent des ressources de la chaîne du Simandou et le gouvernement guinéen.

Principaux résultats

Une grande diversité d'habitats terrestres et de taxons a été observée durant cette étude RAP. Les types d'habitats comprennent des prairies de montagne, des forêts de montagne (à la fois des forêts galeries et des forêts de ravins), des forêts semi-décidues de basse altitude (forêts primaires et secondaires), des savanes, des cours d'eau de montagne, des habitats buissonnants d'altitude, des plantations pérennes (café, cacao, banane) et des jachères. Les études réalisées sur les habitats ont porté sur les plantes (ligneuses et herbacées), les invertébrés (sauterelles), les amphibiens, les reptiles, les oiseaux, les petits mammifères (y compris les chauves-souris) et les grands mammifères (avec une attention particulière portée aux primates).

Au total, l'équipe du RAP a répertorié 804 espèces. Plusieurs d'entre elles sont nouvelles pour la science, dont au moins cinq invertébrés (dont deux sont aussi de nouveaux genres) et trois amphibiens. L'équipe du RAP a observé une extension de la zone de répartition de certaines espèces. Plusieurs espèces sont nouvelles pour la Guinée dont 11 invertébrés, 3 amphibiens, 7 oiseaux, 3 chauves-souris et 1 musaraigne. La Forêt Classée du Pic de Fon héberge de nombreuses espèces d'importance internationale pour la conservation, parmi lesquelles un arbre (*Neolemonniera clitandrifolia*) classé par l'UICN comme « Menacé d'extinction », ainsi que 16 arbres classés comme « Vulnérable » et un arbre classé comme « Presque Menacé » ; un amphibien (*Bufo superciliaris*) et quatre espèces de reptiles (*Chamaeleo senegalensis, Varanus ornatus, Python sebae,* et *Osteolaemus tetraspis*) qui sont protégés sur le plan international (CITES) ; deux espèces de primates classées par l'UICN comme « Menacé d'Extinction » : *Pan troglodytes verus* (Chimpanzé d'Afrique Occidentale) et *Cercopithecus diana diana* (Cercopithèque diane), deux espèces classées comme « Presque Menacé » : *Cercocebus atys atys* (Cercocèbe fuligineux) et *Procolobus verus* (Colobe de Van Beneden) ; huit oiseaux d'importance mondiale pour la conservation, y compris trois classés par l'UICN comme « Vulnérable » : *Schistolais leontica,* le Prinia de Sierra Leone, *Criniger olovaceus,* le Bulbul à barbe jaune et *Lobotos lobatus* l'Echenilleur à barbillons; et sept espèces de chauves-souris classées par l'UICN, une comme « Vulnérable » (*Epomops buettikoferi*) et buettikoferi six espèces comme « Presque Menacé » (*Rhinolophus alcyone, R. guineensis, Hipposideros jonesi, H. fuliginosus, Kerivoula cuprosa, Miniopterus schreibersii*). Les chasseurs locaux et les guides ont confirmé la présence du micropotamogale (*Micropotamogale lamottei*) classé par l'UICN comme « Menacé d'extinction ».

Nombre d'espèces recensées

Plantes	409
Sauterelles (Orthoptera)	40
Amphibiens	32
Reptiles	12
Oiseaux	233
Chauves-souris (Chiroptera)	21
Petits mammifères (non-volant)	17
Grands mammifères	39
Primates	13

Nouvelles espèces découvertes

Sauterelles (Orthoptera)	Au moins 4 esp.
	Thyridorhoptrum sp.
	Afrophisis sp.
	2 esp. d'un nouveau genre de Meconematinae
	peut-être *Ruspolia* sp.
Blatte cavernicole (Blaberidae: Oxyhaloinae)	*Simandoa conserfariam* Gen. et sp. n
Amphibiens	*Arthroleptis/Schoutedenella* sp.
	Amnirana sp.
	Bufo ? sp.

Nouveaux inventaires pour la Guinée

Sauterelles (Orthoptera)	*Phaneroptera minima*
	Phaneroptera ? *maxima*
	Zeuneria melanopeza
	Weissenbornia praestantissima
	Ruspolia jaegeri
	Thyridorhoptrum sp. n.
	Meconematinae Gen. n., sp. n. 1
	Meconematinae Gen. n., sp. n. 2
	Afrophisis sp. n.
	Anodeopoda cf. *lamellate*
Blattes cavernicoles (Blaberidae: Oxyhaloinae)	*Simandoa conserfariam* Gen. et sp. n
Amphibiens	*Arthroleptis/Schoutedenella* sp.
	Amnirana sp.
	Bufo ? sp.
Oiseaux	*Bubo poensis*
	Neafrapus cassini
	Indicator willcocksi
	Smithornis capensis
	Phyllastrephus baumanni
	Cercotrichas leucosticta
	Vidua camerunensis
Chauves-souris	*Myotis welwitschii*
	Kerivoula phalaena
	Kerivoula cuprosa
Petits mammifères non-volants	*Crocidura grandiceps* (musaraigne)

Recommandations pour la conservation

Les résultats de cette étude RAP montrent que la Forêt Classée du Pic de Fon abrite des espèces terrestres et aquatiques uniques ; des connexions intactes entre l'habitat, la flore et la faune ; et l'un des derniers habitats montagneux intacts, à la fois en Guinée et dans les montagnes de Haute Guinée. A l'issue de cette étude, nous recommandons fortement que la Forêt Classée du Pic de Fon reçoive un statut de protection accrue et que les réglementations actuelles concernant ses statuts de classement soient contrôlées, appliquées et renforcées au besoin. Les limites de la zone protégée devraient être clairement définies et appliquées en collaboration avec les communautés locales, avec l'appui de programmes d'éducation et de sensibilisation, ainsi que d'études scientifiques et socio-économiques supplémentaires. Une importante proportion d'espèces végétales et animales recevrait une protection adéquate si la Forêt Classée du Pic de Fon était incorporée dans un système efficacement géré de zones protégées comprenant les forêts, les prairies, des zones représentatives des différentes variations de micro-habitats avec une attention particulière aux habitats critiques et aux espèces endémiques et menacées. L'équipe du RAP a trouvé peu d'espèces invasives durant cette étude, et des mesures devraient être prises afin d'empêcher l'introduction de telles espèces.

Nos résultats soulignent également le fait que toute exploitation potentielle des ressources minérales dans la zone du Pic de Fon, si elle est entreprise, nécessitera davantage de recherches sur les impacts hydrologiques potentiels et les impacts possibles sur les espèces dépendant des habitats de haute altitude. Les habitats de prairie de montagne sont extrêmement rares en Afrique de l'Ouest. Aussi, toutes les activités détruisant ou endommageant irréversiblement ce type d'habitat devraient-elles être considérées avec précaution.

Les 24 villages environnants dépendent de cette forêt pour nombre de services fournis par l'écosystème (dont la nourriture, l'eau et les produits médicinaux) et tout devrait être mis en œuvre pour garantir que la forêt continuera à fournir ces services. Les besoins de la communauté doivent être pris en compte afin que chaque zone protégée reste viable, tandis que les empiètements illégaux (et la destruction subséquente de la flore et de la faune) doivent être empêchés pour en assurer la viabilité à long terme, à la fois dans l'intérêt des communautés locales et pour protéger l'une des dernières forêts intactes d'Afrique de l'Ouest. La propriété foncière est un sujet extrêmement sensible, et s'il est crucial de placer la région du Pic de Fon sous une forme renforcée de protection légale le plus rapidement possible, le succès de cette zone protégée repose sur la discussion et une étroite collaboration avec les communautés et autres parties prenantes dans la région. Ces dialogues faciliteront l'entente entre le gouvernement, les groupes de conversation et les communautés. L'intention est d'atteindre les objectifs de conservation en utilisant une variété de mesures pour répondre aux besoins en terres agricoles, en bois de chauffe, en gestion durable de l'environnement et en autres services fournis par la forêt.

Résumé

INTRODUCTION

Située en Afrique de l'Ouest, la forêt guinéenne est l'un des 25 écosystèmes terrestres les plus riches en biodiversité mais aussi l'un des plus menacés. La forêt de Haute Guinée, l'un des deux principaux blocs forestiers de l'écosystème guinéen, s'étend sur six pays ouest africains, du sud de la Guinée à la Sierra Leone, en passant par le Liberia, la Côte d'Ivoire, le Ghana et le Togo à l'est. On estime que l'écosystème forestier de Haute Guinée couvrait à l'origine une superficie d'au moins 420 000 km², mais des siècles d'activité humaine ont entraîné la perte de près de 70% du couvert forestier original (Bakarr et al. 2001). Seules subsistent de la forêt originelle quelques parcelles isolées qui abritent la flore et la faune uniques de la région. Ces vestiges de forêt renferment des communautés écologiques d'une diversité exceptionnelle, une flore et une faune uniques et une mosaïque de types forestiers qui servent de refuge à de nombreuses espèces endémiques.

La topographie de l'Afrique de l'Ouest est en général relativement uniforme et plate, à part quelques régions montagneuses, comme les zones de haute altitude de Haute Guinée, les régions de haute altitude du Togo/Volta, le plateau de Jos au Nigeria et les montagnes du Cameroun. Les habitats montagneux d'Afrique de l'Ouest ont ainsi une étendue très limitée et sont particulièrement menacés de disparition. La chaîne de montagnes du Simandou, qui fait partie des zones de haute altitude de Haute Guinée, est située au sud-est de la Guinée et s'étend sur 100 km de Komodou au nord à Kouankan au sud. La partie sud du massif du Simandou appartient à l'écosystème forestier de Haute Guinée. La Forêt Classée du Pic de Fon se trouve à l'extrême sud de ce massif. La Forêt Classée du Pic de Fon a été créée en 1953. Avec une superficie de 25 600 hectares, elle est la troisième plus importante forêt de la région forestière guinéenne (Guinée Forestière). Située dans la zone de transition entre la savane et la forêt, elle offre des types d'habitats variés allant de la forêt pluviale à la savane guinéenne humide. Par ailleurs, l'altitude de la Forêt Classée du Pic de Fon varie entre 600 à 1 600 mètres au-dessus du niveau de la mer (le Pic de Fon est le point le plus haut du massif à 1 656 mètres et est le second sommet le plus élevé de la Guinée). On y trouve un type d'habitat de plus en plus rare pour la zone forestière de la Haute Guinée qui est la prairie de montagne.

Très peu de données publiées existent sur la chaîne de Simandou, et plus particulièrement sur la Forêt Classée du Pic de Fon. Cependant, le Pic de Fon est apparu comme une zone potentiellement riche en biodiversité compte tenu de ses habitats montagneux, du gradient altitudinal et de ses forêts relativement intactes. Le Pic de Fon a été désigné comme zone prioritaire pour l'évaluation de la biodiversité au cours de l'Atelier pour la Sélection des Priorités de Conservation International (CI) qui s'est tenu à Elmina au Ghana en 1999.

Etendue du projet

La chaîne du Simandou abrite également une richesse minéralogique potentielle, notamment des gisements importants de minerai de fer. Rio Tinto and Exploration Limited (RTM&E) mène actuellement des activités d'exploration sur quatre sites contigüs de la chaîne du Simandou. Rio Tinto a été l'instigateur de cette étude RAP. Les résultats seront incorporés aux études initiales de préfaisabilité environnementale et sociale et contribueront aux études de base et aux évaluations plus détaillées nécessaires à l'Etude d'Impact Environnemental et Social (ESIA) qui sera menée en parallèle avec l'étude globale de faisabilité, si le projet se poursuit au-delà de l'actuelle phase d'exploration.

Compte tenu de la richesse potentielle en biodiversité du Pic de Fon, Rio Tinto a entériné un accord avec Conservation International pour évaluer la biodiversité régionale ainsi que les menaces potentielles et les opportunités pour la conservation dans la Forêt Classée du Pic de Fon. Ce partenariat vise à réaliser des avancées importantes pour la conservation de la biodiversité et l'industrie, ainsi que pour le gouvernement et le peuple de Guinée.

La collaboration initiale a porté sur un ensemble d'activités bénéfiques aux deux parties : un inventaire biologique de type RAP et une évaluation des menaces et des opportunités socio-économiques. Ces deux activités ont permis l'élaboration du Plan d'Action Initial pour la Biodiversité du Pic de Fon. Par la suite, CI et RTM&E, en collaboration avec d'autres parties prenantes locales et régionales, ont constitué une alliance d'un mandat plus étendu pour la mise en oeuvre des recommandations présentées dans ce rapport.

L'alliance a pour objectif le renforcement des capacités des communautés locales et des autorités régionales, pour un développement économique durable et pour une gestion efficace de la biodiversité et des ressources naturelles de la région. Les organisations partenaires et les communautés locales vont travailler ensemble pour comprendre les sources des menaces sur la biodiversité et pour répondre au besoin d'une gestion locale des ressources naturelles et des moyens de subsistance alternatifs. La Forêt Classée du Pic de Fon est la zone prioritaire pour ces actions mais les partenaires tâcheront d'en faire bénéficier toute la région de Nimba, y compris la Réserve Naturelle du Mont Nimba, la Forêt Classée de Diécké, la Réserve de la Biosphère de Ziama, la Forêt Classée de Béro-Tétini, la Forêt Classée de Tibé et la Forêt Classée du Milo. La planification régionale de la conservation aura une approche corridor qui prendra en considération les caractéristiques de la biodiversité, les processus des écosystèmes (bassins versants et disponibilité en ressources) et l'utilisation des terres.

Aperçu général et objectifs de l'expédition RAP

Le Programme d'Evaluation Rapide (RAP) de Conservation International, un département du Center for Applied Biodiversity Science (CABS), a été créé en 1990 en réponse à la perte croissante de la biodiversité dans les écosystèmes tropicaux. C'est un programme novateur d'inventaires biologiques, conçu pour fournir des informations scientifiques afin de catalyser l'action de conservation dans les zones tropicales soumises à une menace immédiate de conversion de l'habitat naturel. En collaboration avec le Center for Environmental Leadership in Business (CELB) et le programme Afrique de l'Ouest de CI, une expédition RAP a été organisée en novembre 2002 dans la Forêt Classée du Pic de Fon, pour mieux appréhender la diversité biologique de cette zone. L'objectif principal de cette expédition RAP était de collecter des données scientifiques sur la diversité et les statuts des espèces de la Forêt Classée du Pic de Fon, afin de produire des recommandations sur leur conservation et leur gestion. De plus, les données collectées par l'équipe du RAP ont été utilisées pour développer un Plan d'Action Initial pour la Biodiversité de la région du Pic de Fon (voir pages 134-135).

Les objectifs spécifiques de l'expédition étaient les suivants :

- Dresser un aperçu bref, mais complet, de l'intégrité et de la diversité existante de l'écosystème du Pic de Fon et évaluer son importance relative sur le plan de la conservation;

- Réaliser une évaluation des menaces sur la biodiversité de la zone d'étude;

- Fournir une formation «sur le terrain» complémentaire aux biologistes guinéens sous la direction et la supervision d'écologistes de terrain expérimentés ;

- Formuler des recommandations préliminaires sur la gestion et la recherche pour l'écosystème du Pic de Fon, ainsi que des recommandations sur les priorités de conservation pour une meilleure politique de conservation à l'échelle locale, nationale et internationale, pour le Pic de Fon et la chaîne du Simandou en particulier et la région d'altitude de Haute Guinée en général, et

- Mettre à la disposition des décideurs et du grand public en Guinée et ailleurs les données du RAP pour pouvoir identifier et décrire la biodiversité et les caractéristiques uniques de l'écosystème du Pic de Fon pour une meilleure sensibilisation à l'égard de cet écosystème et sa conservation.

L'expédition était composée d'une équipe pluridisciplinaire de 13 chercheurs, y compris des représentants de deux départements du gouvernement guinéen, la Direction Nationale des Eaux et Forêts (DNEF) et le Centre Forestier de N'Zérékoré (voir section sur les participants) et de deux organisations locales, le Centre de Gestion de l'Environnement des Monts Nimba et Winrock International. Cette équipe réunissait des chercheurs internationaux et nationaux spécialistes des écosystèmes et de la biodiversité terrestres

L'équipe du RAP a examiné des groupes taxinomiques sélectionnés pour déterminer la diversité biologique de la zone, le niveau d'endémisme et le caractère unique de l'écosystème. Les expéditions RAP étudient des groupes taxinomiques spécifiques et des espèces indicatrices avec comme objectif le choix de taxons dont la présence permet d'identifier un type d'habitat ou sa condition. Nous avons décidé de mener des inventaires sur les plantes, les orthoptères, les amphibiens et les reptiles, les oiseaux, les micro-mammifères y compris les chauves-souris et les grands mammifères.

Zone d'étude

L'expédition RAP s'est déroulée du 26 novembre au 7 décembre 2002, au début de la saison sèche, et a couvert deux sites. Le premier site était situé entre les prairies et les forêts de montagne connectées directement à des forêts plus vastes à plus basse altitude. Ce site variait de 1 000 à 1 600 mètres d'altitude nm. Le campement était situé à 1350 mètres nm (08°31'52"N, 08°54'21"W) et à 3,5 km (3461 m) du Pic de Fon.

Le deuxième site se composait de forêts primaires et secondaires et de quelques petites parcelles de savane à 600 mètres d'altitude nm environ. Le campement (08°31'29"N, 08°56'12"W) était situé à 6 km environ du village de Banko (en direction de la crête de Simandou) et à 1,5 km environ à l'intérieur de la forêt classée. Le campement était à proximité de la forêt de plaine, avec de nombreux cours d'eau de différentes tailles s'écoulant du massif et des collines recouvertes de végétation de savane secondaire. Cette forêt de plaine contient en particulier la *Parinari excelsa*, plante qui est dispersée par la roussette, ce qui est une caractéristique frappante des forêts de montagne à l'ouest du Dahomey Gap. Sur des surfaces plus plates vers le village, la majeure partie de la forêt a été convertie en plantations de cacao et de café. Cette zone contient également quelques parcelles de forêt secondaire. Des plantations de bananiers et de plantes comestibles et des cultures d'épices, notamment de poivre, se trouvaient à l'intérieur des limites de la forêt classée, avec une transition abrupte avec la zone de forêt qui commençait juste au pied de la montagne.

RÉSUMÉ DES RÉSULTATS

Lors du RAP, les zones prioritaires pour la conservation sont identifiées à travers l'étude des groupes taxinomiques à partir des critères suivants : la richesse en espèces, le niveau d'endémisme, le statut rare ou menacé des espèces et les habitats critiques. L'évaluation de la richesse spécifique peut être utilisée pour comparer plusieurs zones dans une région donnée pour le nombre d'espèces par zone . Les mesures du niveau d'endémisme indiquent le nombre d'espèces endémiques à une zone déterminée, et fournissent une indication à la fois du caractère unique de la zone et des espèces qui seront menacées par une altération de leur habitat (ou au contraire, des espèces susceptibles d'être conservées par des aires protégées). L'évaluation des espèces rares et/ou menacées au sein d'une zone donnée indique la proportion de richesse en espèces classées dans la Liste rouge de l'UICN (UICN 2002) confirmées ou possibles dans la zone. Plusieurs de ces espèces répertoriées dans la Liste rouge des espèces menacées disposent d'un statut légal de protection renforcée, ce qui confère une importance et un poids supplémentaires aux décisions portant sur la conservation. Décrire le nombre

d'habitats ou de niches écologiques critiques d'une zone permet d'identifier les habitats rares ou peu connus au sein d'une région qui contribuent à la variété des milieux et par conséquent à la diversité des espèces. Ce qui suit est un résumé, basé sur ces critères, des principaux résultats de nos deux semaines de travail sur le terrain.

Importance globale

Espèces au statut de conservation préoccupant :

- Nous avons recensé 17 espèces végétales listées par l'UICN – une espèce classée comme « En danger » (*Neolemonniera clitandrifolia*), quinze espèces classées comme « Vulnérable » et une « Quasi menacée ».

- Un amphibien (*Bufo superciliaris*) et quatre espèces de reptiles (*Chamaeleo senegalensis, Varanus ornatus, Python sebae* et *Osteolaemus tetraspis*) ont été identifiés dans la réserve et sont protégés par la législation internationale (espèces CITES).

- Nous avons enregistré huit espèces d'oiseaux présentant de l'importance sur le plan de la conservation mondiale – deux dans les plateaux, sept dans les plaines. La plus importante de ces espèces s'appelle *Schistolais leontica* (Prinia de Sierra Leone) (citée comme « Vulnérable » dans la Liste rouge de l'UICN). C'est un oiseau très peu répandu dans les plateaux ouest africains et dont la présence a été notée seulement dans trois autres sites au monde.

- Nous avons enregistré sept espèces de chauves-souris, soit 33,3 % d'un total de 21 espèces enregistrées, qui figurent dans la liste des espèces menacées de l'UICN : un espèce citée comme « Vulnérable » (*Epomops buettikoferi*) et six espèces citées comme « Presque Menacée » (*Rhinolophus alcyone, R. guineensis, Hipposideros jonesi, H. fuliginosus, Kerivoula cuprosa, Miniopterus schreibersii*)

- Des chasseurs et guides locaux ont confirmé la présence du micropotamogale de Lamotte (*Micropotamogale lamottei*) (classifiée comme « En danger » par l'UICN).

- Nous avons trouvé de nombreuses traces de présence de plusieurs espèces de grands mammifères répertoriées par l'UICN : une espèce, *Syncerus caffer* (buffle d'Afrique) classée comme « Dépendant de la conservation » et cinq espèces, *Cephalophus maxwellii* (céphalophe de Maxwell), *Cephalophus niger* (céphalophe noir), *Cephalophus dorsalis* (céphalophe à bande dorsale noire), *Cephalophus silvicultor* (céphalophe à dos jaune), et *Neotragus pygmaeus* (antilope royale), classées comme « Quasi Menacées ».

- Nous avons observé directement deux espèces de primates qui figurent dans la liste des espèces « En danger » de l'UICN: *Pan troglodytes verus* (Chimpanzé d'Afrique occidentale) et *Cercopithecus diana diana* (Cercopithèque Diane) et une espèce considérée comme « Presque menacée » par l'UICN : *Cercocebus atys atys* (mangabey fuligineux). Une des espèces indirectement confirmées, *Procolobus verus* (colobe vert olive), est considérée « presque menacée ».

Espèces nouvelles ou endémiques:

- Un nouveau crapaud qui est probablement une espèce endémique à la chaîne de Simandou.

- Nous avons enregistré au moins deux nouvelles espèces d'amphibiens, probablement endémiques: une espèce *Arthroleptis* à peau verruqueuse et une nouvelle espèce *Amnirana*.

- Nous avons recueilli quatre nouvelles espèces d'orthoptères, probablement endémiques, ainsi qu'un nouveau genre de blatte trouvée dans une grotte lors de l'étude sur les chauves-souris.

Sur la base des critères d'identification des zones prioritaires pour la conservation pris en compte lors des RAP, la Forêt Classée du Pic de Fon s'est révélée un site important à tous points de vue. Dans tous les groupes étudiés, nous avons relevé un ensemble riche et unique d'espèces de forêts et de prairies, confirmant que la Forêt Classée du Pic de Fon est l'une des plus diverses des régions étudiées en Afrique.

En ce qui concerne l'endémisme et les espèces rares et/ou menacées d'extinction, ce site remplit les conditions définies par BirdLife International pour être une Zone d'Importance pour la Conservation des Oiseaux (Important Bird Area - IBA) sur la base du nombre d'espèces menacées et la présence de nombreuses espèces à aire de répartition et biome restreints (Fishpool and Evans 2001).

La plupart des espèces d'amphibiens trouvées ont une répartition très restreinte dans les habitats forestiers et montagneux de Haute Guinée et ne sont connues que d'un petit nombre de localités. Autre découverte notable de cette étude, nous avons recensé la chauve-souris *Myotis welwitschii* pour la première fois en Afrique de l'Ouest, ce qui étend son aire de répartition de plus de 4 400 km. La liste des espèces menacées ou endémiques recensées au cours de cette courte étude souligne à la fois la nécessité d'une protection renforcée de la forêt et le besoin d'études supplémentaires.

La population de chimpanzés, en particulier, justifie la conduite d'études supplémentaires. Les chimpanzés sont classés « En danger » dans la Liste rouge UICN des espèces menacées, et sont listés en Annexe I de la CITES (Espèces les plus gravement menacées d'extinction). Parmi les trois sous-espèces de chimpanzés, la sous-espèce d'Afrique occidentale *Pan troglodytes verus* est la plus menacée par la destruction de son habitat, par la pression de la chasse, par le commerce de la viande de brousse et des animaux de compagnie, ainsi que par la transmission de maladies. Le nombre des chimpanzés dans toute l'Afrique de l'Ouest est estimé actuellement entre 25 000 et 58 000 individus dont la plupart vit dans des zones non protégées (Kormos and Boesch 2003). La protection des populations encore existantes de chimpanzés et l'établissement d'aires protégées supplémentaires sont d'une importance vitale. La Forêt Classée du Pic de Fon semble abriter un à quatre communautés distinctes de *Pan troglodytes verus*. La densité des chimpanzés dans cette forêt est plus élevée que leur densité moyenne en forêt dégradée, ce qui semble indiquer que cette zone contient encore un nombre important de chimpanzés, et qu'elle constitue un habitat important en terme de conservation de cette espèce en Guinée. Le fait que nous n'ayons pas trouvé de nids en dessous de 900 m d'altitude indique que les chimpanzés ont probablement leur principale aire de distribution ou le coeur de leur domaine dans les zones d'altitude.

Enfin, en ce qui concerne les habitats critiques, une grande partie de la forêt étudiée au cours de ce RAP était une forêt primaire intacte de grande qualité. Il ne reste que 4,1% de la forêt à canopée fermée originelle en Guinée (Sayer et al. 1992) et la protection de ce reliquat est cruciale. En outre, la prairie d'altitude du Site 1 représente un milieu rare et menacé en Afrique de l'Ouest (Fahr et al. 2002).

Plantes

Au cours de l'étude RAP, nous avons recensé 409 espèces végétales. La prairie d'altitude du Pic de Fon a été façonnée par des facteurs liés à la géologie, à l'altitude et au climat, et à l'exception des zones adjacentes à la route, la plus grande partie de sa végétation ne résulte pas d'activités anthropiques. La composition des espèces de forêts galeries et de ravins du Site 1 et d'une partie de la forêt classée du Site 2 est aussi représentative d'une forêt pratiquement intacte de qualité. Dans ces forêts montagneuses nous avons recensé 18 espèces d'arbres répertoriées par l'UICN – une espèce « En danger » (*Neolemonniera clitandrifolia*), 16 espèces classées « Vulnérable » (*Albizia ferruginea, Antrocaryon micraster, Cordia platythyrsa, Cryptosepalum tetraphyllum, Drypetes afzelii, D. singroboensis, Entandrophragma candollei, E. utile, Gorcinia kola, Guarea cedrata, Khaya grandifoliola, Lophira alata, Milicia regia, Nauclea diderrichii, Neostenanthera hamata,* et *Terminalia ivorensis*) et une espèce classée « Quasi menacé ». L'agriculture empiète sur cette forêt, particulièrement sur le Site 2, et nous avons observé plusieurs pratiques agricoles non soutenables (comme l'abattage des arbres pour en récolter les fruits, plutôt que de simplement les cueillir) dans les forêts secondaires et les plantations.

Insectes

Cette étude représente la première enquête sur les orthoptères de la famille des Tettigoniidés dans la zone du Pic de Fon. Ces insectes semblables aux sauterelles et aux criquets se rencontrent en grand nombre dans les écosystèmes de forêt et de savane. Ce sont des herbivores importants pour la structure de la végétation. Nous avons collecté quarante espèces, dont au moins quatre sont nouvelles pour la science, et dix sont nouvelles pour la Guinée (augmentation de 16%). Ces quarante espèces représentent au moins 24 genres et 6 sous-familles. La population d'*Anoedopoda lamellata* dans la savane peut être distincte du reste de la population africaine. La faune d'orthoptères de la famille des Tettigoniidés dans la zone du Pic de Fon se caractérise par un potentiel élevé d'endémisme et devrait faire l'objet d'autres enquêtes. En outre, un nouveau genre de blatte a été découvert dans une grotte explorée pour l'étude des chauves-souris.

Amphibiens

Au total, nous avons relevé 32 espèces d'amphibiens, dont trois au moins étaient nouvelles pour la science et probablement endémiques à la chaîne du Simandou. Nous avons continuellement ajouté de nouvelles espèces à la liste. Notre enquête s'étant déroulée au début de la saison sèche, nous pouvons raisonnablement estimer que le nombre total d'espèces de la zone est au moins le double du nombre d'espèces recensées. Sur les espèces observées, seules 10 d'entre elles (28,6%) sont trouvées ailleurs, 71,4% des espèces appartiennent uniquement à l'Afrique de l'Ouest. Plus de la moitié des espèces observées (19 espèces soit 54.3) sont endémiques à la forêt pluviale de Haute Guinée. Onze des espèces sont endémiques à la région occidentale de la zone forestière de Haute Guinée. Plusieurs de ces espèces endémiques ont une distribution très limitée dans la région montagneuse de Haute Guinée et sont connues uniquement de quelques localités. La faune d'amphibiens de toute la région était composée d'un ensemble unique d'amphibiens de savane et de forêt, allant d'espèces qui nécessitaient un habitat de forêt pluviale primaire intacte à des espèces ayant besoin d'habitats de savane plus secs. Un nouveau crapaud est très vraisemblablement une espèce endémique à la chaîne de Simandou. En outre, nous avons recensé une espèce d'*Arthroleptis* à peau verruqueuse (comme l'espèce endémique *A. crusculus* du Mont Nimba) et une nouvelle espèce d'*Amnirana*. Une espèce d'amphibien recensée (*Bufo superciliaris*) est protégée par la législation internationale (CITES) et est importante sur le plan de la conservation. Compte tenu des résultats de cette enquête, nous estimons que le Pic de Fon présente l'une des plus grandes diversités en amphibiens en Afrique de l'Ouest.

Reptiles

Pendant notre enquête, nous avons constaté la présence de 12 espèces de reptiles ou obtenu des preuves de leur présence. Cependant, la faune d'ophidiens à elle seule comprend probablement plus de 60 espèces dans cette zone (hypothèse basée sur des enquêtes effectuées dans des zones voisines, se caractérisant par des types d'habitat semblables ou même moins divers (Angel et al. 1954a, b; Barbault 1975; Rödel et al. 1995, 1997, 1999; Rödel et Mahsberg 2000; Ernst et Rödel 2002; Ineich 2002; Branch et Rödel 2003). Sur les espèces que nous avons recensées, quatre reptiles (*Chamaeleo senegalensis*, *Varanus ornatus*, *Python sebae* et *Osteolaemus tetraspis*) sont importants sur le plan de la conservation et sont donc protégés par la législation internationale (CITES).

Oiseaux

Le nombre total d'espèces d'oiseaux recensées jusqu'a présent sur l'ensemble de la Guinée se situe aux alentours de 625. Nous avons enregistré 233 espèces dont 131 dans la région des plateaux et 198 dans la région des plaines. En comparaison, 287 espèces ont été constatées dans la Forêt Classée de Ziama et 141 dans la Forêt Classée de Diécké – les deux autres sites du sud-est de la Guinée qui ont été étudiés ; 383 espèces sont connues du versant libérien du Mont Nimba, mais ceci est le résultat de nombreuses années d'études intensives. Nous avons enregistré huit espèces présentant de l'importance sur le plan de la conservation – deux dans la région des plateaux, sept dans la région des plaines. L'espèce la plus importante est *Schistolais leontica* (Prinia de Sierra Leone), une espèce à distribution très restreinte dans les zones d'altitude ouest-africaines et classée comme «Vulnérable» par l'UICN. Cette espèce semble vivre uniquement dans la végétation dense à la lisière de la forêt et le long des cours d'eau à plus de 700 mètres d'altitude et pourrait être particulièrement vulnérable à la modification des habitats à plus haute altitude. Nous avons observé six des 15 espèces d'oiseaux qui déterminent la Zone d'Endémisme d'Oiseaux de la Forêt de Haute Guinée – c'est à dire des espèces terrestres dont l'aire de reproduction mondiale est inférieure à 50 000 km2. Sur les 153 espèces du biome guinéo-congolais qui se trouvent en Guinée, nous en avons recensé 104 (68%) dans le cadre de cette étude. De plus, nous avons enregistré sept espèces pour la première fois en Guinée. Ce site est considéré comme une Zone d'Importance pour la Conservation des Oiseaux, d'après la définition de BirdLife International qui se base sur le nombre d'espèces menacées et la présence en nombre à la fois d'espèces à distribution restreinte et d'espèces limitées à leur biome. Nous notons que la Forêt Classée du Pic de Fon présente de nombreuses similarités avec le Mont Nimba par l'altitude et par la diversité d'habitats rapprochés. Il est donc probable que beaucoup d'autres espèces pourront être découvertes dans la forêt du Pic de Fon.

Chauves-souris

Nous avons trouvé des chauves-souris appartenant à 6 familles, 14 genres et 21 espèces, représentant un tiers des espèces de chauves-souris actuellement connues en Guinée. Les roussettes (Pteropodidae) représentaient la plus grande proportion des espèces au niveau des familles (33,3%). Sept espèces de chauves-souris sur les 21 trouvées figurent dans la liste des espèces menacées de l'UICN: un espèce est citée comme «Vulnérable» (*Epomops buettikoferi*) et six espèces comme «Presque Menacée» (*Rhinolophus alcyone, R. guineensis, Hipposideros jonesi, H. fuliginosus, Kerivoula cuprosa, Miniopterus schreibersii*). Nous n'avons exploré qu'une seule grotte de chauves souris, habitée par le *Rousettus aegyptiacus*, mais de nombreuses autres grottes existent certainement dans la zone. Les espèces *Hipposideros jonesi* et *Rhinolophus guineensis* trouvées pendant l'inventaire sont endémiques à la région de Haute Guinée et dépendent de la présence de grottes. Nous avons recensé *Miniopterus schreibersii*, une espèce d'occurrence isolée dans la région de Haute Guinée et qui vit également dans les grottes. Les deux espèces *H. jonesi* et *M. schreibersii* ne sont pas connues dans la région du Mont Nimba. L'observation de *Myotis welwitschii* est la première non seulement pour la région forestière de la Guinée et de Haute Guinée, mais pour toute l'Afrique de l'Ouest, prolongeant ainsi l'aire de distribution connue de l'espèce de 4 400 km à l'ouest. Cette espèce vit dans les régions montagneuses de son aire de distribution et elle est extrêmement rare dans tous les endroits où sa présence a été relevée (Fahr et Ebigbo en cours de publication). Nous avons trouvé trois espèces de *Kerivoula*, un genre forestier très rare. Deux de ces espèces, *K. phalaena* et *K. cuprosa*, sont de nouvelles espèces pour la Guinée. Durant l'inventaire, des espèces ont continuellement été ajoutées à la liste, ce qui indique que celle-ci est loin d'être complète. L'importante diversité topographique de cette zone et sa proximité avec le Mont Nimba (où 49 espèces ont été recensées) et le Massif du Ziama, laissent penser que la chaîne du Simandou est susceptible de contenir 50 à 65 espèces, incluant potentiellement des espèces menacées qui ont été décrites dans les zones voisines.

Petits mammifères non volants

Nous avons capturé trois espèces de musaraignes et sept espèces de rongeurs et observé six autres espèces d'écureuils. Le nombre d'espèces potentiel est certainement plus élevé à condition de disposer de plus longues périodes d'échantillonnage sur un plus grand nombre de sites dans la forêt et la savane. Les résultats étaient caractéristiques de la forêt de montagne à feuillage semi-caduc avec de plus hauts niveaux de biomasse de petits mammifères à une altitude plus basse, suggérant ainsi une productivité forestière accrue en raison de l'abondance d'eau, de la grande diversité en plantes et en micro-habitats et, dans le cœur de la forêt, de faibles niveaux de perturbation. Les chasseurs et les guides locaux ont confirmé la présence du micropotamogale de Lamotte (*Micropotamogale lamottei*), une espèce menacée, sur la base de photos et de descriptions de l'animal, mais nous n'avons pas été en mesure d'observer cette espèce pendant le RAP.

Grands mammifères

Nous avons observé, identifié au bruit ou photographié 31 espèces de grands mammifères au Site 1 et 34 espèces au Site 2 pour un total global de 39 espèces de grands mammifères. A l'aide de pièges photographiques, nous avons quatre photos de deux espèces, trois de l'athérure d'Afrique *Atherurus africanus* et une photo de l'espèce en danger *Pan troglodytes verus* (chimpanzé d'Afrique occidentale). Nous n'avons pas observé de léopards ou d'oryctéropes, mais des braconniers locaux ont signalé que ces espèces étaient toujours présentes dans la forêt du Pic de Fon. Des traces, des excréments et les observations de chasseurs locaux nous ont permis de confirmer la présence de *Syncerus caffer* (buffle africain), répertorié par l'UICN comme « Dépendant de la conservation », ainsi que plusieurs céphalophes, *Cephalophus*

maxwelli (céphalophe de Maxwell), *Cephalophus niger* (céphalophe noir), *Cephalophus dorsalis* (céphalophe à bande dorsale noire), *Cephalophus silvicultor* (céphalophe à dos jaune), et *Neotragus pygmaeus* (antilope royale) classée par l'UICN comme « Quasi Menacé ».

Primates

Sur les quinze espèces de primates qui vivent probablement dans la région du Pic de Fon, nous avons pu confirmer directement la présence de huit espèces (par des observations visuelles ou acoustiques) et confirmer indirectement la présence de cinq autres espèces. Sur les huit espèces observées directement, deux espèces sont citées comme en danger par l'UICN : *Pan troglodytes verus* (chimpanzé d'Afrique occidentale), *Cercopithecus diana diana* (cercopithèque Diane), et une espèce est citée comme presque menacée : *Cercocebus atys atys* (mangabey fuligineux). Sur les espèces confirmées indirectement, une espèce, *Procolobus verus* (colobe vert olive) est citée comme presque menacée. Même si la diversité des primates est assez importante dans la région, leur abondance est faible ; la plupart des espèces ont été peu rencontrées, à l'exception du galago de Demidoff (*Galagoides demidoff*) que nous avons entendu régulièrement et des chimpanzés que nous avons entendus chaque jour au Site 2. Nous estimons que la population moyenne de chimpanzés au Pic de Fon (en nous basant sur une estimation de 10 000 hectares de zone forestière sur les 25 600 hectares de forêt classée) se situe aux environs de 75 (avec une variation de 21 à 246). Les communautés de chimpanzés regroupant habituellement entre 10 et 100 individus, le Pic de Fon abriterait un à quatre groupes communautaires distincts. La densité de chimpanzés dans cette forêt est plus élevée que la densité moyenne pour une forêt dégradée. Cette région semble toujours abriter un nombre important de chimpanzés et constitue un habitat important pour la protection des chimpanzés en Guinée.

EVALUATION SOCIOÉCONOMIQUE DES MENACES ET DES OPPORTUNITES

Aperçu et objectifs

Conservation International (CI) a effectué une évaluation socioéconomique des menaces et des opportunités du Pic de Fon en novembre et décembre 2002. Les objectifs de cette évaluation étaient les suivants:

- Identifier et confirmer les principales parties prenantes de la région;

- Identifier et caractériser les menaces sur la biodiversité et les obstacles auxquels se heurte la conservation dans la région;

- Dresser une liste prioritaire initiale des menaces sur la biodiversité, et

- Identifier les opportunités et les solutions possibles pour atténuer ces menaces.

Pour la première étape de cette évaluation, CI a supervisé une étude approfondie de la littérature grise et publiée, incluant deux études préliminaires commandées par Rio Tinto. Cette revue de la littérature n'a produit que très peu de données socio-économiques spécifiques à la zone du Pic de Fon, et a révélé un manque général de relevés historiques et d'études échelonnées dans le temps, ce qui rend pratiquement impossible l'analyse des tendances socioéconomiques. A la suite de cet examen, CI a organisé un atelier à Conakry les 12 et 13 décembre 2002, au terme du RAP, pour obtenir des informations d'un large éventail de parties prenantes et d'experts sur les menaces et les opportunités pour la conservation de la biodiversité dans la région du Pic de Fon. Pour compléter la revue de la littérature, des informations étaient apportées par les personnes et les structures engagées dans la conservation de la zone immédiate, de la Guinée et de la région au sens large. Cet exercice a aussi permis de rassembler une documentation supplémentaire qu'il n'aurait pas été possible d'obtenir par les recherches préliminaires. Les participants ont été sélectionnés de manière à représenter une vaste gamme de perspectives et d'expertise en matière d'efforts de conservation dans toute la Guinée, dans l'ensemble du massif du Mont Nimba et dans la région du Pic de Fon. Au total, 33 participants de 17 organisations, incluant le gouvernement national, des bailleurs multilatéraux, des bailleurs bilatéraux, des ONG, des centres de recherche scientifique et Rio Tinto, ont assisté à l'atelier.

RÉSUMÉ DES MENACES ACTUELLES ET POTENTIELLES SUR LA BIODIVERSITÉ

Chasse

La chasse pour la viande de brousse a un impact presque immédiat sur les espèces de grands mammifères, y compris les primates. Les résultats du RAP confirment que la faune de la forêt est soumise à des pressions intensives de chasse. L'équipe RAP a trouvé 11 cartouches de fusils de chasse, deux pièges et plusieurs campements de chasse près du Site 1, ainsi que 36 cartouches de fusils de chasse, 38 pièges au collet ou pièges à filet et a entendu des coups de fusil tous les jours et au moins quatre fois par nuit près du Site 2. L'évaluation des menaces et des opportunités, à travers la revue de la littérature et l'atelier, a confirmé que la chasse pour la viande de brousse constitue une menace importante pour la conservation.

Empiétement agricole

Le PGRR estime que près de 35 à 40% de la Forêt Classée du Pic de Fon est empiétée par l'agriculture à la fois de subsistance et de rentes pratiquée par les communautés voisines. La plantation de cultures pérennes, comme le café, nécessite un investissement important de la part des exploitants agricoles, ce qui indique que ces agriculteurs ne semblent pas craindre des réactions de la part des autorités par rapport aux cultures sur le long-terme à l'intérieur du périmètre du Pic de Fon.

Feux de brousse

Les feux de brousse incontrôlés, qu'ils soient naturels ou provoqués pour défricher des terres, préparer des pâturages ou chasser le gibier, sont une menace constante. Le recours aux feux de brousse pour préparer des pâturages pour le bétail a été identifié comme une menace régionale dans la littérature et à l'atelier. Nous n'avons pas trouvé de références spécifiques sur l'empiètement par l'élevage sur le Pic de Fon. Cependant les 24 villages qui l'entourent pratiquent tous des activités d'élevage. Des feux de brousse sont également allumés pour rabattre le gibier pendant la chasse.

Exploitation forestière

Les documents examinés pendant l'étude de la littérature et l'atelier mentionnent l'exploitation du bois comme une menace actuelle à la biodiversité du Pic de Fon. L'exploitation commerciale se répand en Guinée et la demande en charbon de bois est en train de dépasser rapidement la capacité de reproduction des ressources. L'équipe du RAP a observé que la zone qui s'étend de l'entrée de la forêt classée au pied de la montagne près du Site 2 est dénudée d'arbres (quelques squelettes subsistent), en raison probablement de l'utilisation du bois pour la construction de maisons, de meubles, la collecte du bois de chauffe, ainsi que du défrichement à des fins agricoles.

Pêche excessive et collecte de produits forestiers non ligneux

L'exploitation non réglementée de ressources telles que le poisson et les produits forestiers non ligneux du Pic de Fon a été identifiée par la revue de la documentation et durant l'atelier. Cependant, ni l'enquête ni l'atelier n'ont fourni d'informations concernant les espèces de poisson qui sont habituellement pêchées ou l'ampleur de la collecte des produits forestiers non ligneux habituellement ramassés, sans mentionner les taux d'extraction ou l'impact général de ces activités. L'équipe du RAP a observé quelques pratiques de collecte non soutenables de produits forestiers non ligneux, en particulier des méthodes de cueillette qui consistaient à abattre des arbres entiers pour en recueillir les fruits au lieu de les cueillir sur l'arbre.

Extraction minière artisanale

Plusieurs sources ont mentionné des activités d'extraction illégales à l'intérieur des limites du Pic de Fon et les participants à l'atelier de Conakry ont également identifié cette forme d'extraction comme une menace. En particulier, comme il était également noté dans la littérature, les participants ont fait observer l'impact destructeur de l'extraction artisanale ou à petite échelle d'or et de diamant dans les lits des ruisseaux et des cours d'eau. L'équipe RAP a observé cette pratique au moins une fois et a identifié l'extraction artisanale comme une menace probable. Comme pour la pêche excessive et la collecte de produits forestiers non ligneux, la nature et l'ampleur exactes de cette menace n'ont pas encore été déterminées et nous recommandons vivement la mise à jour des informations sur ces activités.

Construction routière

Les routes récemment aménagées pour faciliter les activités d'exploration minière, ont un impact potentiel sur la biodiversité, notamment en exacerbant la menace représentée par la chasse pour la viande de brousse. Les routes offrent un accès plus facile à des zones auparavant isolées du Pic de Fon. De plus, les améliorations apportées aux routes existantes peuvent offrir un meilleur accès à des marchés plus distants pour la viande de brousse. Une autre menace potentielle liée au développement routier est l'introduction ou la prolifération d'espèces allogènes et envahissantes. De plus, les impacts potentiels sur les espèces de l'altération de l'habitat, ou de la perturbation sonore liée aux routes, méritent d'être examinés. L'équipe du RAP a observé une érosion sédimentaire le long des routes ; l'érosion peut entraîner un envasement accru des cours d'eau, ce qui pourrait avoir un effet négatif sur les communautés aquatiques et riveraines. Cependant, l'impact actuel de l'érosion des routes sur les espèces du Pic de Fon est encore à considérer.

Fragmentation de l'habitat

La fragmentation de l'habitat, résultant de la conversion et de l'empiètement continu sur la Forêt Classée du Pic de Fon, représente une future menace potentielle pour la région et a été observée dans la forêt de plaine près du Site 2. Il existe peu de sites en Afrique de l'Ouest qui couvrent de tels gradients d'altitudes, au voisinage immédiat les uns des autres, et de tels sites sont importants pour un grand nombre d'espèces. Les oiseaux, par exemple, ont parfois besoin d'un habitat à une altitude différente suivant les saisons. Certaines espèces endémiques peuvent être répandues à travers la Forêt Classée du Pic de Fon et nécessiter seulement une protection plus générale par type d'habitat. D'autres espèces peuvent être restreintes à une grotte particulière.

Activités minières commerciales

Les résultats préliminaires d'un programme d'exploration de RTM&E indiquent la présence d'un dépôt de minerai de fer d'une concentration exceptionnelle sur la chaîne du Simandou. Les impacts possibles d'opérations minières commerciales au sein de la Forêt Classée du Pic de Fon représentent une préoccupation majeure. Une altération de la topographie dans des zones d'altitude de la Forêt Classée du Pic de Fon pourrait avoir un effet important sur l'habitat rare que représente la prairie de montagne et pourrait également affecter les espèces endémiques et menacées (comme les chimpanzés qui nidifient à plus de 900 m, les espèces de chauves-souris menacées qui dépendent des grottes, les espèces d'amphibiens endémiques et *Schistolais leontica* (Prinia de Sierra Leone). L'altération topographique pourrait aussi avoir un impact potentiel plus large sur les fonctions climatiques et hydrologiques du Pic de Fon et des zones environnantes, au sein des bassins versants qui en dépendent. Le développement minier peut enfin avoir pour conséquence potentielle une augmentation du taux de déboisement lié à la migration de travailleurs dans une zone donnée.

L'extraction à ciel ouvert est particulièrement préoccupante. Lorsqu'elle fut employée dans le passé, en particulier par des sociétés ayant peu de considération pour les impacts environnementaux et sociaux, cette méthode s'est traduite par la destruction d'habitats rares, ainsi que par de graves impacts sur les villages situés à proximité immédiate des sites miniers (Colston et Curry Lindahl 1986, Gatter 1997, Toure et Suter 2001, 2002). Bien que les pratiques des compagnies minières internationales diffèrent souvent de façon significative de celles de la plupart des compagnies nationales, les dommages potentiels pour l'habitat du Pic de Fon de l'exploitation à ciel ouvert devraient être examinés avec attention.

La plupart de ces sources de pression sur la biodiversité vont très probablement s'intensifier car la croissance et la migration de la population humaine augmentent la demande en viande de brousse, en terres cultivables et en autres ressources. Cette région présente le taux de croissance le plus élevé de la population en Guinée, ce qui renforce l'urgence d'efforts concertés de conservation. D'autres facteurs, qui n'ont pas fait l'objet d'un examen détaillé, confèrent une dimension encore plus complexe à ces menaces : le manque de capacités au niveau des acteurs régionaux (communautés, gouvernement, ONG et communauté scientifique), les troubles civils et les conflits fonciers.

La Guinée est considérée comme l'un des pays les plus pauvres au monde, et cependant l'un des plus riches en terme de ressources minières, ce qui est peut-être l'aspect le plus décourageant pour la planification de la conservation dans la région. Le Rapport sur le Développement Humain du Programme des Nations Unies pour le Développement (PNUD) a classé pendant plusieurs années la Guinée en queue du peloton de quelques 170 pays (UNDP 2003). L'économie guinéenne n'est pas très diversifiée et son développement potentiel est restreint par un manque général d'infrastructures de base.

Les habitants des 24 villages entourant le Pic de Fon dépendent de cette forêt pour la nourriture, l'eau, le combustible, les produits médicinaux et d'autres services fournis par l'écosystème. Tous les efforts doivent être mis en œuvre pour garantir que la forêt continuera à fournir ces services, ou que des alternatives appropriées pourront être trouvées quand les pratiques actuelles ne sont pas soutenables (comme dans le cas de la chasse pour la viande de brousse ou de la conversion de la forêt en terres cultivées). La plupart des menaces détaillées ci-dessus ne peuvent être contrées que par le biais d'une planification attentive, d'une coopération avec ces communautés et de l'assistance de partenaires nationaux et internationaux. Dans cette optique, les besoins communautaires doivent être pris en compte pour que toute zone protégée demeure viable, et les empiètements illégaux (et la destruction conséquente de la flore et de la faune au sein de la forêt) doivent être contrôlés pour assurer une viabilité à long terme, à la fois dans l'intérêt des communautés locales et pour protéger l'une des dernières forêts intactes en Afrique de l'Ouest.

Un important travail de terrain complémentaire, qui dépasse la portée de l'effort actuel en terme de budget et de durée, est nécessaire pour développer une solide connaissance quantitative des menaces actuelles et futures et des opportunités dans la zone du Pic de Fon. Cette information supplémentaire sera d'une importance cruciale pour élaborer une stratégie de conservation appropriée pour la région.

ACTIONS DE CONSERVATION RECOMMANDÉES

Les résultats de cette étude RAP montrent que l'écosystème du Pic de Fon contient des espèces terrestres et aquatiques uniques ; des connections intactes entre le milieu, la flore, la faune ; et l'un des derniers habitats de montagne relativement intacts, à la fois en Guinée et sur les Plateaux de Haute Guinée. A la suite de cette étude, nous recommandons fortement que la Forêt Classée du Pic de Fon reçoive un statut de protection plus élevé et que la réglementation actuelle concernant son statut classé soit contrôlée, mise en application, et renforcée si nécessaire. Les limites de la zone protégée devraient être définies et imposées en collaboration avec les communautés locales, et des programmes d'éducation et de sensibilisation, ainsi que des études scientifiques et socioéconomiques supplémentaires, devraient être menés. La conservation de la biodiversité nécessitera un investissement important en terme de développement des capacités.

Il faut aussi prendre en compte les implications liées à une mise en application plus stricte du contrôle de l'accès des communautés locales à la forêt. La propriété foncière est un problème extrêmement délicat, et s'il est capital de placer au plus vite la région du Pic de Fon sous une forme de protection légale renforcée, le succès de cette zone protégée reposera sur un dialogue ouvert et une étroite collaboration avec les communautés de la région. Ces dialogues faciliteront l'entente entre le gouvernement, les groupes de conservation et les communautés. L'intention est d'atteindre les objectifs de conservation en utilisant une variété de mesures pour répondre aux besoins en terres agricoles, en bois de chauffe, en gestion durable de l'environnement et en autres services fournis par la forêt.

Si la Forêt Classée du Pic de Fon est intégrée dans système de zones protégées, qui inclut à la fois les forêts et les savanes et qui est représentatif des nombreuses variations des micro-habitats, une forte proportion des espèces végétales et animales de cette région seront protégées.

Par ailleurs, la chaîne du Simandou fait partie des Hauts Plateaux du Nimba, qui incluent le Mont Nimba, Diécké, la chaîne de Wonegizi-Ziama et le Mont Béro-Tétini. La planification des investissements de conservation doit donc être faite à un niveau régional, car la plupart des menaces les plus pressantes relèvent de dynamiques régionales. Dans l'immédiat, un obstacle important auquel la stratégie de conservation de la région est confrontée est le manque de données concrètes, en particulier sur l'évolution des facteurs socio-économiques. Par conséquent, la conduite de recherches complémentaires ayant pour objectif la collecte de données de base est l'une des priorités.

Les actions spécifiques recommandées sont les suivantes:
Actions immédiates:

- Recueillir des images satellite à l'aide du système d'information géographique (SIG) et des données par télédétection pour surveiller le couvert forestier du Pic de Fon, afin d'évaluer son évolution.

- Clarifier les lois existantes régissant la chasse et l'utilisation des terres dans la Forêt Classée du Pic de Fon en réalisant et en publiant une étude sur les lois existantes qui assurent la protection des espèces en danger et menacées contre la chasse et les autres menaces dans la région du Pic de Fon, et en identifiant toute discordance et contradiction possibles.

- Organiser et mettre en œuvre une campagne de sensibilisation pour dissuader les pratiques de chasse insoutenables.

- Effectuer des études socioéconomiques de référence.

- Effectuer d'autres inventaires biologiques de référence dans toute la chaîne du Simandou afin de déterminer les priorités de conservation et les menaces futures possibles, y compris des inventaires des prairies de montagne et du grand bloc forestier situé au sud des sites actuels du RAP.

- Prévenir l'introduction d'espèces colonisatrices (noter que très peu ont été trouvées au cours de cette étude).

Actions à long terme:

- Renforcer la capacité des acteurs nationaux, régionaux et locaux à faire respecter les lois en vigueur régissant la chasse et autre exploitation des ressources du Pic de Fon; continuer à définir les limites de la forêt du Pic de Fon et instaurer et renforcer les capacités d'un système de surveillance et d'application des lois.

- Améliorer la capacité des acteurs régionaux à effectuer des recherches et à gérer la Forêt Classée du Pic de Fon.

- Créer des partenariats entre ONG, organismes publics et institutions de recherche locales afin d'élaborer de meilleurs plans de gestion à long terme et de mettre en œuvre des protocoles de suivi pour le Pic de Fon.

- Etablir des accords de conservation basés sur des mesures incitatives qui compenseraient les coûts occasionnés par la conservation des ressources en question, et qui produiraient une impulsion économique pour réduire la pauvreté et améliorer le niveau social par l'apport d'activités rémunératrices.

- Consolider les lois guinéennes sur la chasse et la protection de la faune et la flore.

Etudes scientifiques recommandées
Voir chapitres individuels pour plus d'informations

Hydrologie
Nous recommandons vivement la conduite d'une enquête sur l'hydrologie de la chaîne de Simandou et du Pic de Fon et sur les conséquences climatologiques possibles d'activités minières éventuelles qui auront pour effet d'abaisser la crête de la chaîne de montagnes. La crête de la chaîne de Simandou sert très vraisemblablement de barrière contre les vents du sud-ouest, ce qui se traduit par des niveaux de précipitations plus élevés que le niveau habituel pour les régions de cette latitude. L'abaissement de la crête se traduira probablement par une réduction des taux de précipitations. En outre, toute la chaîne de montagnes sert probablement de réservoir d'eau. Une réduction de ce réservoir, liée à la réduction des précipitations, se traduira presque certainement par la disparition de la forêt le long de la chaîne de montagnes. Ce changement non seulement affectera les faunes aquatique et forestière, mais pourrait aussi avoir de graves conséquences pour la population humaine locale et nécessite donc une évaluation plus approfondie. Nous recommandons aussi de surveiller la qualité de l'eau et les niveaux de minéraux dans les cours d'eau afin d'évaluer les effets de l'érosion et des activités de prospection minière.

Faune herpétologique
Pendant notre enquête, nous avons observé trois nouvelles espèces probablement endémiques à cette région, un crapaud nain, une espèce d'*Arthroleptis* et une espèce d'*Amnirana*, toutes deux à peau verruqueuse. Ces espèces se trouvent uniquement le long de la crête. Des études supplémentaires sur ces espèces et la taille de leur population respective sont nécessaires afin de définir des zones protégées à accès restreint dans leur aire de distribution. Ces espèces risquent en effet d'être fortement menacées par les activités de prospection minière actuelles ainsi que par l'extraction éventuelle de minerai de fer. Pendant la saison des pluies, d'autres inventaires devront être réalisés dans d'autres zones de la chaîne de Simandou afin d'acquérir une connaissance plus complète sur la diversité de la faune herpétologique du Pic de Fon.

Primates
Afin de garantir une protection adéquate des primates du Pic de Fon, nous recommandons la réalisation d'autres inventaires à des saisons différentes et sur des périodes plus longues afin de confirmer les caractéristiques de distribution des primates, leurs modes territoriaux et l'abondance de leurs populations. Ces inventaires sont particulièrement nécessaires aux mesures de protection pour les quatre espèces menacées du Pic de Fon.

Grands mammifères
Nous recommandons la mise en place d'un protocole peu onéreux de surveillance des grands mammifères par des pièges photographiques dans la forêt. Les chercheurs et les responsables de la conservation pourront ainsi obtenir des informations constantes sur les ressources locales de faune sauvage et pourront prendre des mesures de conservation plus précises.

Chauves-souris
Sur la base des recherches préliminaires, il apparaît que la zone du Pic de Fon présente une richesse exceptionnelle en chauves-souris, grâce à la grande variété d'habitats et au gradient d'altitude. Des inventaires supplémentaires sont nécessaires. Une caractéristique frappante de la faune de chauves-souris des plateaux de Haute Guinée est le nombre disproportionné d'espèces qui sont soit endémiques à la Haute Guinée soit représentées dans cette région par des populations isolées. Certaines espèces sont non seulement endémiques à la Haute Guinée, mais ont une aire de distribution infime. Nous insistons sur l'urgence de réaliser d'autres inventaires pour évaluer la présence dans la Forêt Classée du Pic de Fon d'espèces globalement menacées qui ont été trouvées dans des habitats similaires et ou voisins (Fahr et al. 2002).

Petits mammifères non volants
Il est recommandé de poursuivre la réalisation d'inventaires biologiques, en employant diverses méthodes pendant des périodes plus longues et à divers endroits du Pic de Fon, pour vérifier en particulier la présence d'autres espèces de musaraignes, notamment l'endémique crocidure du Mont Nimba et le micropotamogale de Lamotte, observé à la fois au Mont Nimba et à Ziama.

Oiseaux
Des inventaires complémentaires devront être menées sur différents sites de la forêt et pendant des saisons différentes (par exemple au début de la saison des pluies, en avril-mai, quand un plus grand nombre d'espèces sera probablement en train de nicher et donc plus facile à détecter par le chant), ainsi que dans des habitats que nous n'avons pas pu couvrir. Il est probable que d'autres espèces menacées et des espèces migratrices en provenance de l'Europe seront découvertes, et que l'on observera la migration altitudinale de certaines espèces à l'intérieur de la forêt.

Plantes
Nous recommandons d'avoir recours aux images obtenues par SIG et aux données obtenues par télédétection pour surveiller le couvert forestier du Pic de Fon, ce qui permettra d'évaluer son évolution. En outre, des inventaires de la flore sont nécessaires sur le versant oriental plus sec du Pic de Fon afin d'étudier les espèces endémiques éventuelles.

Invertébrés
D'autres études s'avèrent nécessaires dans la chaîne du Simandou afin d'estimer la distribution et la taille des populations d'espèces nouvellement découvertes pendant ce RAP. Aussi, la collecte d'informations supplémentaires s'impose au début de la saison des pluies lorsque la plupart des espèces est en grand nombre et au stade adulte. L'énigmatique ensemble des espèces du genre *Ruspolia* peut inclure plusieurs espèces non décrites et une étude génétique et bioacoustique devra donc être conduite sur les espèces de ce genre dans la chaîne de Simandou.

Poissons

Nous recommandons un inventaire des poissons d'eau douce de la région afin de compléter les données recueillies pendant le RAP et d'aborder d'autres questions liées à la pêche soulevées pendant l'évaluation des menaces et des opportunités. Un autre volet pourrait consister à évaluer la qualité de l'eau et les niveaux de minéraux dans les cours d'eau pour pouvoir déterminer les effets de l'érosion et des activités de prospection minière.

RÉFÉRENCES

Angel, F., J. Guibé and M. Lamotte. 1954a. La réserve naturelle intégrale du mont Nimba. Fascicule II. XXXI. Lézards. Memoirs de l'Institut fondamental d'Afrique noire, sér. A, 40: 371-379.

Angel, F., J. Guibé, M. Lamotte and R. Roy. 1954b. La réserve naturelle intégrale du mont Nimba. Fascicule II. XXXII. Serpents. Bulletin de l'Institut fondamental d'Afrique noire, sér. A, 40: 381-402.

Bakarr, M., B. Bailey, D. Byler, R. Ham, S. Olivieri, and M. Omland (eds.). 2001. From the Forest to the Sea: Biodiversity Connections from Guinea to Togo. Washington DC: Conservation International.

Barbault, R. 1975. Les peuplements de lézards des savane de Lamto (Côte d'Ivoire). Annales de l'Université d'Abidjan, sér. E, 8: 147-221.

Böhme, W. 1999. Diversity of a snake community in a Guinean rain forest (Reptilia, Serpentes). In: Rheinwald, G. (ed.): Isolated Vertebrate Communities in the Tropics. Pp. 69-78. Proceedings of the 4th International Symposium, Bonner zoologische Monographien, 46.

Branch, W.R. and M.-O. Rödel. 2003. Herpetological survey of the Haute Dodo and Cavally forests, western Ivory Coast, Part II: Trapping results and reptiles. Salamandra, Rheinbach 39(1) 21-38.

Colston, P.R. and K. Curry-Lindahl. 1986. The Birds of Mount Nimba, Liberia. British Museum (Natural History). London. 129 pp.

Ernst, R. and M.-O. Rödel. 2002. A new Atheris species (Serpentes: Viperidae) from Taï National Park, Ivory Coast. Herpetological Journal 12: 55-61.

Fahr, J. and N.M. Ebigbo. 2003. A conservation assessment of the bats of the Simandou Range, Guinea with the first record of Myotis welwitschii (GRAY, 1866) from West Africa. Acta Chiropterologica 5(1): 125-141.

Fahr, J., H. Vierhaus, R. Hutterer and D. Kock. 2002. A revision of the Rhinolophus maclaudi species group with the description of a new species from West Africa (Chiroptera: Rhinolophidae). Myotis 40: 95-126.

Fishpool, L.D.C. and M.I. Evans (eds.). 2001. Important Bird Areas in Africa and Associated Islands: Priority sites for conservation. BirdLife Conservation Series No. 11. Pisces Publications and BirdLife International. Newbury and Cambridge, UK.

Gatter, W. 1997. Birds of Liberia. Pica Press, The Banks, Mountfield. 320 pp.

Hatch & Associates, Inc. 1998. "Preliminary Environmental Characterisation Study: Simandou Iron Ore Project Exploration Programme, South-Eastern Guinea". Montreal, Canada.

Hutson, A.M., S.P. Mickleburgh and P.A. Racey (comp.). 2001. Microchiropteran Bats: Global Status Survey and Conservation Action Plan. IUCN/SSC Chiroptera Specialist Group. Gland, Switzerland: IUCN. x + 258 pp.

Ineich, I. 2002. Diversité spécifique des reptiles du Mont Nimba. Unpublished manuscript.

IUCN. 2002. 2002 IUCN Red List of Threatened Species. www.redlist.org.

Kormos, R. and C. Boesch. 2003. Status survey and conservation action plan. The West African Chimpanzee. IUCN/SSC Action Plan.

Nicoll, M.E., G.B. Rathbun and IUCN, SSC Insectivore Tree-Shrew and Elephant-Shrew Specialist Group. 1990. African insectivora and elephant-shrews: an action plan for their conservation. IUCN, Gland, Switzerland, iv+53 pp.

Rödel, M.-O. and D. Mahsberg. 2000. Vorläufige Liste der Schlangen des Tai-Nationalparks / Elfenbeinküste und angrenzender Gebiete. Salamandra 36: 25-38.

Rödel, M.-O., K. Grabow, C. Böckheler and D. Mahsberg. 1995. Die Schlangen des Comoé-Nationalparks, Elfenbeinküste (Reptilia: Squamata: Serpentes). Stuttgarter Beiträge zur Naturkunde, Stuttgart, Serie A, Nr. 528: 1-18.

Rödel, M.-O., K. Grabow, J. Hallermann and C. Böckheler. 1997. Die Echsen des Comoé-Nationalparks, Elfenbeinküste. Salamandra, 33: 225-240.

Rödel, M.-O., K. Kouadio and D. Mahsberg. 1999. Die Schlangenfauna des Comoé-Nationalparks, Elfenbeinküste: Ergänzungen und Ausblick. Salamandra 35: 165-180.

Sayer, J.A., C.S. Harcourt and N.M. Collins (Eds.). 1992. The Conservation Atlas of Tropical Dorests: Africa. Macmillan. 288 pp.

Toure, M. and J. Suter. 2001. Workshop report of the 1st trinational meeting (Côte d'Ivoire, Guinea, Liberia), 12-14 September 2001, Man, Côte d'Ivoire. Initiating a Tri-national Programme for the Integrated Conservation of the Mount Nimba Massif. Fauna & Flora Int., Conservation International & BirdLife Int., Abidjan. 56 pp. http://www.fauna-flora.org/around_the_world/africa/mount_nimba.htm

Toure, M. and J. Suter. 2002. Workshop report of the 2nd trinational meeting (Côte d'Ivoire, Guinea, Liberia), 12-15 February 2002, N'Zérékoré, Guinea. Initiating a Tri-national Programme for the Integrated Conservation of the Nimba Mountains. Fauna & Flora Int., Conservation International & BirdLife Int., Abidjan. 82 pp.

[UNDP] United Nations Development Programme. 2003. Human Development Report 2003. Oxford University Press, Oxford. 375 pp.

Plan d'Action Initial pour la Biodiversité de la Forêt Classée du Pic de Fon, Guinée

Ce Plan d'Action Initial pour la Biodiversité (IBAP) est basé sur les données collectées durant l'expertise biologique terrestre du Programme d'Evaluation Rapide (RAP) et l'évaluation des menaces et des opportunités socio-économiques pour la Forêt Classée du Pic de Fon. L'objectif de cet IBAP est de catalyser les efforts de conservation de la biodiversité de la forêt classée du Pic de Fon, et aussi de promouvoir les actions de conservation dans la Grande Chaîne du Nimba, car nombre de menaces à la biodiversité notées pour la Forêt Classée du Pic de Fon relèvent de dynamiques régionales. Ce plan évoluera et se consolidera sans doute au fur et à mesure que les dynamiques biologiques et socio-économiques du Pic de Fon et de la Grande Chaîne du Nimba seront mieux comprises, et à cette fin CI et RTM&E, ainsi que d'autres parties prenantes locales et régionales, forment maintenant une alliance plus étendue dans la région pour faire progresser les recommandations initiales présentées dans ce rapport. Toute réaction à ce plan est la bienvenue.

Le contexte régional

La Forêt Classée du Pic de Fon est l'une des composantes de la Grande Chaîne du Nimba, qui comprend le Mont Nimba, Diécké, la chaîne de Wonegizi-Ziama et la zone du Mont Béro-Tetini. De nombreuses menaces pèsent sur la biodiversité de la Forêt Classée du Pic de Fon, telles que décrites dans le chapitre sur l'évaluation des menaces et des opportunités socio-économiques de ce rapport. La chasse, l'empiètement agricole, les feux de brousse, l'exploitation forestière, la pêche excessive et la collecte des produits forestiers non ligneux, l'exploitation minière artisanale et le développement routier ont tous été identifiés comme des menaces existantes. De plus, la fragmentation des habitats et les opérations minières commerciales à grande échelle, si elles commencent effectivement, ont été identifiées comme étant de futures menaces importantes sur la biodiversité.

La plupart de ces sources de pression sur la biodiversité vont très probablement s'intensifier car la croissance et la migration de la population humaine augmentent la demande en viande de brousse, en terres cultivables et en autres ressources. Cette région présente le taux de croissance le plus élevé de la population en Guinée, ce qui renforce l'urgence d'efforts concertés de conservation. D'autres facteurs, qui n'ont pas fait l'objet d'un examen détaillé, confèrent une dimension encore plus complexe à ces menaces : le manque de capacités au niveau des acteurs régionaux (communautés, gouvernement, ONG et communauté scientifique), les troubles civils et les conflits fonciers.

La planification des investissements de conservation doit se poursuivre à un niveau régional. Dans l'immédiat, le manque de données concrètes, particulièrement sur l'évolution des tendances socio-économiques, représente un obstacle à l'identification des impacts de la conservation et à la mise en place d'une stratégie régionale de conservation. Par conséquent, la conduite de recherches supplémentaires centrées sur la collecte de données de base figure parmi les principales priorités.

Objectifs, étapes marquantes et résultats clés de la conservation

Les résultats de cette étude RAP montrent que l'écosystème de la Forêt Classée du Pic de Fon contient des espèces terrestres et aquatiques uniques ; des connexions intactes entre le milieu, la flore et la faune ; et l'un des derniers habitats de montagne relativement intacts, aussi bien en Guinée que sur les Plateaux de Haute Guinée. L'évaluation des menaces et des opportunités socio-économiques indique que la conservation de la biodiversité de la Forêt Classée du Pic de Fon et de la Grande Chaîne du Nimba subit de nombreuses menaces. Sur la base des résultats de cette étude, l'objectif global de conservation de ce Plan d'Action Initial pour la Biodiversité est d'améliorer la gestion des ressources naturelles dans la Grande Chaîne du Nimba en luttant contre les menaces sur la biodiversité, en promouvant des pratiques saines de production et en développant des sources alternatives et durables de revenus pour les communautés qui dépendent largement des forêts et des ressources naturelles de la région. Nous avons identifié cinq étapes clés pour la réalisation des objectifs de conservation, et plusieurs résultats pour chacune d'entre elles:

1. Les données et les analyses sur la biodiversité pour la Grande Chaîne du Nimba, et en particulier pour la Forêt Classée du Pic de Fon, seront disponibles au public.

- Une analyse de l'imagerie satellite existante et des données de télédétection sur les Plateaux de Haute Guinée (y compris la Forêt Classée du Pic de Fon) ;

- Des suivis biologiques, couvrant plusieurs saisons, des taxons identifiés au cours du RAP, incluant des inventaires dans le grand bloc forestier au sud des sites d'étude du RAP ;

- Une évaluation scientifique de l'importance de la biodiversité de la Forêt Classée du Pic de Fon par rapport à des zones voisines comparables dans la Grande Chaîne du Nimba ; et

- Une étude de l'hydrologie de la Forêt Classée du Pic de Fon.

2. Protocole de suivi de la biodiversité établi pour la Forêt Classée du Pic de Fon.

- Pièges photo mis en place dans la Forêt Classée du Pic de Fon pour suivre la distribution et les comportements territoriaux des grands mammifères, y compris les primates, et les activités de chasse ;

- Données de base acquises par imagerie satellite et détection des changements, utilisées pour suivre les modifications dans le temps et pour mettre à jour la stratégie de conservation de la biodiversité pour la Forêt Classée du Pic de Fon ;

- Un protocole établi pour suivre la qualité des eaux et le niveau de minéralisation dans les torrents afin de déterminer l'impact des activités humaines dans la Forêt Classée du Pic de Fon ; et

- Des partenariats entre les ONG, les agences gouvernementales et les institutions de recherche locales établis, pour développer des plans de gestion à long terme et mettre en œuvre des protocoles de suivi dans la Forêt Classée du Pic de Fon.

3. Un programme communautaire d'utilisation outenable des ressources pour la Forêt Classée du Pic de Fon.

- Des études socio-économiques de base sur les communautés à l'intérieur et autour de la Forêt Classée du Pic de Fon sont conduites et rendues publiquement disponibles, avec une attention particulière pour les stratégies de subsistance et leurs impacts sur la biodiversité ;

- Une analyse des routes à l'intérieur de la Forêt Classée du Pic de Fon pour mieux comprendre leur impact sur la biodiversité ; et

- Basé sur les études socio-économiques de base, un plan d'utilisation soutenable des ressources est développé et mis en oeuvre avec les communautés à l'intérieur et autour de la Forêt Classée du Pic de Fon. Un tel plan devrait inclure une variété de mesures visant les besoins liés à la production agricole, aux réserves de bois pour l'énergie, à la gestion soutenable de la faune et de la flore, et à d'autres services produits par la forêt, et pourrait inclure des accords de conservation basés sur des mesures incitatives qui compenseraient les coûts occasionnés par la conservation des ressources en question, et qui produiraient une impulsion économique pour réduire la pauvreté et améliorer le niveau social par l'apport d'activités rémunératrices.

4. Une sensibilisation et la capacité à traiter des problèmes de conservation de la biodiversité sont développées dans les agences gouvernementales clés, les institutions scientifiques nationales et régionales, et les communautés locales intéressées par la conservation de la biodiversité dans la Forêt Classée du Pic de Fon

- Un cadre local de scientifiques et techniciens formés pour effectuer des recherches et suivre l'évolution de la biodiversité dans la Forêt Classée du Pic de Fon et suffisamment de fonds disponibles pour supporter cette capacité ; et

- Une campagne de sensibilisation mise en oeuvre pour contrer les pratiques de chasse destructives dans la Forêt Classée du Pic de Fon.

***5. Une structure légale et physique est mise en place pour
la protection de la biodiversité dans la Forêt Classée du
Pic de Fon et la Grande Chaîne du Nimba.***

- Une révision des lois existantes concernant la chasse et la
protection de la faune sauvage en Guinée et en particulier dans la Forêt Classée du Pic de Fon ;

- Une stratégie est développée avec les communautés de la
Forêt Classée du Pic de Fon et le gouvernement guinéen
pour développer et renforcer les lois guinéennes régissant la chasse et la protection de la faune sauvage ; et

- Un réseau d'aires protégées est développé dans la
Grande Chaîne du Nimba, qui protège formellement
la biodiversité régionale sur tout le gradient altitudinal, depuis les sommets jusqu'aux forêts de piedmont,
basé sur les résultats de l'évaluation scientifique de
l'importance de la biodiversité de la Forêt Classée du
Pic de Fon par rapport à des zones comparables dans la
Grande Chaîne du Nimba.

Etudes scientifiques recommandées – voir résumé analytique
p. 133.

Chapitre 1

Revue de l'écologie et de la
conservation de la Forêt Classée du Pic
de Fon, chaîne du Simandou, Guinée

INTRODUCTION

Les caractères actuels de biodiversité ainsi que l'endémisme animal et végétal du bloc forestier de Haute Guinée datent de l'époque du Pléistocène, 15 000-250 000 ans av. JC. Les conditions de sécheresse qui s'ensuivirent sous les tropiques créèrent des refuges isolés, et avec les expansions et contractions répétées de la forêt originelle, les faunes et flores qui en résultèrent, évoluant dans de nouveaux habitats, ont connu de considérables spéciations (Lebbie 2001). A l'origine, il a été estimé que l'écosystème forestier de Haute Guinée couvrait jusqu'à 420 000 km², mais des siècles d'activité humaine ont entraîné la perte de presque 70% du couvert forestier originel (Bakarr et al. 2001). Le bloc forestier de Haute Guinée restant se limite à un certain nombre d'îlots jouant le rôle de refuges pour les espèces uniques de faune et de flore de la région. Ces forêts restantes hébergent des communautés écologiques d'une diversité exceptionnelle, une faune et une flore distinctes et des habitats de forêts mosaïques qui offrent refuge à de nombreuses espèces endémiques.

Des études de sites spécifiques, menées dans différents pays du bloc forestier de Haute Guinée montrent un haut niveau d'endémisme local pour les espèces végétales, probablement représentatif de l'ensemble de la région (Bakarr et al. 2001). Parmi les 1300 espèces connues du Parc National Taï en Côte d'Ivoire (4550 km²) presque 700 sont confinées à l'écosystème forestier de Haute Guinée et 150 sont endémiques à Taï (Davis et al. 1994). Au Mont Nimba, à la frontière de la Guinée, de la Côte d'Ivoire et du Liberia, 2000 espèces de plantes sont supposées présentes dans une zone de seulement 480 km², dont 13 espèces endémiques à ce site.

Si le bloc forestier de Haute Guinée abrite un assemblage unique d'écosystèmes, il recèle aussi une incroyable richesse minéralogique, le tout dans une région confrontée à une extrême pauvreté, à une augmentation de la densité de population humaine, à une faible implication du gouvernement dans l'environnement, et à des troubles civils périodiques mais persistants. Déterminer comment utiliser au mieux les ressources biologiques et géologiques, à l'avantage des communautés locales et de la biodiversité régionale, représente un challenge complexe auquel sont confrontés les industries d'extraction, les gouvernements d'Afrique de l'Ouest, les organisations de développement, les ONG et les habitants locaux.

Géographie

La chaîne du Simandou est l'un des composants des plateaux de Haute Guinée, qui comprennent aussi le Mont Nimba, Diécké, la chaîne du Wonegisi-Ziama, le Mont Béro-Tetini, les montagnes de Loma-Tingi et le Fouta Djalon. La chaîne du Simandou est formée de pics de haute altitude et de plateaux qui s'étendent à travers la Guinée sud-orientale, de Komodou au nord à Kouankan et Boola au sud. Le plus haut point, le Pic de Fon, est à 1656 m d'altitude. La chaîne du Simandou se situant à la transition entre des zones de savane et de forêt, elle offre une variété de types d'habitats allant de la savane guinéenne humide aux forêts pluviales de plaine de Guinée Occidentale, aux forêts galeries et de montagne guinéennes, et jusqu'au type d'habitat rare et exceptionnel que constitue la prairie de montagne. Les milieux de montagne ouest-africains ont une étendue très limitée et sont par conséquent particulièrement menacés, et la chaîne du Simandou est l'un des rares sites de toute l'Afrique de l'Ouest abritant ce type d'habitat. Par endroits, la chaîne du Simandou présente une inclinaison abrupte et couvre une échelle d'altitude allant de 500 à plus de 1600 m, ce qui a pour résultat de placer tous ces différents types de milieux à grande proximité les uns des autres.

A l'extrémité sud de la chaîne du Simandou s'étend la Forêt Classée du Pic de Fon (comprenant le Pic de Fon, le point le plus élevé de la chaîne et second sommet de la Guinée). La République de Guinée compte 113 forêts classées nationales, dont la plupart ont été classées au milieu du XXè siècle par le gouvernement colonial français pour des raisons commerciales et environnementales. La Forêt Classée du Pic de Fon, créée en 1953, est la troisième plus grande dans la région de Guinée Forestière et couvre approximativement 25 600 ha. A l'origine, la classification des forêts en Guinée visait la conservation de la forêt et du sol, qui, dans le cas du Pic de Fon, devient instable lorsqu'il est cultivé en raison de l'escarpement des pentes. De plus, il est reconnu que la chaîne du Simandou agit comme une barrière naturelle contre les feux de savane et régule le débit des cours d'eau qui prennent leur source au sein de la chaîne (incluant les rivières Diani, Loffa et Milo) (Hatch 1998).

Climat

Comparée à plusieurs des zones majeures de forêts pluviales dans le monde, la forêt pluviale guinéenne est plus sèche et la saisonnalité y est plus prononcée (WWF et IUCN 1994). La plus grande partie de la région guinéo-congolaise reçoit entre 1600 et 2000 mm de pluie par an, et cette pluviosité est moins également distribuée que dans d'autres forêts pluviales (White 1983). Dans le bloc oriental principal de la région guinéo-congolaise, les précipitations présentent en général deux pics, séparés par une période sèche relativement prononcée et une autre moins prononcée (White 1983). Dans toute la région guinéo-congolaise, la température mensuelle moyenne est à peu près constante tout au long de l'année (White 1983).

La chaîne du Simandou traverse deux des quatre régions naturelles de Guinée. La partie nord de la chaîne est située dans la région de Haute Guinée ou du plateau Mandingue, caractérisée par un climat soudano-guinéen chaud et sec, alors que la partie sud de la chaîne s'étend dans la région de Guinée Forestière ou dorsale sud-guinéenne, caractérisée par un climat guinéen forestier, avec des précipitations annuelles plus élevées et une courte saison sèche de trois mois.

Végétation et diversité d'habitat

Bien que les études biologiques antérieures soient pour la plupart non disponibles, on estime que la diversité et l'endémisme au sein de cette forêt sont relativement importants. Le Pic de Fon est escarpé et il présente un couvert de végétation dense dans sa partie sud occidentale, ce qui rend l'accès aux forêts de haute altitude relativement difficile.

Les zones de haute altitude au sein de la Forêt Classée du Pic de Fon contiennent deux types d'habitats : de la forêt de montagne, limitée aux plus hautes élévations à des forêts galeries de ravins courant en bandes étroites le long des lits des torrents (ces lits de rivières sont généralement au fond des ravins situés entre les pentes), et de la prairie de montagne. Les forêts de montagne et les prairies d'altitude sont toutes deux des végétations typiques originaires des montagnes du Simandou. Cependant, comme il y a peu de montagnes de cette altitude en Afrique de l'Ouest, ces habitats sont rares et exceptionnels pour la région. Un autre facteur augmentant le potentiel de forte diversité et d'endémisme est la présence de ces différents habitats à grande proximité les uns des autres ainsi que de la forêt pluviale guinéenne de piedmont trouvée à plus basse altitude dans la forêt classée.

Etant donné que le Pic de Fon agit en tant que barrière physique entre l'harmattan (vent sec et chargé de poussière venant du nord) et les alizés du sud-ouest chargés d'humidité, les pentes occidentales escarpées du Pic de Fon sont plus humides que les pentes orientales et donc couvertes de forêt plus denses. Sur le coté occidental du Pic de Fon, la forêt est limitée à de fines bandes de galeries forestières courant le long des rivières qui descendent juste en-dessous de la crête. La face occidentale de la crête présente des bandes de galeries forestières plus vastes et plus denses et d'importants îlots de forêts de ravin qui descendent de la crête et rejoignent les forêts de piedmont à l'ouest. Cet effet de barrière explique le fait que le Pic de Fon reçoive plus de précipitations que la normale, comparé aux zones de basse altitude sous des latitudes similaires. La chaîne montagneuse du Simandou constitue un réservoir d'eau d'importance majeure pour toute la région environnante et est la source d'un système hydrographique complexe qui approvisionne le système de drainage du Niger, de la Loffa, du Makona et du Diani (Hatch 1998).

Géologie

Dans presque toute la région guinéo-congolaise, l'altitude est inférieure à 1000 mètres. En Afrique de l'Ouest, la quasi-totalité de la région guinéo-congolaise est issue de la roche Précambrienne. Le paysage est formé de plateaux relativement bas et de plaines interrompues par des inselbergs résiduels et de petits plateaux plus élevés. Les plus importants de ces derniers sont le Fouta Djalon, les plateaux de Haute Guinée et la chaîne Togo-Atacora (White 1983). Les plateaux de Guinée atteignent 1752 m au Mont Nimba et 1947 m dans les Monts Loma. Contrastant avec le Fouta Djalon, les plateaux de Guinée montrent peu de surfaces plates et les collines y sont arrondies (White 1983).

Contrairement à la grande majorité de l'Afrique de l'Ouest, la partie australe de la chaîne du Simandou s'élève au-dessus de la plaine à quelques 600 à 800 m avec une crête principale atteignant 1250 à 1650 m. Sur la face orientale de la crête principale du Simandou, du moins dans sa portion australe, les pentes les plus douces correspondent à des talus de type « canga » qui se présentent comme des cônes de déjection des torrents, le plus grand ayant une étendue d'1 km². Dans la même zone, les pentes les plus élevées faisant face à l'est sont escarpées et entièrement couvertes d'affleurements minéralisés, de matériaux éboulés constitués de roches similaires et de « canga » formant des brèches. Les zones sommitales sont communément couronnées de roches plus ou moins dispersées de minéralisation massive ou semi-massive. Ce caractère est particulièrement développé au sommet du Pic de Fon, où des blocs de minerai de fer massif sont plus ou moins étroitement amalgamés dans une zone boisée (Hatch 1998).

Système hydrologique

Les régions montagneuses de Guinée, au sein de l'écosystème forestier de Haute Guinée qui recouvre le Pic de Fon et sa forêt classée, a été décrit comme le « château d'eau de l'Afrique de l'Ouest » (MMGE 2002) en raison de son rôle essentiel dans l'hydrologie de toute la région. La chaîne de montagnes du Simandou apparaît comme une zone réservoir d'eau cruciale pour toute la région environnante et la source d'un système hydrographique complexe qui alimente les systèmes de drainage du Niger, de la Loffa, de la Makonna et du Diani (Hatch 1998).

Description des sites d'étude du RAP

L'expédition RAP a eu lieu du 26 novembre au 7 décembre 2002, au début de la saison sèche et les suivis étaient basés sur deux sites, l'un proche du sommet du Pic de Fon et l'autre à moins haute altitude près du village de Banko (voir Carte). Les emplacements des deux camps ont été choisis en fonction de leur proximité avec les sites d'échantillonnage voulus, mais la sélection des camps a été limitée en fonction de leur accessibilité. Les camps permettaient l'accès à tous les principaux types de milieux de la région explorée, offrant à l'équipe une première impression assez complète. Cependant ils ne sont sans doute pas totalement représentatifs de la forêt entière qui contient potentiellement des zones plus riches et diverses, mais moins faciles d'accès.

Le premier site (Site 1) se trouvait entre de la prairie de montagne et des îlots de forêt montagneuse (comprenant des forêts de ravin et des forêts galeries) directement reliés à de plus vastes forêts pluviales de plaine à plus basse altitude. Ce site couvrait une échelle d'altitude comprise entre 1000 et 1600 m. Le camp était situé à 1350 m d'altitude (08°31'52"N, 08°54'21"W) et à 3,5 km (3461 m) du Pic de Fon (voir Carte).

Le second site (Site 2) comprenait des forêts secondaires et primaires et de petits îlots savanicoles à environ 600 m d'altitude. Le camp (08°31'29"N, 08°56'12"W) était situé à approximativement 6 km du village de Banko (en direction de la crête du Simandou) et se trouvait à environ 1,5 km à l'intérieur des limites de la forêt classée (voir Carte). Le camp était situé près d'une forêt pluviale de plaine, et présentait de nombreux cours d'eau de montagne de différentes tailles provenant de la chaîne montagneuse et quelques collines couvertes d'une végétation de savane dérivée. Cette forêt de plaine est caractérisée par *Parinari excelsa,* un arbre dont les

graines sont dispersées par les chauves-souris frugivores, ce qui constitue un trait remarquable des forêts de montagnes situées à l'ouest du Dahomey Gap. Sur un terrain plus plat et dans la direction du village, la plupart des forêts ont été converties en plantations de cacao et de café, et cette zone contient également de la forêt secondaire. Des plantations de bananiers et des jachères, entremêlées d'épices cultivées telles que le poivre, se trouvaient à l'intérieur des limites de la forêt classée avec une transition claire et soudaine avec le début de la forêt juste au pied de la montagne.

Profil de la population

Actuellement, la densité de population dans le quart sud-ouest de la Guinée, la région de Guinée Forestière, demeure relativement basse, estimée de 9 à 16 habitants au km². Toutefois, la densité de population est peut-être plus élevée sur les contreforts du sud de la région du Pic de Fon (Konomou et Zoumanigui 2000).

Un afflux massif de réfugiés a eu lieu en Guinée Forestière depuis les guerres en Sierra Leone et au Liberia. Il y a plus de réfugiés en Guinée que dans tout autre pays africain. A la fin de 1996 il a été estimé que près de 650 000 réfugiés en provenance du Liberia et de Sierra Leone se trouvaient en Guinée (UNHCR 1997). Konomou et Zoumanigui (2000) mentionnent que la population de réfugiés atteint 629 275 personnes, soit près de 40 % de la population en Guinée Forestière. Ces afflux spectaculaires de réfugiés du Liberia et de la Sierra Leone (et récemment de Côte d'Ivoire) expliquent sans doute les taux de croissance de population élevés enregistrés dans les préfectures de Kérouané, Beyla, et Macenta (voir chapitre 10 pour plus de détails).

Statuts de protection légaux

La Guinée est l'un des 150 pays membres de la CITES et elle a ratifié la Convention concernant la protection du patrimoine mondial culturel et naturel (WHC, Paris, 1972) et la Convention relative à la coopération en matière de protection et de mise en valeur du milieu marin et des zones côtières de la région de l'Afrique de l'Ouest et du Centre (WACAF, Abidjan, 1981). La Guinée a signé mais non ratifié la Convention africaine relative à la conservation de la nature et des ressources naturelles (ACCN) (Barnett et Prangley 1997).

En Guinée, les organes du gouvernement responsables de la faune et de la flore sont le Ministère de l'Agriculture et la Direction Nationale des Eaux et Forêts (DNEF). La loi qui régit l'utilisation de la faune sauvage est le «Code de la Protection de la Faune Sauvage et Réglementation de la chasse» (République de Guinée, 1988). Dans ce code, rédigé en 1988, adopté en 1990 et amendé en 1997, les espèces sont classées soit comme (1) intégralement protégées, (2) partiellement protégées ou (3) autres espèces. Les espèces qui sont intégralement protégées ne peuvent pas être chassées, capturées, détenues ou exportées (excepté si un permis scientifique est obtenu du gouvernement). La chasse, la capture et la détention d'une espèce intégralement protégée sont passibles d'une peine de six mois à un an de prison et d'une amende de 40 000 à 80 000 FG, ou l'une de ces deux pénalités. Pour les espèces qui ne sont pas spécialement protégées, les chasseurs doivent obéir à la réglementation de la chasse, doivent avoir un permis de chasse et ne peuvent chasser que du 13 décembre au 30 avril et uniquement du lever au coucher du soleil (Kormos et al. 2003).

Dans la région du Pic de Fon, un bloc forestier d'environ 25 600 ha a été désigné comme forêt classée – la Forêt Classée du Pic de Fon. Les forêts classées de Guinée Forestière couvrent un total d'environ 323 000 ha, Ziama et Diécké étant les deux plus importantes avec une étendue respective de 116 170 ha et 59 000 ha (Robertson 2001). La Forêt Classée du Pic de Fon est la troisième en taille (Konomou et Zoumanigui 2000). Le statut de forêt classée est censé fournir une protection en tant que propriété du gouvernement, mais pas nécessairement pour la conservation de la biodiversité. Il existe six types d'aires protégées en Guinée : parc national, réserve naturelle intégrale, réserve naturelle gérée, réserve spéciale ou sanctuaire de faune, zone d'intérêt cynégétique et zone de chasse. Une forêt classée peut recevoir l'un de ces statuts d'aire protégée, mais en elle-même elle n'indique pas de statut de conservation.

RÉFÉRENCES

Bakarr, M., B. Bailey, D. Byler, R. Ham, S. Olivieri and M. Omland (eds.). 2001. From the Forest to the Sea: Biodiversity Connections from Guinea to Togo. Washington DC: Conservation International.

Camara, W. and K. Guilavogui. 2001. "Identification des villages riverains autour des Forêts classées du Pic de Fon et Pic de Tibé". PGRR/Centre Forestier N'Zérékoré, Mesures Riveraines.

Davis, S.D., V.H. Heywood and A.C. Hamilton. (eds.). 1994. Centers of Plant Diversity: A Guide and Strategy for Their Conservation (Volume 1). Gland, Switzerland: World Wide Fund for Nature and IUCN-The World Conservation Union.

Fahr, J. and N.M. Ebigbo. 2003. A conservation assessment of the bats of the Simandou Range, Guinea, with the first record of Myotis welwitschii (GRAY, 1866) from West Africa. Acta Chiropterologica 5(1): 125-141.

Fahr, J., H. Vierhaus, R. Hutterer and D. Kock. 2002. A revision of the Rhinolophus maclaudi species group with the description of a new species from West Africa (Chiroptera: Rhinolophidae). Myotis 40: 95-126.

Hatch & Associates, Inc. 1998. "Preliminary Environmental Characterisation Study: Simandou Iron Ore Project Exploration Programme, South-Eastern Guinea." Montreal, Canada.

Konomou, M. and K. Zoumanigui. 2000. "Zonage Agro-Ecologique de la Guinée Forestière". Centre de Recherche Agronomique de Sérédou. Ministère de l'Agriculture et des Eaux et Forêts: Guinée.

Kormos, R., Boesch, C., Bakarr, M.I. and Butynski, T. (eds.). 2003. West African Chimpanzees. Status Survey and Conservation Action Plan. IUCN/SSC Primate Specialist Group. IUCN, Gland, Switzerland and Cambridge, UK.

Lebbie, A. 2001. Distribution, Exploitation and Valuation of Non-Timber Forest Products from a Forest Reserve in Sierra Leone. PhD Dissertation, University of Wisconsin-Madison, USA.

Mittermeier, R.A., N. Myers, C.G. Mittermeier and P.R. Gil. 1999. Hotspots: Earth's Biologically Richest and Most Endangered Terrestrial Ecoregions. CEMEX.

MMGE [Ministry of Mines, Geology, and the Environment]. 2002 (January). National Strategy and Action Plan for Biological Diversity; Volume 1: National Strategy for Conservation Regarding Biodiversity and the Sustainable Use of these Resources. Guinea/UNDP/GEF: Conakry.

[PGRR] Projet de Gestion des Ressources Rurales. 2001. "Pic de Fon – Situation Actuelle, Analyse et Recommandations (Proposition d'axes de collaboration)". Centre Forestier N'Zérékoré.

[PGRR] Projet de Gestion des Ressources Rurales. 2002a. "Proposition de stratégie pour la surveillance du Pic de Fon – Attente PGRF". Centre Forestier N'Zérékoré, Section Conservation – Biodiversité.

[PGRR] Projet de Gestion des Ressources Rurales. 2002b. "Rapport d'Activites – Délimitation du versant sud-ouest de la Forêt Classée du Pic de Fon". Centre Forestier N'Zérékoré (Antenne Béro).

Robertson. 2001. In: Fishpool, L.D.C. and M.I. Evans (eds.). Important Bird Areas in Africa and Associated Islands. Pisces Publications. 1162 pp.

White, F. 1983. The vegetation of Africa: A descriptive memoir to accompany the Unesco/AETFAT/UNSO vegetation map of Africa. Paris, Unesco. 356 pp.

WWF and IUCN. 1994. Centers of plant diversity. A guide and strategy for their conservation. Vol 1. Europe, Africa, South West Asia and the Middle East. 3 Volumes. IUCN Publications Unit, Cambridge, U.K.

Chapitre 2

Evaluation botanique rapide de la Forêt Classée du Pic de Fon, Guinée

Jean-Louis Holié et Nicolas Londiah Delamou

RÉSUMÉ

Une évaluation rapide de la flore de la Forêt Classée du Pic de Fon a été menée durant 11 jours du 27 novembre au 7 décembre 2002. Des données ont été récoltées sur deux sites en utilisant des transects linéaires au sein de différents types d'habitats, pour élaborer une liste des espèces présentes sur l'ensemble de la forêt classée et afin de décrire ses habitats. Les espèces recensées sur les deux sites suggèrent que les forêts galeries de ravin et les prairies de montagne du Site 1 étaient relativement peu dégradées alors que la forêt pluviale semi-sempervirente de basse altitude du Site 2 était soumise à une pression plus importante du fait de l'agriculture des résidents locaux. Nous avons recensé 409 espèces dont 18 espèces d'arbres classées par l'UICN, une comme « Menacé d'extinction », 16 comme « Vulnérable » et une « Presque Menacé ».

INTRODUCTION

L'étendue approximative de la couverture forestière tropicale humide en Guinée, comprenant les forêts de plaine, de montagne, de marais, et de mangrove, était à l'origine de 185 800 km². Il a été estimé qu'il ne reste aujourd'hui que 7655 km² de ce couvert forestier (ou 4,1% de la couverture forestière originelle ; Sayer et al. 1992) avec une perte moyenne annuelle de 1,8% de la couverture forestière en Guinée entre 1981 et 1985 (WRI 1992).

Les régions forestières sud-orientales de Guinée, y compris la Forêt Classée du Pic de Fon, sont inclues dans le Centre d'Endémisme Régional (RCE) guinéo-congolais, une zone d'environ 2,8 millions de km² en Afrique Centrale et de l'Ouest. Sur l'ensemble du RCE guinéo-congolais, il a été estimé que 12 000 espèces sont présentes (WWF et IUCN 1994), environ 80% de ces espèces et 20% des genres endémiques sont présents dans la région. La portion occidentale du RCE guinéo-congolais, la forêt pluviale guinéenne, couvre environ 420 000 km² et est centrée autour des régions forestières de Guinée, de la Sierra Leone, du Liberia, du sud de la Côte d'Ivoire et du Ghana, avec une chaîne de montagnes au nord (WWF et IUCN 1994).

La flore de Guinée est faiblement connue dans son ensemble et d'importantes zones du pays restent encore à étudier. En Guinée, seule la flore du Mont Nimba est relativement bien étudiée. La majorité des travaux sur la flore du Nimba sont ceux de Shnell (1952) et Leclerc et al. (1955). Plus de 200 espèces ont été décrites du Mont Nimba et environ 16 d'entre elles pourraient être endémiques (Adam 1971-1983). La région a été identifiée comme un centre de diversité botanique par le Programme de Conservation des Végétaux de l'UICN-WWF (UICN/WWF 1988). La chaîne du Simandou fait partie de la Grande Chaîne du Nimba et la végétation de la Forêt Classée du Pic de Fon peut être considérée comme ressemblant à celle du Mont Nimba, compte tenu des similarités de la géologie, du climat et de l'altitude. Cependant, antérieurement à cette expertise, les études publiées sur la végétation de la chaîne du Simandou consistent uniquement en une étude du Pic de Fon par Schnell (1961).

SITE D'ÉTUDE ET MÉTHODES

Conséquence de l'orientation sud-ouest de la chaîne de montagne, la Forêt Classée du Pic de Fon couvre trois types d'habitats primordiaux. Les montagnes jouent un rôle de barrière physique entre les vents secs venant du nord-est et les vents humides du sud-ouest. En conséquence, les pentes occidentales du Pic de Fon sont plus humides et couvertes de forêts plus denses que les pentes orientales. Alors que la forêt des pentes orientales est principalement restreinte à de petits îlots de forêts galeries de ravin courant le long des nombreux torrents et rivières, sur les pentes occidentales les îlots de forêts d'altitude sont directement liés à une ceinture continue de forêt sempervirente de plaine. Les résidents des 24 villages qui entourent la forêt classée ont activement utilisé cette forêt, au moins durant les 25 dernières années, et aux altitudes les plus faibles, près des villages, ne subsiste qu'une petite zone de forêt primaire. Dans les zones riveraines extérieures à la forêt classée, particulièrement à proximité du village de Banko, la majeure partie de l'espace a été convertie en plantations pérennes ou en champs de riz, alors que d'autres zones ont été mises en jachère depuis un à cinq ans. Les plaines adjacentes à l'est de la forêt classée sont dominées par une savane arborée, avec de petites forêts galeries le long des ruisseaux qui jaillissent de la chaîne du Simandou.

Nous avons mené 11 jours de travail de terrain, quatre jours dans les hauteurs à des altitudes de 900 à 1500 m (Site 1, 27-30 novembre) et sept jours sur le piedmont à 550-800 m (Site 2, 1-7 décembre). Notre objectif initial était d'identifier les types d'habitats. Nous avons utilisé des transects pour réaliser l'inventaire des espèces, collectant des spécimens de plantes qui étaient impossible à identifier sur le terrain. Dans les prairies, pour l'inventaire des lianes et des graminées, nous avons recensé les plantes jusqu'à 2 m de chaque coté du transect. Pour l'inventaire des buissons, nous avons recensé les plantes jusqu'à 20 m de part et d'autre du transect. Pour les forêts galeries de ravin du Site 1, qui étaient souvent encaissées dans les vallées, nous avons méthodologiquement choisi de les traverser en suivant le cours des rivières qui prenaient généralement leur source dans la montagne. Nous avons alors suivi un transect perpendiculaire au sens de l'eau. Pour la forêt pluviale semi-sempervirente de basse altitude, qui est plus dense, nous avons utilisé la même méthode que pour les forêts de montagne.

Sur chaque site inventorié, nous avons fait une description complète de la végétation. La détermination des espèces était faite progressivement. Les espèces non identifiées *in situ* étaient collectées et identifiées ultérieurement à l'aide de clefs d'identification. Les spécimens ont été déposés au Centre Forestier de N'Zérékoré, en Guinée. En parallèle à ces inventaires, nous avons fait des observations sur l'état, l'histoire et l'impact anthropique sur les sites.

L'accès à la forêt sur le Site 1 était plus difficile du fait du manque d'accès par la route et du peu de chemins existants. Le Site 2, près du village de Banko et à l'intérieur de la forêt classée, était plus aisément accessible à pied du fait des nombreux chemins menant du village à l'entrée de la forêt classée et grâce aux nombreux et anciens chemins de chasse parcourant la forêt classée.

RÉSULTATS

La Forêt Classée du Pic de Fon recèle trois principaux types d'habitats : les prairies montagneuses de haute altitude (Site 1); les forêts de montagne qui consistent en forêts galeries et de ravin dans les vallées et les lits des rivières du Pic de Fon (Site 1) ; et la forêt pluviale semi-sempervirente de basse altitude (Site 2). Dans nos inventaires de ces trois types d'habitats de la forêt classée, nous avons catalogué 409 espèces de plantes (voir Appendice 1).

La végétation de la prairie de montagne est liée aux conditions édaphiques (sol latéritique avec une concentration de fer élevée) et est dominée principalement par des *Fabaceae* et des *Poaceae*. La prairie de montagne couvre la majorité des terres des parties orientales et septentrionales du Pic de Fon. Les espèces de graminées les plus typiques trouvées dans la prairie de montagne, sur le Site 1, comprennent *Bouteloua gracilis (Poaceae)*, *Dolichos nimbaensis (Fabaceae)*, *Droogmansia scaettaiana (Fabaceae)*, et *Protea angolensis (Proteaceae)*, parmi tant d'autres. Les espèces de buissons telles que *Craterispermum laurinum* et *C. caudatum (Rubiaceae)* et spécialement *Ficus eriobotryoides (Moraceae)* se distinguent à cet endroit parmi les graminées.

Un certain nombre de graminées et de buissons typiques des prairies secondaires ont aussi été trouvés dont *Pennisetum purpureum (Poaceae)*, *Imperata cylindrica (Poaceae)* et *Hyparrhenia rufa (Poaceae)* et des buissons dont *Crossopteryx febrifuga (Rubiaceae)* et *Parkia biglobosa (Mimosaceae)*. Cependant, comme les Fabaceae telles que *Dolichos nimbaensis* (endémique à la prairie d'altitude du Mont Nimba) et *Rhynchospora corymbosa (Cyperaceae)* sont associées à des prairies édaphiques et sont plus typiques des prairies de montagne du Pic de Fon, nous pensons que cet habitat représente un climax édaphique.

La forêt de montagne du Site 1, le long de lit de torrents, est dominée par les espèces riveraines *Cryptosepalum tetraphyllum (Caesalpiniaceae)*, *Teclea verdoorniana (Rutaceae)*, *Garcinia polyantha (Clusiaceae)*, et *Tarenna vignei subglabrata (Rubiaceae)*. A plus haute altitude nous avons noté la présence de *Strychnos spinosa (Loganiaceae)*, *Balanites wilsoniana (Balanitaceae)*, *Morus mesozygia (Moraceae)*, et *Stereospermum acuminatissimum (Bignoniaceae)*. Les zones inventoriées sont pour la plupart d'étroites bandes le long de cours d'eau entre les montagnes avec des flancs abruptes qui sont difficiles d'accès et d'orientation. Ces ravins et forêts galeries constituent un habitat intact avec très peu de traces de dégradation.

La forêt pluviale semi-sempervirente du Site 2 est dominée par un grand nombre d'arbres, de lianes et de graminées non trouvées dans d'autres sites alentour. Les familles les plus représentées sont les *Rubiaceae, Sapotaceae, Clusiaceae, Ulmaceae, Moraceae, Euphorbiaceae* et *Caesalpiniaceae*. Les espèces de grands arbres recensées étaient *Neolemonniera clitandrifolia (Sapotaceae), Xylia evansii (Mimosaceae), Sterculia oblonga (Sterculiaceae), Guarea cedrata (Meliaceae), Khaya grandifoliola (Meliaceae), Celtis Adolfi –Friderici (Ulmaceae), Mansonia altissima (Sterculiaceae), Parkia bicolo*r *(Mimosaceae)* et d'autres espèces listées en Annexe 1. Les lianes inventoriées comprennent *Strychnos aculeatea (Loganiaceae), S. spinosa, Dalbergia hostilis (Fabaceae), Hippocratea* spp. *(Hippocrateaceae)* and *Calycobolus africanus (Convolvulaceae)*. Pour les graminées, nous avons noté la présence de *Geophila* spp. *(Rubiaceae), Solanum nigrum (Solanaceae), Trema orientalis (Ulmaceae), Marantochloa* spp. *(Marantaceae), Leptaspis cochleata (Poaceae), Olyra latifolia (Poaceae)*, et *Cyathula prostrata (Amaranthaceae)*.

La Forêt Classée du Pic de Fon recèle à la fois des unités forestières primaires non exploitées et fortement exploitées. Dans les zones les moins dégradées, des arbres tels que *Cryptocepalum tetraphyllum (Caesalpiniaceae), Parinarim excelsa (Chrysobalanaceae), Teclea verdoorniana (Rutaceae), Morus mesozygia (Moraceae), Celtis* spp. *(Ulmaceae), Funtumia* spp. *(Apocynaceae)* et *Guarea cedrata (Meliaceae)* étaient typiques. Dans les zones exploitées, des arbres, des graminées, des herbes de colonisation récente tels que *Solanum incanum, S. verbascifolium (Solanaceae), Ageratum conyzoides (Asteraceae), Physalis angulata (Solanaceae), Trema orientalis (Ulmaceae), Anthocleista nobilis (Loganiaceae)* et *Harungana madagascariensis (Clusiaceae)* étaient présents.

DISCUSSION

En raison de la composition spécifique, nous pensons que la prairie de montagne du Site 1, de peuplement antérieur à l'activité d'exploration minière, a subi peu d'interférence humaine, et mis à part les perturbations issues de la création des routes, les peuplements d'espèces de prairie de montagne sont relativement intacts. Nous avons noté la présence d'un certain nombre d'espèces associées à la prairie secondaire, mais en raison de la prédominance d'espèces associées à la prairie édaphique, nous pensons que la végétation représente ici une végétation climacique pour cet habitat.

De même, les forêts galeries de ravin du Site 1 constituent aussi un habitat intact avec très peu de traces de dégradation, et une composition spécifique ressemblant à la fois à la forêt d'altitude de *Parinari excelsa* et à la forêt périphérique, semi sempervirente guinéo-congolaise plus sèche telle que décrite par White (1983).

Nous souhaitons noter que la topographie de l'Afrique de l'Ouest est en général relativement uniforme et plate avec seulement quelques régions montagneuses, et que les habitats de la chaîne du Simandou est l'un des rares endroits dans toute l'Afrique de l'Ouest qui recèle des prairies de montagne et donc l'un des seuls lieux ou des espèces adaptées à ce type d'habitat peuvent vivre. La carte (pg. 18) indique à quel point les habitats d'altitude sont rares au sein de toute l'Afrique de l'Ouest.

Un facteur accroissant le potentiel de la Forêt Classée du Pic de Fon pour un haut niveau de diversité et d'endémisme est que ces types d'habitats rares sont situés à grande proximité les uns des autres ainsi que des forêts de plaine trouvées à faible altitude, à l'intérieur des limites de la forêt classée.

Espèces importantes pour la conservation
Parmi les 409 espèces recensées, nous avons relevé 18 espèces d'arbres répertoriées dans la Liste Rouge de l'UICN, une comme « Menacé d'Extinction », 16 comme « Vulnérable » et une « Presque Menacé ». Voir Tableau 2.1 pour la liste et les explications des catégories de menaces. Parmi ces espèces d'arbres, toutes ont été trouvées dans la forêt de plaine, et une seule (*Milicia excelsa* – nt) a aussi été trouvée dans la forêt de montagne. En outre, pour la forêt de plaine, les espèces *Teclea verdoorniana, Schefflera barteri* et *Balanites wilsoniana* constituent un groupe rare qui devrait être protégé.

Espèces utilisées pour l'alimentation, la médecine et le commerce international
Parmi les espèces recensées, nous avons identifié 98 espèces connues pour la valeur qu'elles représentent pour l'alimentation, la médecine ou le commerce international. Parmi celles-ci, 76 espèces sont utilisées pour l'alimentation, 24 sont communément utilisées en Afrique pour leurs propriétés médicinales, et 16 espèces sont commercialisées au niveau international, et économiquement importantes (dont *Elaeis guineensis*, palmier à huile, une espèce originaire d'Afrique de l'Ouest). Quelques unes des plus importantes espèces, telle que *Rauvolfia vomitoria*, sont utilisées localement pour l'alimentation et la médecine, alors qu'elles sont aussi internationalement exportées, dans le cas de *R. vomitoria* pour l'industrie pharmaceutique. Pour une liste complète des plantes identifiées durant notre étude et utilisées pour l'alimentation, la médecine et le commerce, voir Annexe 2.

Utilisation locale des ressources forestières
Les résidents des villages autour de la Forêt Classée du Pic de Fon ont activement utilisé cette forêt durant les 25 dernières années. Par conséquent, aux faibles altitudes, près des villages, ne subsiste qu'un faible pourcentage de forêt primaire. Nous avons noté que dans les alentours de la forêt classée, particulièrement à proximité du village de Banko (et probablement aussi à proximité des autres villages) plusieurs zones avaient été converties en plantations pérennes ou en champs de riz, tandis que d'autres zones sont en jachère (la plupart des jachères observées dataient d'entre un et cinq ans). Nous avons noté un certain nombre de plantations pérennes, par exemple de café et de bananes,

Tableau 2.1. Espèces de plantes globalement menacées recensées durant l'expertise RAP

Famille	Genre/espèce	Statuts de Menace
Anacardiaceae	*Antrocaryon micraster*	VU A1cd
Annonaceae	*Neostenanthera hamata*	VU A1c, B1+2c
Boraginaceae	*Cordia platythyrsa*	VU A1d
Caesalpiniaceae	*Cryptosepalum tetraphyllum*	VU A1c, B1+2c
Clusiaceae	*Garcinia kola*	VU A1cd
Combretaceae	*Terminalia ivorensis*	VU A1cd
Euphorbiaceae	*Drypetes afzelii*	VU A1c, B1+2c
Euphorbiaceae	*Drypetes singroboensis*	VU A1c
Meliaceae	*Entandrophragma candollei*	VU A1cd
Meliaceae	*Entandrophragma utile*	VU A1cd
Meliaceae	*Guarea cedrata*	VU A1c
Meliaceae	*Khaya grandifoliola*	VU A1cd
Mimosaceae	*Albizia ferruginea*	VU A1cd
Moraceae	*Milicia excelsa*	nt
Moraceae	*Milicia regia*	VU A1cd
Ochnaceae	*Lophira alata*	VU A1cd
Rubiaceae	*Nauclea diderrichii*	VU A1cd

Statuts de Menace (Liste Rouge des Espèces Menacées de l'UICN 2002)

EN : Menacé d'Extinction

  B) Etendue d'occurrence estimée à moins de 5000 km² ou aire d'occupation estimée à moins de 500 km² et estimations indiquant l'une des deux situations suivantes:
  1) Sévèrement fragmentée ou connue pour ne pas exister dans plus de cinq sites
  2) Déclin continu, déduit, constaté ou prévu, dans l'un des cas suivants:
    c) superficie, étendue et/ou qualité de l'habitat

VU: Vulnérable

  A) Réduction de la taille de la population de l'une des façons suivantes:
  1) Une réduction constatée, estimée, déduite ou supposée d'au moins 20% dans les 10 dernières années ou trois générations selon la plus longue des deux périodes, en se basant sur l'un des éléments suivants (à préciser) :
    c) Réduction de la zone d'occupation, de la zone d'occurrence, et/ou de la qualité de l'habitat
    d) Niveaux d'exploitations réels ou potentiels
  B) Etendue d'occurrence estimée inférieure à 20 000 km² ou aire d'occupation estimée en dessous de 2000 km², et estimations indiquant l'un des deux cas suivant :
  1) Population sévèrement fragmentée et présente dans 10 localités au plus
  2) Déclin continu constaté, déduit ou prévu dans l'un des cas suivants:
    c) superficie, étendue et/ou qualité de l'habitat

nt: Presque Menacé
Taxons près d'être qualifiés de Vulnérable

indiquant que les villageois ne pensent pas que les cultures pérennes à l'intérieur des limites de la forêt classée risquent de provoquer de réaction de la part des autorités. Des études antérieures menées par le PGRR (Projet de Gestion des Ressources Rurales), un projet du Centre Forestier de N'Zérékoré, ont estimé qu'entre 35 et 40% de la forêt classée a subi d'une façon ou d'une autre un impact causé par l'empiètement agricole (PGRR 2001). Nous avons aussi observé un certain nombre de modes de prélèvement non soutenables, en particulier la méthode de récolte des fruits qui consiste à abattre l'arbre, plutôt que de récolter les fruits sur l'arbre sans le détruire afin qu'il puisse de nouveau fructifier à la saison suivante.

RÉFÉRENCES

Adam, J.G. 1971-1983. Flore descriptive des Monts Nimba. Vols. 1 – 6. Mémoires du Muséum d'Histoire Naturelle B.20: 1-527; 22: 529-908.

Cunningham, A.B. 1993. African Medicinal Plants: setting priorities at the interface between conservation and primary health care. Working paper 1. UNESCO, Paris.

[FAO] Food and Agriculture Organization of the United Nations. 1990. Appendix 3 - Commonly Consumed Forest and Farm Tree Foods in the West African Humid Zone. The Major Significance of 'Minor' Forest Products: The Local Use and Value of Forests in the West African Humid Forest Zone. FAO, Rome.

[FAO] Food and Agriculture Organization of the United Nations and the US Department of Health, Education and Welfare. 1968. Appendix 4 - Index of Scientific Names of Edible Plants. Food Composition Table for Use in Africa.

IUCN. 2002. 2002 IUCN Red List of Threatened Species. Downloaded on 13 October 2003. www.redlist.org.

IUCN/WWF. 1988. Centres of Plant Diversity: a guide and strategy for their conservation. IUCN-WWF Plants Conservation Programme/IUCN Threatened Plants Unit. 40 pp.

Leclerc, J.C., M. Lamotte, J. Richard-Molard, G. Rougerie and P. Porteres. 1955. La Réserve Naturelle Intégrale du Mont Nimba. La chaine du Nimba: essai géographique. Mémoires de l'Institut Française d'Afrique Noire 43: 1-256.

MacKinnon, J. and K. MacKinnon. 1986. Review of the proteced areas system in the Afrotropical Realm. IUCN, Gland, Switzerland, in collaboration with UNEP. 259 pp.

Marshall, N.T. 1998. Searching for a Cure: Conservation of Medicinal Wildlife Resources in East and Southern Africa, TRAFFIC International.

PGRR. 2001. Pic de Fon – Situation Actuelle, Analyse et Recommandations (Proposition d'axes de collaboration). Centre Forestier N'Zérékoré.

Safowora, A. 1982. Medicinal Plants and Traditional Medicine in Africa. John Wiley and Sons Limited, Chichester.

Sayer, J.A., C.S. Harcourt and N.M. Collins (Eds.). 1992. The Conservation Atlas of Tropical Forests: Africa. Macmillan. 288 pp.

Schnell, R. 1952. Végétation et flore de la region montagneuse du Nimba. Mémoires de l'Institut Française d'Afrique Noire, 22, p. 1-604.

Schnell, R. 1961. Contribution à l'étude botanique de la chaîne de Fon (Guinée). Bull. Jard. Bot. État Brux., 31, p. 15-54.

White, F. 1983. The vegetation of Africa: a descriptive memoir to accompany the Unesco/AETFAT/UNSO vegetation map of Africa. Natural Resources Research XX. Unesco, Paris. 356 pp.

[WRI] World Resources Institute. 1992. World Resources 1992-93: a guide to the global environment. Oxford University Press, New York. 385 pp. (Prepared in collaboration with UNEP and UNDP.)

WWF and IUCN. 1994. Centres of plant diversity. A guide and strategy for their conservation. 3 volumes. IUCN Publications Unit, Cambridge, U.K. Vol. 1, p. 119.

Chapitre 3

Etude préliminaire des sauterelles (Insecta: Orthoptera: Tettigoniidae) de la Forêt Classée du Pic de Fon, Guinée

Piotr Naskrecki

RÉSUMÉ

Cette étude est la première évaluation des sauterelles (Tettigoniidae) de la région du Pic de Fon. 40 espèces ont été collectées, dont 4 sont nouvelles pour la science et 10 nouvellement inventoriées pour la Guinée (16 % d'augmentation). Les populations savanicoles de *Anoedopoda* cf. *lamellate* doivent être distinguées du reste de la population africaine. Les sauterelles du Pic de Fon ont un fort potentiel d'endémisme et devraient être étudiées de façon plus approfondie.

INTRODUCTION

Le niveau de diversité spécifique pris en considération lors des prises de décision en matière de conservation tropicale est globalement limité à de grandes espèces bien étudiées (charismatiques) de vertébrés tels que les primates, les éléphants ou les grands prédateurs. La diversité des oiseaux, des reptiles ou des amphibiens, est plus rarement prise en compte dans le choix des aires protégées. Mais jusqu'ici, il n'y a jamais eu un seul cas dans lequel les données sur les invertébrés aient influencé à elles seules la décision d'attribuer un statut de conservation à une zone ou un site tropical (bien que de tels cas deviennent plus fréquents en Europe, Nouvelle Zélande et Amérique du Nord ; Murphy et Weiss 1988; Sherley et Hayes 1993; Ortiz 2001). Plusieurs facteurs sont responsables de cette situation. En premier lieu, l'état de nos connaissances sur les communautés invertébrées tropicales est rudimentaire ou inexistant pour de nombreuses régions. En conséquence, il est pratiquement impossible d'évaluer l'abondance ou même la présence ou l'absence d'individus spécifiques d'invertébrés.

Cette absence presque totale de données sur la diversité des invertébrés sous les tropiques a pour tragique conséquence que des espèces disparaissent ou sont sérieusement menacées sans que personne n'en connaisse même l'existence. Il y a très peu de cas documentés d'extinction d'espèces invertébrées (St. Helena earwig, California dune neduba ; Rentz 1977), et pratiquement tous ne sont dévoilés que lorsqu'une espèce d'un groupe connu pour son haut niveau d'endémisme et de fidélité à son habitat est découverte par un taxinomiste dans une collection muséologique, collectée d'un site déjà altéré ou détruit par les activités humaines.

Dans les sciences biologiques, les obstacles taxinomiques rendent le traitement et l'identification du matériel invertébré et des données collectées en zones tropicales extrêmement difficiles. Alors qu'il y a des centaines de spécialistes des vertébrés travaillant sur la faune tropicale, et que l'identification des espèces n'est pas un problème, dans la grande majorité des groupes d'invertébrés, l'identification est problématique et souvent impossible dans les conditions de terrain. De plus, le nombre important d'espèces encore non décrites (dans certains groupes d'invertébrés, plus de 90% des espèces attendent toujours d'être reconnues et nommées) rend la collection et la comparaison des données de zones tropicales très difficile.

Ironiquement, il n'y a pas mieux que les invertébrés pour servir d'outils de conservation et d'indicateurs de l'état des habitats. De nombreuses espèces sont extrêmement sensibles aux changements même mineurs de l'humidité, de la température, de la concentration en métaux lourds dans le sol et l'eau, ou de la composition des communautés de plantes, et peuvent potentiellement servir de premier système d'alarme vis-à-vis des changements qui pourraient éventuellement affecter les communautés de vertébrés. D'autre part, plusieurs espèces d'invertébrés constituent la première source d'alimentation d'autres animaux et rendent des services écologiques incommensurables, comme la pollinisation, la production du sol ou la décomposition. Plusieurs petites espèces, en raison de leurs besoins oligotypiques et de leurs aptitudes de dispersion limitées, sont plus vulnérables que les animaux de plus grande taille. Pourtant, elles échappent régulièrement à l'attention des autorités de la conservation et sont irrémédiablement perdues.

Durant l'expertise du présent Programme d'Evaluation Rapide (RAP) de la partie sud occidentale de la chaîne du Simandou, dans l'est de la Guinée, un groupe d'invertébrés, les sauterelles (Tettigoniidae) ont été échantillonnées dans le but de déterminer leur biodiversité et leur potentialité en tant qu'indicateurs du caractère unique des deux sites. Les membres de ce groupe présentent un potentiel remarquable comme indicateurs environnementaux (Samways 1997) et plusieurs espèces ont des aires de répartition remarquablement réduites, résultant en un haut degré d'endémisme dans tout le groupe. Exception faite d'une unique étude basée sur le matériel collecté au Mont Nimba (Chopard 1954) et de quelques recensements individuels éparpillés dans les révisions taxinomiques (Brunner von Wattenwyl 1895; Beier 1962, 1965) la faune des sauterelles de Guinée est pratiquement inconnue. Il existe ici un fort potentiel, partiellement démontré par cette étude, de découverte de nouveaux taxons dans ce pays, aussi bien au niveau générique que spécifique. Chopard (1954) a recensé 60 espèces de la région du Mont Nimba et Beier (1965) a ajouté une espèce supplémentaire à la liste.

MÉTHODES

Trois méthodes de collecte se sont montrées d'une grande efficacité pour récolter les sauterelles dans des habitats mixtes comme la zone investiguée : la lumière noire (lumière UV), la recherche visuelle et le balayage au filet. Malheureusement, la méthode de la lumière noire n'était pas utilisable lors de cette étude, réduisant potentiellement les chances de collecter des espèces nocturnes volantes, comme de nombreux représentants des Phaneropterinae. Toutefois, la possibilité d'employer d'autres sources lumineuses (lumière fluorescente sur le Site 1, lampes à pétrole sur le site 2) a permis de déterminer que pratiquement aucune sauterelle nocturne volante n'était présente dans la zone. Aussi, le fait de n'avoir pas utilisé la lumière noire comme méthode de collecte n'a-t-il probablement eu qu'un impact minimal sur le comptage final des espèces collectées.

Le balayage au filet a été employé dans des milieux savanicoles, dans les strates inférieures de la forêt et dans des habitats de lisière buissonnants adjacents à la forêt.Cette méthode s'est avérée très efficace pour récolter des sauterelles se nourrissant de graines dans les hautes herbes, ainsi que de nombreuses sauterelles arboricoles qui se fixent à l'envers sur la surface inférieure des feuilles. Le balayage a été standardisé en réalisant une série de six balayages consécutifs avant que le contenu du filet ne soit inspecté, et quatre séries de balayage étaient réalisées sur chaque transect de 100 m dans les habitats savanicoles. Au total 30 transects ont été échantillonnés.

La recherche visuelle de nuit s'est révélée être de loin la plus efficace des méthodes de collecte, en terme de nombre d'espèces récoltées, plutôt que du simple nombre de spécimens collectés. La plus grande partie de la collecte était effectuée entre 8 h du soir et 2 h du matin, quand l'activité de pratiquement toutes les espèces de sauterelles est la plus importante. Cette méthode a été uniquement standardisée en tenant compte du temps passé à collecter.

Des représentants de toutes les espèces rencontrées ont été collectés et des spécimens échantillons ont été conservés dans de l'alcool a 95 % et sous forme de spécimens secs épinglés. Ces spécimens seront déposés dans les collections du Muséum de Zoologie Comparée de l'Université de Harvard et de l'Académie de Sciences Naturelles de Philadelphie (cette dernière deviendra aussi le dépositaire officiel des holotypes de plusieurs nouvelles espèces rencontrées au cours de cette étude d'après leur description formelle).

En plus de la collection physique de spécimens, la stridulation des espèces sonores a été enregistrée à l'aide d'un magnétophone professionnel Sony WM-D6C Walkman et d'un microphone unidirectionnel Sennheiser. Ces enregistrements sont essentiels pour établir l'identité de plusieurs espèces discrètes du genre *Ruspolia* et *Thyridoropthrum*, dont les caractères morphologiques seuls ne suffisent pas pour identifier l'espèce.

RÉSULTATS

En dépit des conditions météorologiques défavorables (début de saison sèche) un nombre de sauterelles relativement élevé a été collecté. Au total, 40 espèces ont été récoltées, représentant au moins 24 genres (52 % de la diversité générique d'Afrique de l'Ouest) et 6 sous-familles. La seule sous-famille connue pour la région, mais non collectée durant cette étude, est Hetrodinae, qui comprend des espèces typiques des habitats de plaine xériques, aussi son absence parmi les taxons collectés n'est-elle pas surprenante. Au moins quatre des espèces récoltées sont nouvelles pour la science, incluant au moins un nouveau genre, mais il est possible que de nouveaux taxons supplémentaires se trouvent parmi les spécimens encore non identifiés du genre graminicole *Ruspolia*. Ce dernier a grand besoin d'une révision taxinomique, qui devrait être basée en grande partie sur les propriétés acoustiques des appels spécifiques à l'espèce. La Figure 3.1 présente les signatures digitales des «espèces type» de *Ruspolia* enregistrées pendant l'étude. Le Tableau 3.1 présente la liste complète des espèces récoltées durant cette étude.

Sur le premier site, ouvert et particulièrement sec (Site 1: 08°31'52"N; 08°54'21"W) la majorité des espèces récoltées étaient des éléments savanicoles. La plupart des sauterelles étaient des granivores nocturnes, avec seulement deux espèces prédatrices (*Thyridorhoptrum*). Les ravins forestiers du site hébergeaient très peu d'espèces purement sylvicoles. Trois îlots forestiers distincts ont été explorés : deux îlots à une altitude de 1200- 1300 m, et un à une altitude approximative de 1000 m (08°32.9'N; 08°53.9'W). Les seuls taxons connus pour être exclusivement associés à des milieux forestiers étaient *Mormotus* sp. 1 et *Afrophisis* sp. n. Le reste des espèces récoltées dans les forêts était constitué d'éléments savanicoles qui pénètrent apparemment les îlots de forêt relativement étroits et secs. La présence dans l'un des îlots de l'espèce strictement graminivore *Plastocorypha nigrifrons* était particulièrement surprenante. Les taxons de sauterelles normalement associés à la litière végétale comme *Afromecopoda* ou *Euthypoda* étaient totalement absents. Ceci est peut-être simplement une erreur d'échantillonnage due au cours laps de temps passé à collecter sur le site, mais la couche de litière dans les îlots forestiers était très sèche, et n'abritait qu'un faible nombre des autres insectes (ex. Collembola, Dermaptera, Formicidae) qui se trouvent normalement dans un tel milieu. Détail intéressant, les espèces afro-tropicales *Anoedopoda lamellata*, trouvées en milieu forestier dans l'ensemble de leur aire de répartition, n'était présentes sur ce site qu'en savane ouverte. Des études supplémentaires devraient être menées afin de déterminer

si la population du Pic de Fon est conspécifique aux populations forestières en Afrique Centrale et Australe. Mais à partir des caractéristiques de son appel et des différences morphologiques mineures du veinage des ailes, nous pouvons supposer sans risque que la population du Pic de Fon représente une nouvelle espèce du genre.

Au moins deux espèces supplémentaires récoltées sur le premier site sont nouvelles pour la science. *Aphrophisis* sp. n. représente un petit genre d'insectivores spécialisés, connus jusqu'à présent seulement du Togo (1 espèce) et du Cameroun (1 espèce). Il est très probable que cette espèce soit endémique de la région, car la plupart des espèces de son groupe (Phisidini) sont connues pour avoir des aires de répartition très limitées (Jin and Kevan 1992). La présence de *Thyridorpthrum* sp. n. indique l'existence d'un complexe auparavant inconnu d'espèces discrètes au sein du genre. Bien qu'étant morphologiquement très similaire de l'espèce largement distribuée *T. senegalense*, la nouvelle espèce trouvée dans la savane et en lisière de forêt sur le premier site a un appel d'avertissement très distinct et caractéristique (Fig. 3.2). Cependant, la nouvelle espèce ayant de bonnes capacités à voler, ce taxon n'est sans doute pas exclusivement lié à la zone, et on doit probablement le trouver sur d'autres sites de la chaîne du Simandou. En plus des espèces nouvelles pour la science, au moins 4 espèces récoltées sur le premier site sont nouvellement recensées en Guinée.

Le second site (Site 2: 08°31'29"N; 08°56'12"W) était situé à une altitude de 590 m et il présentait une

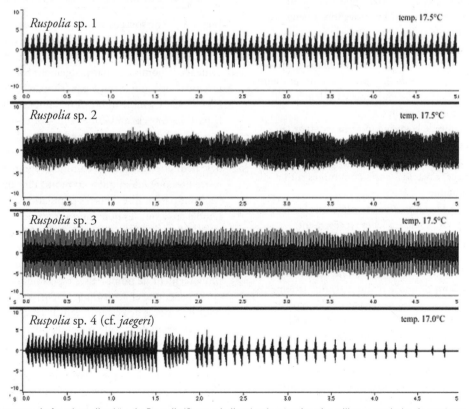

Figure 3.1. Oscillogramme de 4 espèces discrètes de *Ruspolia* (Conocephalinae) présentes dans les milieux savanicoles de montagne africaine du Site 1.

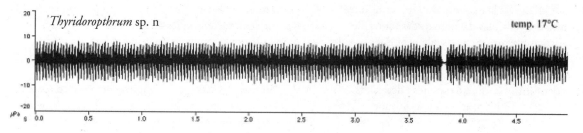

Figure 3.2. Oscillogramme d'une nouvelle espèce de *Thyridoropthrum* (Conocephalinae) des milieux savanicoles de montagne africaine du Site 1.

humidité et un couvert forestier beaucoup plus importants. Dix-neuf des 29 espèces collectées sur ce site étaient des éléments typiquement sylvicoles, trouvés principalement dans la forêt primaire sur le versant occidental de la chaîne. En raison de l'humidité plus élevée et d'une litière végétale plus développée dans la forêt, au moins trois espèces terrestres étaient présentes sur le site (*Afromecopoda frontalis*, *Euthypoda* sp. 1, et *Conocephalus carbonarius*). Au moins 10 des espèces récoltées étaient exclusivement arboricoles et quelques espèces supplémentaires ont été entendues stridulant en hauteur dans la canopée (de telles espèces ne peuvent être capturées que par exploration de la canopée ou après une série de fortes pluies, quand elles sont immobilisées au sol par le vent et l'eau).

Un caractère marquant de la faune des sauterelles de forêt, sur le second site, était la très faible densité de population des espèces observées. En moyenne, à peine un individu était récolté toutes les 30 minutes de recherches dans la forêt. Cette situation était très probablement due au début de la saison sèche, moment où de nombreuses espèces traversent une courte période de diapause. De plus, plusieurs genres (*Euthypoda*, *Mustius*, *Amytta*) étaient seulement présents à l'état de jeunes nymphes.

L'herbe dominante de la savane, *Setaria megaphylla*, abritait plusieurs espèces de *Ruspolia*, *P. nigrifrons*, et *Conocephalus* spp. L'absence de *P. lanceolatus* et *A.* cf. *lamellata*, deux espèces savanicoles très communes sur le premier site, était notable.

Deux espèces de *Meconematinae* récoltées sur le second site sont nouvelles pour la science (elles représentent aussi un nouveau genre qui sera prochainement décrit) (voir pp. 18-21), mais d'autres espèces non décrites sont peut-être présentes parmi les *Ruspolia* non identifiées. Malheureusement, étant donné le manque d'enregistrements sonores comparatifs pour pratiquement toutes les espèces déjà décrites du genre, il serait déraisonnable de prendre des décisions taxinomiques en se basant seulement sur les caractères morphologiques. Une étude supplémentaire des espèces d'Afrique de l'Ouest devrait donc être conduite dès que possible. Les deux nouvelles espèces de *Meconematinae* ont presque certainement une distribution très limitée, car elles sont toutes les deux incapables de voler, et disposent donc d'un faible potentiel de dispersion. Enfin, au moins six

espèces collectées sur le second site sont recensées pour la première fois en Guinée.

De nombreuses espèces collectées au second site représentent les seuls recensements de ces espèces en dehors de la localité du spécimen type. Par exemple, le spécimen de *Weissenbornia praestantissima* collecté ici est seulement le second spécimen connu de cette espèce, décrite à l'origine du Cameroun (Lowry-Kribi Mudung), et la grande population de *Conocephalus carbonarius* trouvée dans les herbes des strates inférieures de la forêt secondaire du site 2 représente les premiers spécimens adultes de cette espèce jamais collectés (cette espèce n'était connue que par un spécimen au stade nymphal provenant de Accra).

RECOMMANDATIONS POUR LA CONSERVATION

Bien qu'elle ne soit pas exhaustive, cette étude a ajouté au moins 10 espèces à la liste de la faune du pays (une augmentation de 16 %). Dans le même temps, elle met en évidence le nombre remarquablement bas d'espèces (10) déjà recensées au Mont Nimba par Chopard (1954) et Beier (1965). Cela semble indicatif de l'important turnover de la part des espèces en Guinée sud – orientale et présente le potentiel de haut degré d'endémisme des sauterelles de cette région. Les nouvelles espèces de sauterelles découvertes devraient être considérées comme géographiquement restreintes ou endémiques, au moins tant que d'autres études n'auront pas localisé d'autres populations de ces espèces. Il semble aussi que les populations savanicoles de *A. lamellate* soient distinctes des populations forestières de cette espèce dans d'autres régions de l'Afrique. De plus le complexe d'espèces discrètes du genre *Ruspolia* comprend probablement de nombreuses espèces non décrites, et une étude génétique et acoustique extensive devrait être conduite pour les espèces de ce genre dans la chaîne du Simandou.

En raison de la forte sensibilité des espèces de sauterelles forestières aux changements de l'humidité et aux compositions des espèces de plantes, toute activité dans la zone qui changerait le régime des eaux et affecterait le climat pourrait potentiellement avoir des effets dévastateurs sur la présente faune. D'autres études sur toute la chaîne du Simandou sont urgemment requises afin d'estimer la distribution et la taille des populations des espèces nouvellement découvertes. Une collecte supplémentaire est aussi nécessaire au début de la saison des pluies, lorsque la plupart des espèces apparaissent en grand nombre et au stade adulte.

Tableau 3.1. Liste complète des espèces Orthoptera collectées durant le RAP au Pic de Fon.

Taxon	Site 1	Site 2	Habitat	Nouveaux inventaires pour la Guinée
Tettigoniidae				
Phaneropterinae				
Phaneroptera minima	+	+	S/RF/BP	Oui
Phaneroptera ? maxima	+		S	Oui
Phaneroptera sp. 1	+	+	S/PF	
Eurycorypha sp. 1	+		S/RF	
Eurycorypha sp. 2		+	S	
Preussia lobatipes		+	BP	
Ducetia fuscopunctata		+	S	
Arantia ? ovalipennis		+	S	
Arantia sp. 1		+	S	
Zeuneria melanopeza		+	S	Oui
Weissbornia praestantissima		+	PF	Oui
Conocephalinae				
Ruspolia jaegeri	+		S	Oui
Ruspolia sp. 1	+		S	?
Ruspolia sp. 2	+		S	?
Ruspolia sp. 3	+	+	S/BP	?
Ruspolia sp. 4	+	+	S/BP	?
Ruspolia sp. 5	+	+	S	?
Pseudorhynchus lanceolatus	+		S	
Plastocorypha nigrifrons	+	+	S	
Conocephalus carbonarius		+	PF	
Conocephalus maculatus	+	+	S/BP	
Conocephalus sp. 1	+		S	
Thyridorhoptrum senegalense	+	+	S/RF/BP	
Thyridorhoptrum sp. n.	+		RF/PF	Oui
Meconematinae				
Amytta sp. 1	+	+	RF/PF	
Amytta sp. 2		+	PF	

S – Savane
RF – Forêt de ravin
PF – Forêt primaire
BP – Plantation de bananes

Taxon	Site 1	Site 2	Habitat	Nouveaux inventaires pour la Guinée
Gen. n., sp. n. 1		+	PF	Oui
Gen. n., sp. n. 2		+	PF	Oui
Listroscelidinae				
Afrophisis sp. n.	+		RF	Oui
Mecopodinae				
Anodeopoda cf. lamellata	+		S	Oui
Afromecopoda frontalis		+	PF	
Euthypoda sp. 1		+	PF	
Pseudophyllinae				
Stenampyx annulicornis		+	PF	
Mustius sp. 1	+		RF/PF	
Mustius sp. 2		+	PF	
Mormotus sp. 1		+	PF	
Mormotus clavaticercus		+	PF	
Lichenochrus cf. centralis		+	PF	
Hambrocomes sp. 1		+	PF	
Adapantus brunneus		+	PF	

Nombre d'espèces recensées – 40
Novelles espèces pour la Guinée – 10 (peut-être plus)
Nouvelles espèces decouvertes – 4 (peut-être plus)
Site 1 – 20
Site 2 – 29
Savane – 21 (14 exclusivement)
Forêt – 22 (18 exclusivement)
Plantation de bananes – 6

RÉFÉRENCES

Beier, M. 1962. Pseudophyllinae. Tierreich 73.

Beier, M. 1965. Die afrikanischen Arten der Gattungs-gruppe "Amytta" Karsch. Beiträge zur Entomologie 15: 203-242.

Brunner von Wattenwyl, C. 1895. Monographie der Pseudophylliden. Wien.

Chopard, L. 1954. La reserve naturelle integrale du Mont Nimba III. Orthopteres Ensiferes. Mem. IFAN 40: 25-97.

Gibbs, G.W. 1998. Why are some weta (Orthoptera: Stenopelmatidae) vulnerable yet others are common? Journal of Insect Conservation 2: 161-166.

Jin, X-B. and K. McE. Kevan. 1992. Afrophisis, a new genus and two new species of small orthopteroids from Africa (Grylloptera Tettigonioidea Meconematidae). Tropical Zoology 4: 317-328.

Murphy, D.D. and S.B. Weiss. 1988. A long-term monitoring plan for a threatened butterfly. Conservation Biology 2: 367-374.

Ortiz, F. 2001. Mexico moves to protect Monarch butterfly reserve. Online. Available: http://www.planetark.org/dailynewsstory.cfm/newsid/13455/story.htm. 21 Nov. 2001.

Rentz D.C.F. 1977. A new and apparently extinct katydid from Antioch Sand Dunes (Orthoptera: Tettigoniidae). Entomological News 88: 241-245.

Samways, M.J. 1997. Conservation Biology of Orthoptera. In: Gangwere, S.K. et al. (eds). The Bionomics of Grasshoppers, Katydids and Their Kin. CAB International, pp. 481-496.

Sherley, G.H. and L.M. Hayes. 1993. The conservation of giant weta (Deinacrida n. sp. Orthoptera: Stenopelmatidae) at Mahoenui, King Country: habitat use, and other aspects of its ecology. NZ Entomol. 16, 55–68.

Chapitre 4

Évaluation rapide des amphibiens et des reptiles dans la Forêt Classée du Pic de Fon, Guinée

Mark-Oliver Rödel et Mohamed Alhassane Bangoura

RÉSUMÉ

Nous avons recensé au moins 32 espèces d'amphibiens de la région de la chaîne du Simandou / Pic de Fon et nous estimons que 50 à 60 espèces sont probablement présentes. La Forêt Classée du Pic de Fon est l'une des régions abritant la plus importante biodiversité d'amphibiens de toute l'Afrique de l'Ouest. Un mélange d'espèces forestières et savanicoles, des espèces répandues et localement endémiques, caractérisent la communauté d'amphibiens du Pic de Fon, reflétant l'unique habitat mosaïque de cette région. Plusieurs espèces rencontrées durant l'expertise RAP sont nouvelles pour la science et il est possible qu'elles soient endémiques à la région du Pic de Fon. La communauté d'amphibiens de la région est actuellement soumise à de nombreuses menaces, tell que la destruction de l'habitat. De plus, de probables altérations futures de l'habitat, telles que les activités d'extraction minières et minérales, pourraient avoir de graves impacts négatifs. Des recherches complémentaires sont nécessaires pour évaluer la répartition et les caractéristiques démographiques des amphibiens du Pic de Fon.

INTRODUCTION

Les amphibiens sont hautement diversifiés en Afrique de l'Ouest (Rödel 2000b, Rödel et Schiøtz in Bakarr et al. 2001). Leur taxinomie est relativement bien connue et la plupart des espèces sont étroitement liées à certains types d'habitats qui leur sont spécifiques. De plus, il est relativement facile d'évaluer une part raisonnable et représentative de la faune d'amphibiens d'une zone donnée en un court laps de temps, au moins durant leur phase reproductive, qui est, presque toujours, la saison des pluies. Ceci fait des amphibiens un groupe convenant parfaitement à des programmes d'évaluation biologique rapide centrés sur la biodiversité de zones scientifiquement inconnues. En outre, la présence ou l'absence de certaines espèces sert d'indicateur du statut (intact ou dégradé) des habitats (cf. Rödel et Branch 2002).

Si les amphibiens africains savanicoles sont généralement très répandus, avec des aires de distribution qui s'étendent souvent du Sénégal à l'Afrique du Sud (Schiøtz 1999, Rödel 2000a), ce n'est pas le cas des espèces forestières. Ces dernières sont principalement restreintes à l'Afrique de l'Ouest, dans une zone s'étirant du Sénégal au nord-ouest, au Nigeria au sud-est (Schiøtz 1967, 1999; Rödel 2000b). Plusieurs de ces espèces ont même des aires de répartition plus étroites, confinées aux forêts de Haute Guinée (la zone forestière à l'ouest du « Dahomey Gap ») ou à de petites parties de ces forêts (ex. Rödel et Ernst 2000, 2002b; Rödel et al. 2002, 2003).

Les inventaires d'amphibiens raisonnablement bons sont encore rares. Jusqu'à présent, pas plus de 25 inventaires ont été compilés pour toute la région de Haute Guinée (pour révision voir Rödel et Agyei 2003). Cependant, l'étude de ces zones montre clairement que la diversité d'amphibiens tend à être plus élevée dans les zones forestières que dans les habitats savanicoles, et les zones comprenant une combinaison de biomes différents présentent probablement une plus grande diversité que les biomes uniformes.

Les reptiles n'ont été que rarement rencontrés au cours de cette étude RAP. Le moment auquel le RAP a eu lieu, au début de la saison sèche, n'était pas favorable à un recensement des reptiles à cause des hautes herbes gênant la vue et de l'activité généralement faible de ces espèces. Sur la base de ces résultats et de ceux d'autres études menées dans d'autres zones d'Afrique de l'Ouest (Angel et al. 1954a, b; Barbault 1975; Böhme 1994b, 1999; Rödel et al. 1995, 1997, 1999; Rödel et Mahsberg 2000; Ernst et Rödel 2002; Ineich 2002; Branch et Rödel 2003), le nombre d'espèces de reptiles recensées pourraient ne représenter que 20 % des espèces vivant dans la zone du Pic de Fon. Nous ne présentons donc qu'un résumé sur les reptiles recensés et un commentaire sur les reptiles protégés. Les analyses sont centrées sur les amphibiens.

SITE D'ÉTUDE ET MÉTHODES

En raison de l'orientation sud-ouest de la chaîne de montagne, la FCPF couvre de nombreux types d'habitats différents. Les pentes occidentales de la FCPF sont plus humides que les pentes orientales, et abritent donc une forêt plus dense. La forêt s'étend également plus au nord et à l'ouest que sur les pentes orientales. Alors qu'à l'est de la FCPF la forêt se réduit principalement à de petites zones de forêts galeries courant le long des nombreuses petites rivières et ruisseaux, à l'ouest les zones de forêt à haute altitude sont directement reliées à une ceinture de forêt pluviale de plaine encore très préservée. Les plaines adjacentes à l'est de la FCPF sont dominées par une savane arborée, avec de petites forêts galeries le long des ruisseaux issus de la chaîne du Simandou. Les zones de haute altitude au sein de la FCPF comportent deux habitats distincts : de la forêt montagneuse (de galerie et de ravin) formant des bandes étroites le long des cours d'eau, et de la prairie de montagne, qui est l'un des types d'habitats les plus exceptionnels et les plus menacés en Afrique de l'Ouest. La moyenne des précipitations annuelles varie entre 1 700 et 2 000 mm. La longue saison sèche dure de novembre à avril.

L'étude RAP s'est déroulée du 26 novembre au 7 décembre 2002, au début de la saison sèche. L'équipe du RAP a principalement travaillé sur deux sites. Le premier site, proche du Pic de Fon (Site 1; 08°31'52"N, 08°54'21"W) comportait de la prairie de montagne et des îlots forestiers le long de cours d'eau, à une échelle d'altitude variant entre environ 1 000 et 1 600 m. Les deux types d'habitats paraissaient être à l'état primaire. Toutefois, des routes construites récemment pour faciliter les activités d'exploration minière, ont déjà causé une érosion sédimentaire. Le second site (Site 2; 08°31'29"N, 08°56'12"W) se trouvait à environ 600 m d'altitude, à 6 km de distance du village de Banko. Il était approximativement à 1,5 km à l'intérieur des limites de la forêt classée. Les deux sites étaient distants d'à peu près 3,6 km. Le deuxième site comprenait de la forêt primaire et de

petites zones de savane dans un paysage très vallonné. Dans la direction du village, une partie de la forêt a été convertie en plantations de cacao et de café, et abrite également de la forêt secondaire. Des plantations de bananiers mélangées à des cultures de condiments, tels que le poivre, se trouvaient à l'intérieur des limites de la forêt classée, avec une transition claire et soudaine avec le début de la forêt, juste au pied de la montagne.

Nous avons essayé de prospecter de la même façon tous les types d'habitats présents. Toutefois, en raison de la sécheresse et de la présence de hautes herbes dans la prairie de montagne, cet habitat a probablement été sous échantillonné. En altitude, la plupart des relevés d'amphibiens proviennent de forêts galeries situées le long de ravins et des parties humides des prairies, ces dernières recelant de petites zones d'eau stagnante. Dans la plaine, nous avons prélevé des échantillons aussi bien dans les forêts pluviales intactes que dans les forêts dégradées et les plantations vers le village de Banko. Du fait du vallonnement du paysage et de la présence de forêt primaire, il n'y avait pratiquement pas d'eaux stagnantes. A la place, cette partie de la réserve offrait une très grande variété de cours d'eau lents ou rapides de différentes tailles. Plusieurs petites rivières couraient également dans les forêts dégradées et les plantations. De plus, ces habitats offraient aussi des mares, des flaques, et des marécages. Le Tableau 4.1 dresse la liste complète des habitats prospectés, incluant leur position géographique, la date de prospection, l'effort d'échantillonnage et une description rapide des caractéristiques de l'habitat. Avant l'étude RAP, l'auteur junior a effectué un travail de terrain dans la région du Pic de Fon, du 2 au 9 novembre 2002. Les recensements des deux études de terrain sont inclus dans cette analyse.

Les amphibiens et les reptiles ont principalement été localisés de façon opportuniste, au cours de suivis visuels effectués par quatre personnes au maximum. Les suivis étaient menés de jour et de nuit. Les techniques de recherche incluaient une observation visuelle du terrain et l'examen de refuges (ex. en soulevant des pierres et des rondins, en fouillant la litière). Nous avons aussi mis en place un suivi acoustique dans tous les types d'habitats disponibles (Heyer et al. 1993). Pendant toute la durée des recherches, il n'a plu qu'à trois reprises, deux fois au cours de la première période de terrain et le premier jour de la deuxième période (voir ci-dessous). Pour cette raison, l'activité reproductive des amphibiens (et donc les mâles appelant) était très limitée, et la plupart des relevés provenaient d'observations visuelles. Plusieurs espèces n'ont pu être recensées qu'à l'état juvénile. Nous avons aussi fait usage de filets immergés pour les têtards dans les eaux qui s'y prêtaient, et contrôlé les lignes de pitfalls temporaires mises en place par le groupe d'étude des petits mammifères, sur le second site du RAP. Toutefois, les résultats obtenus pour les pitfalls étaient négligeables. Notre méthodologie n'a produit que des données qualitatives et semi quantitatives. Pour obtenir des données quantitatives, l'expérimentation de marquage- recapture le long de transects standardisés ou sur des parcelles déterminées aurait été nécessaire. La durée de l'étude RAP était trop limitée pour employer cette méthode. Nous avons mesuré notre effort d'échantillonnage en homme-heures passées à prospecter une zone donnée (Annexe 3). Pour récolter des données comparatives, nous avons passé plus de temps à prospecter des habitats complexes et étendus que ceux qui étaient uniformes ou plus petits.

En supposant que l'effort d'échantillonnage était comparable sur les différents habitats, nous avons calculé le nombre total d'espèces d'amphibiens présentes dans la FCPF. Comme nous ne disposions pas de données quantitatives, nous avons utilisé les estimateurs Chao2 et Jack-knife 1, en nous basant sur la présence ou l'absence de données pour tous les habitats (logiciel: EstimateS, Colwell 1994-2000). Les bases de calcul étaient les listes des espèces recensées chaque jour (17 jours) pour 32 espèces d'amphibiens (voir ci-dessous). Pour éviter des effets d'ordre, tous les calculs ont été basés sur 500 tests. Pour une introduction aux méthodes employées, voir Colwell (1994-2000) et la littérature citée dans ce document.

Plusieurs déterminations restent encore à confirmer. Les connaissances taxinomiques actuelles ne permettent pas de déterminer les grenouilles *Arthroleptis/Schoutedenella* au niveau spécifique. La séparation des espèces en appliquant des critères morphologiques (morpho-spécifiques) est donc difficile voire impossible. Les seuls caractères fiables, sur le terrain, sont les appels (Rödel et Branch 2002, Rödel et Agyei 2003). En raison du temps sec, aucune de ces grenouilles n'a été entendue appelant pendant l'étude RAP. Nous avons donc compté les recensements de ce groupe comme une seule espèce. Cependant, il est beaucoup plus réaliste de supposer qu'au moins deux ou trois espèces ont été recensées. La richesse spécifique est donc estimée conventionnellement.

Nous ne décrivons que les espèces qui sont peu connues, ont un statut taxinomique incertain, et/ou n'ont pas été traitées en détail dans d'autres publications récentes (ex. Schiøtz 1999, Rödel 2000a, Rödel et Branch 2002, Rödel et Agyei 2003). Une liste complète des espèces, incluant tous les sites où une espèce particulière a été recensée pendant l'étude RAP, est donnée dans le Tableau 4.2. La nomenclature suit généralement celle de Frost (2002). Pour les exceptions, voir le texte.

Certains spécimens échantillons ont été anesthésiés et tués dans une solution de chlorobutanol puis conservés dans de l'éthanol à 70 %. Les échantillons ont été déposés dans les collections de l'auteur senior (MOR) et de l'auteur junior (MAB, voir Appendice 2). Les spécimens de MOR seront transférés dans les collections de plusieurs musées d'histoire naturelle. Les spécimens du MAB constitueront la base d'une collection guinéenne de référence, et seront ensuite entreposés à l'Université de Conakry. Les échantillons de tissu (extrémités d'orteils) des espèces recensées étaient conservés dans de l'éthanol à 95 %. Ces échantillons sont conservés dans la collection de MOR et à l'Institut de Zoologie de l'Université de Mainz, Allemagne.

Liste des espèces sélectionnées:

Bufo superciliaris Boulenger, 1888 «1887». Ce très gros crapaud africain n'a pas pu être recensé durant le RAP. Toutefois, nous avons eu des descriptions si précises de la part des villageois, concernant la taille, la coloration et les habitudes d'un gros crapaud forestier, que nous ne doutons pas que la présence de cette espèce puisse être tenue pour confirmée. Selon nos guides locaux, il peut être trouvé au pied de grands arbres à contreforts, en forêt profonde. Les populations malinkés locales croient que ce crapaud donne naissance aux arcs-en-ciel. Cette croyance semble être répandue en Afrique de l'Ouest. L'auteur senior a entendu des histoires similaires chez les Oubi et les Guéré de l'Ouest de la Côte d'Ivoire. *B. superciliaris* est connu pour être répandu depuis l'est de la Côte d'Ivoire jusqu'au nord de la République Démocratique de Congo et au Gabon (Frost 2002). Il a récemment été recensé dans la forêt de Ziama, au sud de la Forêt Classée du Pic de Fon, par Böhme (1994a). D'autres localités connues en Afrique de l'Ouest sont le Parc National de Taï, Côte d'Ivoire (Rödel 2000b), le Mont Nimba (Côte d'Ivoire, Liberia, Guinée; Guibé et Lamotte 1958a) et le Ghana (Hughes 1988, sans donner de localisation exacte). Selon Böhm (comm. pers.), le *B. superciliaris* d'Afrique Centrale et de l'Ouest diffèrent considérablement. *B. superciliaris* a été décrit du «Rio del Rey, Cameroun» par Boulenger (1888 «1887»). Il est donc probable que *Bufo chevalieri* Mocquard, 1909 «1908», décrit de la Côte d'Ivoire et donné sans discussion comme synonyme par Tandy et Keith (1972), doive être réhabilité.

Bufo togoensis Ahl, 1924. Cette espèce a été citée comme synonyme de *B. latifrons* Boulenger, 1900 par Tandy et Keith (1972). *B. latifrons* a été décrit du Gabon, *B. togoensis* a été décrit du Togo. Nos spécimens ont des parotides apparentes plus ou moins lisses et droites, en contact avec les paupières (granuleuses et espacées de la paupière chez *B. latifrons,* Perret et Amiet 1971), et un tympan clairement discernable (indistinct chez *B. latifrons,* Perret et Amiet 1971). Leurs verrues varient en taille et la peau sur le dessus de la tête est légèrement plus lisse. Nos spécimens ressemblent donc plutôt à *B. camerunensis* Parker, 1936. Cependant, leur peau sur les flancs et la partie postérieure est moins rugueuse et celle du dessus de la tête est plus lisse que chez *B. camerunensis*. De plus, les mâles provenant de la Forêt Classée du Pic de Fon dépassaient de loin la taille de ceux des deux autres espèces. Quatre mâles appelant mesurés avaient un SVL de 62,5 – 64 mm (*B. camerunensis* mâles: 46-56 mm, *B. latifrons* mâles: 35-45 mm, Perret et Amiet 1971). Nous pensons que les crapauds ouest-africains de ce groupe devraient faire l'objet d'une révision (voir aussi Rödel et Branch 2002), mais comme nos échantillons sont plus proches de la description de *B. togoensis* par Ahl, nous avons décidé de désigner nos spécimens comme faisant partie de cette espèce. Les crapauds du Mont Nimba, désignés comme *B. latifrons* (Guibé et Lamotte 1958a), ne semblent pas être *B. latifrons* mais correspondent au taxon présentement décrit. Une autre espèce de ce groupe, aussi présentée comme synonyme de *B. latifrons* par Tandy et Keith (1972), est *B. cristiglans* Inger et Menzies, 1961, décrits des Monts Tingi en Sierra Leone.

Bufo togoensis était très abondant dans les massifs forestiers occidentaux de la Forêt Classée du Pic de Fon. Alors que la plupart des espèces anoures avaient déjà cessé de se reproduire, nous avons trouvé de très grands groupes de ces crapauds le long de larges rivières (5 – 10 m de large) à proximité du camp 2 (> 100 mâles, un grand nombre de couples et de femelles, une femelle mesurait 72 mm SVL). Les chants les plus importants étaient émis de nuit avec les pics les plus hauts à l'aube et au crépuscule. Cependant, nous avons aussi entendu des mâles solitaires appelant régulièrement dans la journée. Des chants plus faibles et des mâles solitaires appelant ont été entendus sur presque tous les ruisseaux à l'intérieur de la forêt dense. Les mâles étaient installés le long de la rivière, distants de 50 à 200 cm les uns des autres. Les sites d'appel se trouvaient dans de l'eau peu profonde, sur des pierres et éventuellement de la végétation (40 cm de hauteur). Sur un site, *B. togoensis* appelait conjointement avec *B. maculates* Hallowell, 1854. Nous avons observé une seule fois deux couples à 10h du matin, alors qu'ils frayaient dans une partie très peu profonde de la rivière, avec un faible courant. Les chaînes d'œufs étaient englués avec de la vase, ce qui n'a pas permis un comptage précis des œufs.

Les mâles appelant avaient parfois le dos presque jaune mais changeaient de couleur dans les minutes qui suivait leur dérangement, un caractère aussi connu pour d'autres crapauds africains Rödel 2000a, Channing 2001) et même pour les grenouilles *Petropedetes* (Rödel 2003). Les caractères de coloration étaient très variables. Nous avons trouvé des spécimens avec et sans ligne vertébrale distincte, avec des dos uniformément marron, des têtes rouges, présence ou absence de taches noires derrière le cou, etc.

Bufo (?) sp. Ce crapaud minuscule, uniformément marron, a été capturé à 1 350 m d'altitude dans la prairie de montagne, à proximité immédiate d'une forêt galerie. Nous n'avons pas été en mesure de relier cet individu probablement juvénile à une espèce connue d'Afrique de l'Ouest. Il est particulier par l'absence de glandes parotides, un caractère clairement discernable chez les juvéniles de tous les crapauds connus de la région (Rödel non publié). Compte tenu de son apparence générale (ex : forme du corps, structure des pattes antérieures et postérieures), nous sommes tentés de le classer parmi le genre *Bufo*, bien que d'autres spécimens (adultes) soient nécessaires pour clarifier le statut taxinomique de ce crapaud.

Conraua sp. *C. alleni* (Barbour and Loveridge, 1927) est la seule espèce de ce genre décrite pour la forêt pluviale de Haute Guinée. Des recensements sous ce nom ont été publiés du Liberia, de la Côte d'ivoire, de la Guinée et de la Sierra Leone (Lamotte and Perret 1968). Nous avons recensé deux espèces de *Conraua* de la Côte d'Ivoire (Rödel et Branch 2002, Rödel 2003), toutes les deux très clairement différentes de *C. derooi* Hulselmans, 1972 qui a été décrite du Togo. Nos inventaires comprennent donc au moins une

nouvelle espèce. Morphologiquement et génétiquement, les spécimens de la Forêt Classée du Pic de Fon sont plus proches des spécimens de la forêt de la Haute Dodo que de ceux du Parc National du Mont Sangbé (Rödel et Kosuch données non publiées). Cependant les relations entre nos inventaires et la description d'une nouvelle espèce requièrent des analyses approfondies et seront traitées dans une publication séparée.

Les grenouilles du genre *Conraua* étaient présentes dans toutes les rivières et les ruisseaux de la région. Leurs chants se faisaient régulièrement entendre la nuit, augmentant clairement après le crépuscule. L'activité maximum d'appel était atteinte avant l'aube. Les grenouilles adultes mesuraient jusqu'à 70 mm SVL. Elles étaient toutes colorées de marron ou vert olive avec des taches noires sur le dos. Certaines présentaient une ligne latérale noire. Les têtards ne pouvaient être observés que la nuit. Les nombreux têtards étaient actifs dans les parties peu profondes, presque stagnantes des ruisseaux, le plus souvent sur fond vaseux ou sablonneux. Lorsqu'ils étaient dérangés, les adultes et les têtards s'enfonçaient immédiatement dans la vase (Knoepffler 1985, Rödel et Branch 2002). Les têtards avaient un corps rougeâtre, tacheté de points noirs. La dernière moitié de leur queue, y compris la nageoire, était noire.

Amnirana nov. sp. Nous avons trouvé un adulte *Amnirana* mâle (58 mm SVL) qui différait considérablement des autres *Amnirana* d'Afrique de l'Ouest par un tympan extraordinairement grand et une peau complètement couverte de larges granules. Les mâles d' *Amnirana occidentalis* ont une peau plus lisse (Rödel et Branch 2002). Les mâles d'*A. albolabris* ont une peau fine et granuleuse. La taille des granules du spécimen du Simandou est même supérieure à celle du seul autre *Amnirana* connu avec une texture de peau plus granuleuse, *A. assperima* d'Afrique Centrale (Perret 1977). Cette nouvelle espèce d'*Amnirana* sera décrite par ailleurs. Nous ignorons encore si une très grande femelle *Amnirana* (75 mm), avec une peau lisse et des tympans plus petits, capturée au même endroit, appartient aussi à cette nouvelle espèce, ou bien est simplement une femelle *A. albolabris*, qui a aussi été recensé sur la chaîne du Simandou.

Ptychadena spp. En plus de *Ptychadena aequiplicata*, qui est une espèce typiquement forestière répandue de l'Afrique Centrale à l'Afrique de l'Ouest, nous avons recensé deux autres espèces de *Ptychadena* que nous n'avons pas été en mesure d'attribuer avec certitude à des espèces connues. L'une ressemble à *P. aequiplicata* mais diffère par des crêtes dorsales régulières. Nous disposons de preuves génétiques qu'une espèce secrète de *Ptychadena*, proche de *P. aequiplicata,* vit dans la zone de transition ouest-africaine, entre la forêt et la savane (Vences, Kosuch et Rödel, données non publiées). Notre recensement est peut-être celui de cette espèce non décrite.

Tous les autres spécimens appartenaient à une espèce de *Ptychadena* qui montrait des caractères quelque peu intermédiaires entre *P. mascareniensis* et *P. oxyrhynchus*. Ils étaient caractérisés par des pattes extrêmement longues comme *P. oxyrhynchus*, mais les palmures étaient beaucoup plus réduites que chez cette espèce. L'organisation des crêtes dorsales était similaire à celle de *P. mascareniensis*, ils avaient une crête dorso-latérale blanchâtre et continue, et de claires bandes vertébrales oranges ou vertes. Parmi les adultes, le plus grand spécimen était une femelle de 62 mm qui avait un ventre jaune foncé. Nous avons fréquemment trouvé cette espèce dans les prairies de montagne durant la nuit, à proximité de flaques entre les rochers de la prairie, en forêt secondaire le long de petits ruisseaux et dans une zone humide proche du village de Banko. Nous les avons entendus appeler à côté de flaques entre des rochers dans les prairies de montagne, pendant la nuit. Malheureusement, l'appel n'a pas pu être enregistré. L'impression acoustique de l'appel ne ressemblait pas à *P. mascareniensis* de Côte d'Ivoire (Rödel unpubl. data). Nous avons aussi observé de tous petits têtards de *Ptychadena* dans ces mêmes flaques.

Petropedetes natator Boulenger, 1905. Cette espèce, endémique de la partie occidentale du bloc forestier de Haute Guinée, vit exclusivement dans les torrents au sein des forêts (Guibé et Lamotte 1958a, Lamotte 1966, Lamotte et Zuber-Vogeli 1954, Böhme 1994b, Rödel 2003). Dans la FCPF cette espèce n'était pas très abondante, mais presque omniprésente dans tous les ruisseaux et rivières rapides, de 600 à 1400 m d'altitude. Nous n'avons jamais recensé *P. natator* plus loin qu'à 3-5 m de l'eau. Mesurant 37-42.5 mm, les mâles étaient plus petits que ceux du Parc National du Mont Sangbé en Côte d'Ivoire (41-47 mm, Rödel 2003). Le plus petit juvénile recensé mesurait 18 mm. Plusieurs adultes avaient des mites endoparasites sous la peau de la face ventrale des cuisses. Jusqu'ici, ces mites étaient seulement connues pour une autre grenouille *Petropedetes* du genre *Phrynobatrachus* (Spieler and Linsenmair 1999, Rödel 2000a).

Phrynobatrachus alticola Guibé et Lamotte, 1961. Cette petite espèce de litière est restreinte aux forêts de Haute Guinée (Guibé et Lamotte 1961, Lamotte 1966). En raison de son développement direct, elle est capable de survivre dans des zones forestières dépourvues d'eau libre (Rödel et Ernst 2002a). Elle est caractérisée par sa toute petite taille et trois paires de verrues sur le dos. Les grenouilles de la Forêt Classée du Pic de Fon diffèrent des spécimens ivoiriens par le fait qu'elles ont souvent plus de verrues sur les flancs et le dos (voir Rödel et Ernst 2002a, Rödel 2003). Certains spécimens ont des verrues sur le cou, fusionnées en une petite crête. La plupart des spécimens étaient presque uniformément brun, certains avaient un point clair sur le cou, d'autres montraient des bandes vertébrales oranges, jaunes ou vertes et avaient le dos complètement roux. La partie ventrale des cuisses était jaune, la gorge des mâles adultes était pointillée de taches noirâtres à presque noires. L'iris était toujours roux et or. Les palmures étaient complètement réduites, l'extrémité des orteils était légèrement élargie. Contrairement aux spécimens de Côte d'Ivoire, plusieurs avaient des mites endoparasites sur la partie ventrale des cuisses.

Alors que sur des sites connus de Côte d'Ivoire (Rödel et Ernst 2002a, Rödel et Branch 2002, Rödel 2003) *P. alticola* est facile à détecter par ses appels, mais difficile à voir, dans la FCPF cette espèce était de loin la plus abondante des grenouilles sur le sol forestier. Tandis que sur nos sites ivoiriens *P. alticola* était toujours recensé loin de l'eau libre, un comportement normal compte tenu de leur mode de développement direct, les spécimens de la FCPF ont toujours été attrapés à proximité de petites rivières, depuis la forêt du piedmont jusqu'à de petites poches de forêts galeries sur la crête de la montagne. La grande majorité des spécimens recensés étaient juvéniles. Durant la nuit, les individus formaient des agrégats (souvent > 20 individus), sur de petites plantes qui poussaient à la fois dans et près de l'eau, probablement pour échapper à la prédation. Donc, les spécimens de la FCPF ne montraient pas seulement des différences morphologiques et de coloration par rapport aux inventaires ivoiriens, mais différaient aussi dans le choix de leur habitat. Malheureusement nous n'avons jamais entendu d'appel. Pour cette raison, nous ne pouvons déterminer si notre échantillon est une espèce différente de *P. alticola*. Une possibilité serait *P. tokba* qui a été décrite de la Guinée par Chabanaud (1921). L'auteur senior a examiné des spécimens de série type de *P. tokba* et des spécimens attribués à *P. alticola* par Lamotte (spécimen type de *P. alticola*, probablement perdu, A. Ohler comm. pers.). Morphologiquement, les spécimens des muséum de ces deux espèces sont indiscernables (pour les spécimens examinés voir appendice dans Rödel et Ernst 2002b).

Phrynobatrachus fraterculus (Chabanaud, 1921). Cette espèce de litière a été à l'origine décrite à Macenta, en Guinée. Jusqu'à présent, elle est seulement connue du sud-est de la Guinée, du Liberia, de la Sierra Leone et de l'extrême ouest de la Côte d'Ivoire. Nous n'avons pas connaissance d'un recensement effectué en Guinée Bissau (cf. Frost 2002). Cette espèce est caractérisée par la forme svelte de son corps avec un museau relativement pointu, un dos brun avec ou sans rayure vertébrale claire, une remarquable bande latérale d'un noir profond bien délimitée, la palmure réduite, des doigts et orteils aux extrémités élargies. Les femelles adultes et la plupart des mâles avaient un ventre jaune, couverts soit de quelques points noirs ou de plus larges taches noires. La nette délimitation de la bande latérale noire est le caractère le plus important pour distinguer *P. fraterculus* de de *P. gutturosus*. De plus, *P. fraterculus* a toujours une peau très lisse, alors que la peau de *P. gutturosus* varie entre relativement lisse et très verruqueuse, mais montre toujours au moins quelques petites verrues (cf. figures dans Rödel 2000a). Les mâles mesuraient entre 17 et 19 mm (Rödel et Ernst 2002b). Un mâle (MOR T02.21) mesurait 18,9 mm, avait un ventre blanc dépourvu de taches et deux plis de gorge clairement marqués, parallèles aux mandibules. La gorge de ce mâle était grise. Rien n'a été publié sur la biologie de ces espèces. Dans le Parc National de Taï, en Côte d'ivoire, nous avons trouvé cinq spécimens, durant quatre ans d'intense étude de terrain, au voisinage immédiat des rivières et parties marécageuses des forêts primaires et secondaires (Rödel et Ernst non publié). Le plus grand spécimen recensé était une femelle (MOR T02.25) de 24,7 mm. Dans les zones de basse altitude de la Forêt Classée du Pic de Fon, nous avons souvent trouvé plusieurs adultes et juvéniles, posés sur des feuilles ou sur le sol, presque toujours à proximité immédiate d'un ruisseau. Les alentours étaient composés de forêt secondaire ou de plantations (cacao, banane). Les plus petits juvéniles, mesurant 7,5-8 mm, ont été trouvés sur la berge du ruisseau et dans une mare asséchée. Nous ne savons donc toujours pas si cette espèce se reproduit dans les eaux stagnantes ou courantes. Les adultes étaient tous des femelles, mesurant 24-26 mm. Une femelle de la FCPF (MOR F068) et une provenant de Taï contenaient de nombreux tous petits œufs. Nous supposons que *P. fraterculus* se reproduit comme tous ses congénères dans des eaux stagnantes ou légèrement courantes, en déposant un film d'œufs flottant (Wager 1986, Rödel 2000a).

Phrynobatrachus cf. *maculiventris* Guibé et Lamotte, 1958. *P. maculiventris* a été décrite de la Guinée (mare forestière près de Doromou). Plus tard, il lui a été attribué le synonyme de *P. fraterculus* par Guibé et Lamotte (1963). Dans la Forêt Classée du Pic de Fon, nous avons collecté trois petits spécimens de *Phrynobatrachus* (MOR B49-51) dans une zone très humide de la savane d'altitude (1300 m d'altitude), proche d'une petite forêt galerie. Les deux plus grands sont un mâle (14 mm) et une femelle (19,1 mm). Le juvénile et le mâle ont un ventre noirâtre et réticulé, la femelle a un ventre et une gorge densément ponctués de noir. La gorge du mâle est noire et couverte par un repli distinct. Les trois grenouilles ont un dos verruqueux marron. La femelle a une ligne vertébrale jaune irrégulière. Les palmures sont totalement réduites, les extrémités des doigts et des orteils ne sont pas élargies. La femelle porte quelques très gros œufs. Ces spécimens diffèrent de *P. fraterculus* par leur petite taille, l'absence de ligne latérale noire évidente, un dos rugueux, l'absence d'élargissement des extrémités des doigts et orteils, et une coloration ventrale légèrement différente. Ils diffèrent de *P. gutturosus* par leur coloration ventrale et la forme générale de leur corps. Ils diffèrent de ces deux espèces par le nombre d'œufs, moins important mais beaucoup plus gros, et donc très probablement par un comportement reproductif différent. Chez *P. gutturosus*, le male est aussi plus gros que chez *P.* cf. *maculiventris*. Il reste à confirmer si ces spécimens sont vraiment des *P. maculiventris* ou s'ils représentent une espèce encore non décrite.

Arthroleptis/Schoutedenella spp. Comme mentionné dans des publications antérieures, il n'est pas possible, en l'état actuel des connaissances, de déterminer avec certitude les espèces de groupe (Rödel et Branch 2002, Rödel et Agyei *in press*). Les variations intra spécifiques recouvrent largement les variations interspécifiques. Les critères morphologiques (même la longueur des doigts chez les mâles) ou les caractères de coloration (avec ou sans caractère sableux sur le dos, ligne vertébrale ou autre) ne conviennent pas à leur détermination spécifique. Les caractères fiables, qui permettent au moins la définition des limites spécifiques dans une zone donnée, sont les cris d'alerte. Nous n'avons pourtant jamais entendu d'appels de grenouilles de ce groupe au cours du RAP. Nous avons trouvé des grenouilles *Arthroleptis/Schoutedenella* dans les prairies de montagne (en partie dans des zones très sèches), dans les forêts de haute altitude et les forêts primaires et secondaires de basse altitude. Compte tenu de cette importante diversité d'habitats, il est raisonnable de supposer que nos recensements concernent deux ou trois espèces différentes. Au moins une espèce de prairie de montagne peut être différenciée des autres par une peau très rugueuse (cf. Guibé et Lamotte 1958b). La distance museau-anus était modérée pour tous les spécimens rencontrés, variant de 12 à 25 mm.

Hyperolius picturatus Peters, 1875. Selon Schiøtz (1967, 1999) cette grenouille, endémique de la zone des forêts de Haute Guinée, pourrait comprendre deux espèces. Toutefois, des grenouilles provenant de différentes localités d'Afrique de l'Ouest, examinées par nous jusqu'à ce jour, ont montré qu'elles étaient génétiquement identiques (Kosuch et Rödel non publié). Cela s'applique aussi aux espèces de la FCPF, bien que les populations diffèrent largement de par leur coloration et l'utilisation de leur habitat (cf. Schiøtz 1967, 1999; Rödel et Branch 2002; Rödel 2003; Rödel et Ernst, manuscrit soumis pour publication). Dans la FCPF nous avons trouvé que *H. picturatus* était très commune dans la végétation dense le long de ruisseaux ayant un fort courant, aussi bien en basse et haute altitude. De petits groupes appelaient toujours en cœur pendant la périod du RAP. Tous les mâles avaient un ventre granuleux et jaune, une gorge orange, des cuisses rouges avec une rayure longitudinale sur la face dorsale, un dos brun couvert de taches noires et de larges bandes latérales jaunes bordées de noir. Les quelques femelles étaient de coloration très similaire mais avaient un dos uniformément vert olive. Nous avons trouvé des têtards et des juvéniles en cours de métamorphose (11-12 mm) juste en dehors d'une forêt galerie en savane d'altitude, dans une petite mare encaissée avec un fond boueux et quelques plantes aquatiques.

Afrixalus fulvovittatus (Cope, 1861). La confusion taxinomique dans l'utilisation des noms *A. fulvovittatus* et *A. vittiger* (Peters, 1876), tous deux décrits du Liberia, pour deux espèces *Afrixalus* probablement répandues dans les savanes d'Afrique de l'Ouest, a été discutée en détail par Perret (1976), Schiøtz (1999) et Rödel (2000a). Nous avons trouvé un *A. fulvovittatus* typique (selon Perret 1976; Rödel 2000a) ou Type B (selon Schiøtz 1999) à la fois en prairie de haute altitude, dans les zones marécageuses en lisière de forêt dégradée et dans les plantations proches du village de Banko. *A. fulvovittatus* a été récemment recensé en Guinée par Böhme (1994b). Il semble que *A. fulvovittatus*, vivant à proximité des forêts, ait une distribution plus occidentale et australe que *A. vittiger* qui est une espèce typiquement savanicole.

Kassina cochranae (Loveridge, 1941). Jusque recemment, on pensait qu'une seule espèce de *Kassina* tachetée vivait en Afrique de l'Ouest (Schiøtz 1999; Frost 2002), comprenant une sous-espèce occidentale (*K. c. cochranae*) et une sous-espèce orientale (*K. c. arboricola* Perret, 1985). Cependant, nos recherches ont montré que trois sous-espèces étaient dissimulées dans ces anciens inventaires, restreignant ainsi l'aire de répartition de *K. cochranae* à l'extrême ouest de la Côte d'Ivoire, le sud de la Guinée, le Liberia et la Sierra Leone (Rödel et al. 2002b). Les inventaires ci-inclus sont la confirmation de l'étendue de cette aire de répartition. Nous avons recensé *K. cochranae* dans la Forêt Classée du Pic de Fon, dans une petite mare de la prairie de haute altitude et dans les jachères du piedmont, proches du camp 2. Dans la mare, déjà mentionnée dans le recensement de *H. picturatus*, nous avons capturé une grenouille en cours de métamorphose (28,5 mm) qui montrait des caractères de coloration spécifiques. Les têtards typiques de *Kassina* présents dans cette mare, différaient de ceux de *K. schioetzi* Rödel, Grafe, Rudolf et Ernst, 2003 par deux rayures claires allant de l'extrémité du museau à la base de la nageoire (Rödel 2000a, Rödel et al. 2003), ressemblant donc en partie aux têtards de *K. lamottei* Schiøtz, 1967 (Rödel et Ernst 2001). Leur nageoire caudale était tachetée de rouge et de noir, le corps uniformément vert olive ou tacheté de noir. Leur ventre était blanc et se colorait de taches grisâtres vers les flancs et la tête. Les tailles des têtards étaient (longueur du corps/longueur de la queue ; en mm) : 13/32, 19/45 (hauteur de la nageoire : 10 mm), 16/42, 18/44, 15/33, 17/ 38.

Richesse spécifique et composition des communautés:

Nous avons recensé environ 32 espèces d'amphibiens durant la totalité de la période du RAP (Tableau 4.1). Une courbe d'accumulation toujours croissante indique que de nouvelles espèces d'amphibiens restent à confirmer. Nous avons calculé un nombre de 35-39 espèces d'amphibiens dans l'aire d'investigation (Figure 4.1). Nous aurions donc recensé 82-91% de la totalité des espèces régionales. Nous croyons, pour de nombreuses raisons, que cette estimation est trop optimiste (voir discussion).

Seulement 8 (24,8%) de nos espèces d'amphibiens recensées ont une aire de répartition qui dépasse l'Afrique de l'Ouest (Tableau 4.1). La plupart des espèces étaient cantonnées à l'Afrique de l'Ouest. Plus de la moitié des espèces (19 ; 61,3%) étaient endémiques à la zone de forêts pluviales de Haute Guinée. Quinze espèces étaient endémiques à la partie occidentale du bloc forestier de Haute Guinée (48,4%) et au moins trois espèces (9,7%) pourraient être nouvelles pour la science et endémiques de la chaîne du Simandou. La grande majorité des espèces étaient liées aux habitats de forêt ou au moins aux habitats de jachère (Tableau 4.1).

Nous n'avons trouvé, ou relevé des signes évidents de leur présence, que 12 espèces de reptiles durant ce RAP (Annexe 5). Nous ne pouvons donc pas émettre de jugement concernant ce groupe d'animaux. Pourtant, la faune reptilienne est suspectée dépasser 60 espèces, comme estimé pour d'autres régions d'Afrique de l'Ouest (cf. Angel et al. 1954a, b; Barbault 1975; Rödel et al. 1995, 1997, 1999; Böhme 1999; Rödel et Mahsberg 2000; Ernst et Rödel 2002; Ineich 2002; Branch et Rödel 2003).

DISCUSSION

Même si le moment auquel le RAP a eu lieu, au début de la saison sèche, n'était pas favorable à une évaluation des amphibiens, nous avons recensé 32 espèces d'amphibiens et calculé que seules 3 à 7 espèces d'amphibiens avaient échappé à notre recensement. Toutefois, nous supposons que nous n'avons recensé qu'une portion beaucoup plus faible des espèces régionales que calculé. Etant donné que seules quelques espèces appelaient régulièrement, il est probable que nous avons négligé les espèces qui avaient déjà arrêté de se reproduire. En outre, nous avons seulement prospecté deux zones comparativement petites de la totalité de la chaîne de montagne. Durant nos recherches, nous avons continuellement ajouté de nouvelles espèces à notre liste (Figure 4.1). Au cours de seulement deux semaines de travail de terrain, nous avons réussi à recenser plusieurs espèces rares ou nouvelles. Si l'on considère les très vastes zones de la chaîne du Simandou qui n'ont pas encore été prospectées, incluant les parties les plus sèches (les zones savanicoles au nord) et les plus humides (les forêts du sud), il est raisonnable de supposer que beaucoup plus d'espèces d'amphibiens sont présentes sur l'ensemble de la zone du Pic de Fon. Plusieurs espèces peuvent être restreintes aux prairies et/ou endémiques à la chaîne du Simandou.

Dans l'état actuel des connaissances, les sites d'Afrique de l'Ouest présentant la plus riche diversité en terme d'amphibiens sont le Mont Nimba avec 58 espèces d'amphibiens (Guibé et Lamotte 1958a, b, c, 1963) et le Parc National de Taï, Côte d'Ivoire avec 56 espèces d'amphibiens (Rödel 2000b). Le Parc National de Taï est exceptionnel pour le nombre élevé d'amphibiens forestiers hautement spécialisés et endémiques qu'il héberge. Le Mont Nimba abrite également nombre d'espèces endémiques et recèle une forte diversité d'amphibiens grâce à l'exceptionnelle variété de ses habitats. La situation dans la chaîne du Simandou est probablement la même que sur le Mont Nimba. Comme le Mont Nimba, la chaîne du Simandou est particulière de par la composition unique de sa faune. La communauté d'amphibiens de la FCPF couvre un large éventail : des espèces locales endémiques dont l'aire de répartition couvre pratiquement toute l'Afrique, jusqu'à des espèces restreintes aux forêts pluviales primaires et des espèces savanicoles spécialisées. Nous avons recensé toutes les espèces connues des forêts de Haute Guinée qui dépendent des petits cours d'eau forestiers (*Bufo togoensis, Conraua* sp.*, Petropedetes natator, Astylosternus occidentalis, Cardioglossa leucomystax, Hyperolius chlorosteus*) et plusieurs espèces de jachères (ex. *Phrynobatrachus accraensis, Hyperolius fusciventris, H. concolor*). Nous supposons que les espèces qui ont échappé à nos recensements sont principalement les espèces purement savanicoles qui se reproduisent plus tôt dans l'année (ex. plusieurs espèces de *Ptychadena* et *Phrynobatrachus, Kassina senegalensis, Hyperolius nitidulus, H. lamottei, Leptopelis viridis, Phrynomantis microps*), des espèces forestières qui se reproduisent dans les eaux stagnantes (ex.

Tableau 4.1. Espèces d'amphibiens recensées dans la Forêt Classée du Pic de Fon, avec le nombre de sites d'inventaire, les préférences d'habitats et la distribution africaine des espèces. S = savane, FB = Jachères (forêts et zones agricoles dégradées), F = forêt, A = Afrique (présent aussi en dehors de l'Afrique de l'Ouest), WA = Afrique de l'Ouest (du Sénégal au Nigeria oriental), UG = Haute Guinée (zone forestière de l'ouest du Dahomey Gap), WUG = Haute Guinée occidentale (zone forestière de la Côte d'Ivoire occidentale à l'Ouest de cette région), E = endémique au Pic de Fon, * = Recensements basés sur les descriptions des villageois (espèces CITES), ** = Recensements qui comprennent probablement plusieurs espèces, cf. & sp. = déterminations à confirmer ou probables nouvelles espèces.

Espèces	Habitat #	S	FB	F	A	WA	UG	WUG	E
PIPIDAE									
Silurana tropicalis	9		x	x		x			
BUFONIDAE									
Bufo maculatus	6, 7, 13	x	x		x				
Bufo regularis	2	x			x				
Bufo superciliaris *	6			x	x				
Bufo togoensis	6, 7, 7-9			x			x		
Bufo (?) sp.	2	x							x
RANIDAE									
Hoplobatrachus occipitalis	2, 5, 6, 8, 9, 11, 13	x	x		x				
Conraua sp.	2, 6, 7, 12			x				x	
Amnirana albolabris	6, 7, 8, 13		x	x	x				
Amnirana sp.	12			x					x
Ptychadena cf. *aequiplicata*	7			x	x				
Ptychadena sp. aff. *aequiplicata*	6-9			x					?
Ptychadena sp.	2, 5, 6, 9, 6-9, 11	x						x	
PETROPEDETIDAE									
Petropedetes natator	1, 2, 6, 7			x				x	
Phrynobatrachus accraensis	2, 5, 7, 9, 7-9, 11, 13	x	x			x			
Phrynobatrachus alleni	7			x		x			
Phrynobatrachus alticola	1, 2, 5, 6, 7, 8, 9, 14	x	x				x		
Phrynobatrachus fraterculus	5, 7, 8, 9, 7-9	x	x					x	
Phrynobatrachus liberiensis	6, 7, 8			x			x		
Phrynobatrachus phyllophilus	6, 8			x				x	
Phrynobatrachus cf. *maculiventris*	2	x							x
ARTHROLEPTIDAE									
Arthroleptis sp. **	1, 2, 6, 6-9	x	x	x				x	
Cardioglossa leucomystax	6, 7, 6-9			x	x				
ASTYLOSTERNIDAE									
Astylosternus occidentalis	13			x				x	
HYPEROLIIDAE									
Leptopelis hyloides	2, 6, 7, 9, 6-9, 7-9, 13	x	x			x			
Hyperolius chlorosteus	6, 7, 7-9, 13			x				x	
Hyperolius fusciventris	5, 9, 6-9	x	x			x			
Hyperolius concolor	9, 6-9, 7-9	x					x		
Hyperolius picturatus	2, 7, 11	x	x				x		
Afrixalus dorsalis	5, 9, 6-9	x	x	x					
Afrixalus fulvovittatus	2, 5	x						x	
Kassina cochranae	2, 7-9	x						x	

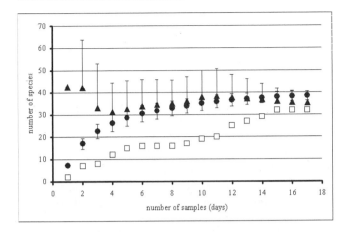

Figure 4.1. Estimation de la richesse spécifique et courbe d'accumulation des espèces d'amphibiens (carrés blancs) de la chaîne du Simandou/région du Pic de Fon, Guinée. Les données sont les valeurs moyennes et l'écart type (sd) de 500 tests de la liste journalière des espèces. Triangles noirs : estimateur Chao2, seulement écart type positif (35,3 ± 4,2 espèces) ; cercles noirs : estimateur Jack-knife 1 (38,5 ± 2 espèces)

Phrynobatrachus gutturosus, P. plicatus, Hyperolius zonatus, Chriromantis rufescens) des espèces fouisseuses qui sont généralement difficiles à étudier (*Geotrypetes seraphini, Hemisus* spp.). Nous estimons par conséquent qu'un total de 50-60 espèces d'amphibiens peut être envisagé pour l'ensemble de la chaîne du Simandou.

La chaîne du Simandou est confrontée à de multiples menaces qui affectent probablement l'herpétofaune de la région. Les feux de brousse et la coupe du bois peuvent affecter les espèces savanicoles et forestières. La plupart des espèces savanicoles semblent être adaptées aux feux ou ont du moins un potentiel de reproduction suffisant pour remplacer la part de population qui meurt durant les feux (voir Rödel 2000a pour une révision).

Par contre, les activités de coupe, à la fois dans un but commercial et pour les cultures à petite échelle, ont un sérieux effet sur le réservoir spécifique des espèces locales et régionales. Les vraies espèces forestières semblent incapables de perdurer dans les forêts dégradées, sans doute à cause d'un microclimat altéré qui offre des conditions qu'elles ne peuvent pas supporter (Ernst et Rödel données non publiées). Les environs de notre second site offraient de larges zones de forêt pluviale primaire de plaine. Les forêts secondaires et les plantations adjacentes hébergeaient une faune d'amphibiens naturellement présente dans la zone de transition entre forêt et savane et dans les perturbations naturelles au sein de la zone forestière, comme les trouées créées par la chute des arbres (Schiøtz 1967, Rödel et Branch 2003). L'ensemble de la faune forestière, y compris

les espèces de jachère, constitue un potentiel élevé pour la conservation, à condition que les forêts restantes soient gérées de façon soutenable et que de vastes parties de la forêt encore préservée demeurent intactes.

Notre premier site d'étude était essentiellement constitué d'une prairie de montagne typique des hautes altitudes d'Afrique de l'Ouest (Guibé et Lamotte 1958a, 1963). Dans ce type d'habitat, les amphibiens ne sont normalement pas très diversifiés, mais comprennent toutefois des espèces tout à fait uniques et à aire de répartition très restreinte. Par exemple les zones les plus élevées du Mont Nimba sont presque exclusivement habitées par seulement deux espèces d'amphibiens, les endémiques du Nimba *Nimbaphrynoides occidentalis* (Angel, 1943) et *Schoutedenella crusculum* (Angel, 1950). Une situation identique existe peut-être sur la chaîne du Simandou où nous avons trouvé un crapaud nain et une espèce de *Arthroleptis/Schoutedenella* dotée d'une peau verruqueuse, les deux étant très probablement nouveaux pour la science. Cette faune risque d'être fortement menacée par toute altération des habitats de haute altitude y compris les activités minières projetées. La distribution de ces espèces le long de la crête de la montagne et la taille respective de leurs populations appelle d'urgence des recherches supplémentaires afin de prédire plus précisément les effets possibles d'une dégradation de l'habitat.

Quatre espèces de reptiles et une espèce d'amphibien sont protégées par les lois internationales et présentent donc un statut de conservation préoccupant. La présence du crocodile nain africain (*Osteolaemus tetraspis*) et du plus gros crapaud africain (*Bufo superciliaris*) en particulier devrait donner lieu à d'autres recherches.

Si les activités minières projetées doivent avoir lieu dans cette zone, il semble urgent d'étudier l'hydrologie de l'ensemble de la chaîne montagneuse et les conséquences climatologiques possibles d'un abaissement de la crête consécutif à certaines pratiques minières. Toutes les espèces forestières seraient potentiellement mises en danger par une altération de la crête de la montagne, en particulier les espèces qui dépendent des cours d'eau permanents. Les montagnes font obstacle aux vents du sud-est, ce qui entraîne des niveaux de précipitation supérieurs à la normale pour les zones à cette latitude. L'abaissement de la crête pourrait causer une diminution du taux de précipitation. Qui plus est, l'ensemble de la montagne agit probablement en tant que réservoir d'eau. Une réduction de ce réservoir, et moins de pluies, auraient presque certainement pour résultat un type de forêt altéré et une progression de la savane vers le sud. Les cours d'eau, qui sont aujourd'hui permanents à 80 %, deviendraient ainsi temporaires. Outre les effets sur les amphibiens et les autres populations de faune aquatique et forestière, une réduction des pluies pourrait aussi avoir des conséquences significatives pour la population humaine locale.

RECOMMANDATIONS POUR LA CONSERVATION

La durée du RAP était trop limitée pour présenter en détails les conséquences pour la conservation. Comme esquissé ci-dessus, un accroissement de la coupe ou des activités d'extraction minière aurait probablement un effet négatif considérable sur l'herpétofaune de la région. Cependant, à l'heure actuelle nous sommes encore loin de présenter un inventaire à peu près complet de ces groupes fauniques. Pour juger plus précisément de l'impact potentiel d'une intensification de l'activité humaine dans la région, nous recommandons fortement un travail de terrain approfondi durant la saison des pluies, incluant d'autres parties de la réserve du Pic de Fon. De telles recherches offriraient non seulement une vue plus complète du réservoir spécifique local, mais permettraient aussi une meilleure compréhension des tailles des populations et des schémas de distribution des espèces dont le statut de conservation est éminemment préoccupant, plus particulièrement les endémiques locaux.

RÉFÉRENCES

Angel, F., J. Guibé and M. Lamotte. 1954a. La réserve naturelle intégrale du mont Nimba. Fascicule II. XXXI. Lézards. Memoires de l'Institut fondamental d'Afrique noire, sér. A, 40: 371-379.

Angel, F., J. Guibé, M. Lamotte and R. Roy. 1954b. La réserve naturelle intégrale du mont Nimba. Fascicule II. XXXII. Serpents. Bulletin de l'Institut fondamental d'Afrique noire, sér. A, 40: 381-402.

Bakarr, M., B. Bailey, D. Byler, R. Ham, S. Olivieri and M. Omland (eds.). 2001. From the forest to the sea: biodiversity connections from Guinea to Togo, Conservation Priority-Setting Workshop, December 1999. Washington, DC: Conservation International. 78 pp.

Barbault, R. 1975. Les peuplements de lézards des savane de Lamto (Côte d'Ivoire). Annales de l'Université d'Abidjan, sér. E, 8: 147-221.

Böhme, W. 1994a. Frösche und Skinke aus dem Regenwaldgebiet Südost–Guineas, Westafrika. I. Einleitung; Pipidae, Arthroleptidae, Bufonidae. Herpetofauna, 16 (92): 11–19.

Böhme, W. 1994b. Frösche und Skinke aus dem Regenwaldgebiet Südost-Guineas, Westafrika. II. Ranidae, Hyperoliidae, Scincidae; faunistisch-ökologische Bewertung. Herpetofauna, 16 (93): 6–16.

Böhme, W. 1999. Diversity of a snake community in a Guinean rain forest (Reptilia, Serpentes). In: Rheinwald, G. (ed.). Isolated Vertebrate Communities in the Tropics. pp. 69-78. Proceedings of the 4th International Symposium, Bonner zoologische Monographien, 46.

Branch, W.R. and M.-O. Rödel. 2003. Herpetological survey of the Haute Dodo and Cavally forests, western Ivory Coast, Part II: Trapping results and reptiles. Salamandra, Rheinbach 39(1): 21-38.

Chabanaud, P. 1919. Énumération des batraciens non encore étudiés de l'Afrique Occidentale Française, appartenant à la collection du muséum. Bull. Mus. Natl. Hist. Nat., 1919: 456–557.

Channing, A. 2001. Amphibians of Central and Southern Africa. New York: Cornell University Press. 470 pp.

Colwell, R.K. 1994-2000. EstimateS, statistical estimation of species richness and shared species from samples. Version 6.0b1. Online. Available: http://viceroy.eeb.uconn.edu/estimates.

Ernst, R. and M.-O. Rödel. 2002. A new Atheris species (Serpentes: Viperidae) from Taï National Park, Ivory Coast. Herpetological Journal, 12: 55-61.

Frost, D.R. 2002. Amphibian Species of the World: an online reference. V2.21. Online. Available: http://research.amnh.org/herpetology/amphibia/index.html. 15 July 2002.

Guibé, J. and M. Lamotte. 1958a. La réserve naturelle intégrale du Mont Nimba. XII. Batraciens (sauf Arthroleptis, Phrynobatrachus et Hyperolius). Memoires de l'Institut fondamental d'Afrique noire, sér. A., 53: 241–273.

Guibé, J. and M. Lamotte. 1958b. Morphologie et reproduction par développement direct d'un anoure du Mont Nimba, Arthroleptis crusculum ANGEL. Bull. Mus. Natl. Hist. Nat., 2e Sér., 30: 125–133.

Guibé, J. and M. Lamotte. 1958c. Une espèce nouvelle de batracien du Mont Nimba (Guinée Française) appartenant au genre Phrynobatrachus: Ph. maculiventris n. sp. Bull. Mus. Natl. Hist. Nat., 2e Sér., 30: 255–257.

Guibé, J. and M. Lamotte. 1961. Deux espèces nouvelles de batraciens de l'ouest africain appartenant au genre Phrynobatrachus: Ph. guineensis n. sp. et. Ph. alticola n. sp. Bull. Mus. Natl. Hist. Nat., 2e Sér., 33: 571–576.

Guibé, J. and M. Lamotte. 1963. La réserve naturelle intégrale du Mont Nimba. XXVIII. Batraciens du genre Phrynobatrachus. Memoirs de l'Institute fondamental d'Afrique noire, sér. A., 66: 601–627.

Heyer, W.R., M.A. Donnelly, R.W. McDiarmid, L.-A.C. Hayek and M.S. Foster. 1993. Measuring and monitoring biological diversity, standard methods for amphibians. Washington, DC: Smithsonían Institution Press. 364 pp.

Hughes, B. 1988. Herpetology in Ghana (West Africa). British Herpetological Society Bulletin, 25: 29-38.

Ineich, I. 2002. Diversite specifique des reptiles du Mont Nimba. unpublished manuscript.

Knoepffler, L.-P. 1985. Le comportement fouisseur de Conraua grassipes (Amphibien anoure) et son mode de chasse. Biologia Gabonica, 1: 239-245.

Lamotte, M. 1966. Types de répartition géographique de quelques batraciens dans l'Ouest Africain. Bulletin de l'Institut fondamental d'Afrique noire, Sér A, 28: 1140–1148.

Lamotte, M. and J.-L. Perret. 1968. Révision du genre *Conraua* Nieden. Bulletin de l'Institut fondamental d'Afrique noire, sér. A., 30: 1603–1644.

Lamotte, M. and M. Zuber-Vogeli. 1954. Contribution à l'étude des batraciens de l'Ouest Africain III. Le développement larvaires de deux espèces rhéophiles, *Astylosternus diadematus* et *Petropedetes natator*. Bull. Inst. fond. Afr. noire, Sér. A, 16: 1222–1233.

Perret, J.-L. 1976. Identité de quelques *Afrixalus* (Amphibia, Salientia, Hyperoliidae). Bull. Soc. neuchat. Sci. Nat., Sér. 3, 99: 19–28.

Perret, J.-L. 1977. Les *Hylarana* (Amphibiens, Ranidés) du Cameroun. Rev. Suisse Zool., 84: 841–868.

Perret, J.-L. 1985. Description of *Kassina arboricola* n. sp. (Amphibia, Hyperoliidae) from the Ivory Coast and Ghana. S. Afr. J. Sci., 81: 196–199.

Perret, J.-L. and J.-L. Amiet. 1971. Remarques sur les *Bufo* (Amphibiens Anoures) du Cameroun. Ann. Fac. Sci. Cameroun, 5: 47-55.

Rödel, M.-O. 2000a. Herpetofauna of West Africa, Vol. I: Amphibians of the West African savanna. Frankfurt/M. (Edition Chimaira), 335 pp.

Rödel, M.-O. 2000b. Les communautés d'amphibiens dans le Parc National de Taï, Côte d'Ivoire. Les anoures comme bio-indicateurs de l'état des habitats. *In:* Girardin, O., I. Koné and Y. Tano (eds.). Etat des recherches en cours dans le Parc National de Taï (PNT), Sempervira, Rapport de Centre Suisse de la Recherche Scientifique, Abidjan, 9: 108-113.

Rödel, M.-O. 2003. The amphibians of Mont Sangbé National Park, Ivory Coast. Salamandra, 39: 91-110.

Rödel, M.-O. and A.C. Agyei. 2003. Amphibians of the Togo-Volta highlands, eastern Ghana. Salamandra, Rheinbach 39(3/4): 207-234.

Rödel, M.-O. and W.R. Branch. 2002. Herpetological survey of the Haute Dodo and Cavally forests, western Ivory Coast, Part I: Amphibians. Salamandra, 38: 245-268.

Rödel, M.-O. and R. Ernst. 2000. *Bufo taiensis* n. sp., eine neue Kröte aus dem Taï-Nationalpark, Elfenbeinküste. Herpetofauna, Weinstadt, 22 (125): 9-16.

Rödel, M.-O. and R. Ernst. 2001. Description of the tadpole of *Kassina lamottei* Schiøtz, 1967. Journal of Herpetology, 35: 678-681.

Rödel, M.-O. and R. Ernst. 2002a. A new reproductive mode for the genus *Phrynobatrachus*: *Phrynobatrachus alticola* has nonfeeding, nonhatching tadpoles. Journal of Herpetology, 36: 121-125.

Rödel, M.-O. and R. Ernst. 2002b. A new *Phrynobatrachus* species from the Upper Guinean rain forest, West Africa, including a description of a new reproductive mode for the genus. Journal of Herpetology, 36 (4): 561-571.

Rödel, M.-O. and D. Mahsberg. 2000. Vorläufige Liste der Schlangen des Tai-Nationalparks/ Elfenbeinküste und angrenzender Gebiete. Salamandra, Rheinbach, 36 (1): 25-38.

Rödel, M.-O., K. Grabow, C. Böckheler and D. Mahsberg. 1995. Die Schlangen des Comoé-Nationalparks, Elfenbeinküste (Reptilia: Squamata: Serpentes). Stuttgarter Beiträge zur Naturkunde, Stuttgart, Serie A, Nr. 528: 1-18.

Rödel, M.-O., K. Grabow, J. Hallermann and C. Böckheler. 1997. Die Echsen des Comoé-Nationalparks, Elfenbeinküste. Salamandra, 33: 225-240.

Rödel, M.-O., K. Kouadio and D. Mahsberg. 1999. Die Schlangenfauna des Comoé-Nationalparks, Elfenbeinküste: Ergänzungen und Ausblick. Salamandra, Rheinbach, 35 (3): 165-180.

Rödel, M.-O., D. Krätz and R. Ernst. 2002a. The tadpole of *Ptychadena aequiplicata* (WERNER, 1898) with the description of a new reproductive mode for the genus (Amphibia, Anura, Ranidae). Alytes, 20: 1-12.

Rödel, M.-O., T.U. Grafe, V.H.W. Rudolf and R. Ernst. 2002b. A review of West African spotted *Kassina*, including a description of *Kassina schioetzi* sp. nov. (Amphibia: Anura: Hyperoliidae). Copeia, Lawrence, 2002 (3): 800-814.

Rödel, M.-O., J. Kosuch, M. Veith and R. Ernst. 2003. First record of the genus *Acanthixalus* Laurent, 1944 from the Upper Guinean rain forest, West Africa, including the description of a new species. Journal of Herpetology, 37: 43-52.

Schiøtz, A. 1967. The treefrogs (Rhacophoridae) of West Africa. Spolia zoologica Musei Haunienses, 25: 1–346.

Schiøtz, A. 1999. Treefrogs of Africa. Frankfurt/M. (Edition Chimaira), 350 pp.

Spieler, M. and K.E. Linsenmair. 1999. The larval mite *Endotrombicula pillersi* (Acarina: Trombiculidae) as a species-specific parasite of a West African savannah frog (*Phrynobatrachus francisci*). Am. Midl. Nat., 142: 152-161.

Tandy, M. and R. Keith. 1972. *Bufo* of Africa. *In:* Blair, W.F. (ed.). Evolution in the genus *Bufo*. Austin & London: University of Texas Press. 119–170.

Wager, V.A. 1986. Frogs of South Africa, their fascinating life stories. Delta, Craighall. 183 pp.

Chapitre 5

Inventaire préliminaire des oiseaux de la Forêt Classée du Pic de Fon

Ron Demey et Hugo J. Rainey

RÉSUMÉ

En 11 jours de travaux sur le terrain dans la Forêt Classée du Pic de Fon, 233 espèces d'oiseaux ont été recensées, dont 131 dans la zone de haute altitude et 198 dans la zone de basse altitude. La protection de huit d'entre elles est d'intérêt mondial (deux observées dans la zone de haute altitude et sept dans la zone de basse altitude), la plus importante étant la Prinia de Sierra Leone *Schistolais leontica*, dont la distribution dans les zones d'altitude d'Afrique de l'Ouest est très limitée. Nous avons trouvé six des 15 espèces à répartition restreinte qui composent la Zone d'Endémisme d'Oiseaux de la forêt de Haute Guinée. Un échantillon significatif des espèces strictement forestières du pays a été rencontré, puisque nous avons identifié 104 des 153 espèces du biome des forêts guinéo-congolaises recensées en Guinée. Sept espèces sont signalées pour la première fois en Guinée. Vu la haute valeur de cette forêt classée pour la conservation, la conduite d'études complémentaires est recommandée afin de compléter la liste des espèces observées. Des recommandations sont également faites pour la gestion de la forêt et la conservation de sa diversité biologique.

INTRODUCTION

Il a été prouvé que les oiseaux sont de bons indicateurs de la diversité biologique d'un site. Leur taxinomie et leur répartition géographique mondiale sont relativement bien connues en comparaison à d'autres taxons (ICBP 1992), ce qui facilite leur identification et permet l'analyse rapide des résultats d'une étude ornithologique. Les statuts de conservation de la plupart des espèces ayant été assez bien évalué (BirdLife International 2000), les résultats et conclusions d'une telle étude peuvent être examinés et appliqués avec efficacité. Les oiseaux font également partie des espèces les plus charismatiques, ce qui peut faciliter la présentation de recommandations pour la conservation à l'intention des décideurs et des parties prenantes.

Des études menées dans certaines forêts existant encore en Afrique de l'Ouest ont démontré que celles-ci sont d'une importance considérable pour la survie de l'avifaune des forêts de Haute Guinée (voir, par exemple, Allport et al. 1989 ; Gartshore et al. 1995 ; Demey et Rainey sous presse). Toutefois, l'avifaune de la majorité des forêts ombrophiles ouest-africaines, dont la superficie diminue rapidement, demeure peu connue. C'est notamment le cas pour l'ensemble de l'avifaune de Guinée, de vastes étendues du pays n'ayant pas encore fait l'objet de recensements (Robertson 2001a). A ce jour, seulement deux zones dans le sud-est de la Guinée ont été explorées (Wilson 1990 ; Halleux 1994). La partie guinéenne du Mont Nimba, site du patrimoine mondial de l'UNESCO, est à peine étudiée, alors que la partie libérienne du massif a fait l'objet d'un suivi à long terme de l'avifaune (Colston et Curry-Lindahl 1986).

Nous avons effectué 11 jours de travail sur le terrain, dont quatre jours dans la zone de haute altitude, entre 900 et 1550 m d'altitude (08°31.52'N, 08°54.21'W) (27 - 30 novembre) et sept jours dans la zone de basse altitude, entre 550 et 800 m (08°31.29'N, 08°56.12'W) (1-7 décembre).

Dans un souci de standardisation, nous avons suivi la nomenclature, la taxinomie et l'ordre de Borrow et Demey (2001).

MÉTHODOLOGIE

La principale méthode utilisée pour cette étude est l'observation des oiseaux en marchant lentement le long des pistes minières et des sentiers forestiers. Des notes ont été prises à la fois sur les observations visuelles et les émissions vocales. Les vocalisations inconnues et celles des espèces rares ont été enregistrées pour permettre leur analyse ultérieure et leur mise en archives de sons. Nous avons essayé de parcourir le plus d'habitats possibles, surtout ceux susceptibles d'abriter des espèces menacées ou peu connues. Cependant, la difficulté d'accès à la plupart des zones de la forêt, en raison des pentes raides et rocheuses, de la végétation dense, et de la rareté ou de l'absence de sentiers, n'ont pas permis de couvrir de grandes superficies.

Dans la zone de haute altitude, les principaux habitats comprennent des prairies situées sur des pentes abruptes et accidentées, des ravins le long de petits ruisseaux, des vallées et dépressions, et des forêts bordées de buissons et de broussailles. Dans la zone de basse altitude, la majeure partie du travail s'est déroulée dans des forêts situées sur des pentes raides, et, aux abords du piedmont, dans une forêt de plaine. Un certain nombre de cours d'eau, bordés de forêt et de végétation basse, parcourent le site. Des formations de savane issues de la dégradation de la forêt caractérisent la zone comprise entre la lisière inférieure de la forêt et la limite de la réserve. Certaines zones en bordure de forêt ont été défrichées pour la culture du café et du manioc et à l'intérieur de la forêt de plaine, se trouvent des plantations de bananiers et quelques cacaoyers. Le travail sur le terrain a débuté peu avant l'aube (vers 06h30) pour se poursuivre jusqu'à 13h00, et de 15h00 jusqu'au coucher du soleil (aux environs de 18h30). Des sorties nocturnes ont également permis de collecter des données sur les hibous et engoulevents, avec enregistrement de leurs vocalisations.

Des captures aux filets japonais ont pu être faites pendant six jours. Les filets étaient constitués de quatre bandes et mesuraient 12 mètres de long sur 3 mètres de haut une fois installés. L'objectif principal était d'obtenir des données sur des espèces discrètes et silencieuses qui peuvent facilement passer inaperçues lors d'observations visuelles ou auditives. Sur les pentes supérieures, les filets ont été installés pendant deux jours pour un total de 131 mètres de filet/heure. Les filets étaient placés dans des zones herbeuses, en lisière de forêt, en forêt, et au-dessus d'un petit cours d'eau forestier. Dans les étages inférieurs, les filets ont été installés pendant quatre jours, pour un total de 199 mètres de filet/heure, en forêt primaire (canopée de 30-40 mètres), en forêt basse (canopée de15-20 mètres), au-dessus de deux petits cours d'eau forestiers et en lisière. Un dispositif spécial a permis d'installer un filet de 12 mètres, dans la canopée, à une hauteur de 20 mètres et en lisière de forêt pendant 17,5 heures.

Chaque jour, une liste exhaustive des espèces identifiées a été établie. Le nombre d'individus ou de groupes a été noté, ainsi que les indications sur la reproduction (par exemple, la présence de jeunes) et des informations concernant l'habitat dans lequel les oiseaux se trouvaient. Ceci nous a permis de produire des indices d'abondance pour chaque espèce, basés sur le taux de rencontre (nombre de jours pendant lesquels l'espèce fut notée et nombre d'individus et de groupes concernés). Des comparaisons peuvent ainsi être établies entre les deux sites et d'autres sites de la région. Les définitions des indices d'abondance sont données dans l'Annexe 6. Le statut de conservation de chaque espèce a été pris en compte dans notre analyse, ainsi que la présence d'espèces à biome restreint et d'espèces dont l'aire de répartition se limite à la Zone d'Endémisme d'Oiseaux des forêts de Haute Guinée. Les statuts de conservation sont définis dans le Tableau 5.1. Un certain nombre d'espèces ont des aires de répartition réduites et se trouvent uniquement dans les forêts de Haute Guinée. De même, de nombreuses espèces se trouvent uniquement dans le biome des forêts guinéo-congolaises et la forte proportion de celles-ci fournit un indice sur la richesse du site. Pour plus de détails, nous renvoyons à Stattersfield et al. (1998), BirdLife International (2000) et Fishpool et Evans (2001).

RÉSULTATS

Au total, 233 espèces d'oiseaux ont pu être recensées. La liste de ces espèces est donnée en Annexe 6 avec mention d'une estimation de l'abondance sur chacun des deux sites et les signes de reproduction observés. Le statut de conservation, l'endémisme au bloc forestier de Haute Guinée, l'appartenance à un assemblage inféodé à un biome et l'habitat sont également indiqués.

Zone de haute altitude

Sur cette zone, 131 espèces ont été recensées (voir Annexe 6), parmi lesquelles deux dont la protection est d'intérêt mondial (BirdLife International 2000; Tableau 5.1) : la Prinia de Sierra Leone *Schistolais leontica*, classée dans la catégorie « Vulnérable » et le Choucador iris *Lamprotornis iris*, considéré comme « Insuffisamment Documenté ». Parmi les 153 espèces du biome des forêts guinéo-congolaises présentes dans le pays, 49 (soit 32%) ont été recensées dans les étages supérieurs du massif.

De plus, un certain nombre d'espèces rares et peu connues en Guinée ou en Afrique de l'Ouest ont été observées. Elles comprennent le Pipit à long bec *Anthus similis*, le Cossyphe à sourcils blancs *Cossypha polioptera*, le Tchitrec à tête noire *Elminia nigromitrata*, le Tisserin de Preuss *Ploceus preussi*, le Sénégali vert *Mandingoa nitidula* et le Sénégali à ventre noir *Euschistospiza dybowskii*. La sous-espèce distincte *henrici* de l'Alouette à nuque rousse *Mirafra africana*, restreinte à quelques zones d'altitude en Sierra Leone et au Mont Nimba, s'est révélée être relativement commune dans les zones herbeuses. Des espèces

Tableau 5.1. Espèces globalement menacées ou dont la protection est d'intérêt mondial recensées pendant l'étude

Espèces		Statut de conservation
Ceratogymna elata	Calao à casque jaune	nt
Lobotos lobatus	Échenilleur à barbillons	VU
Phyllastrephus baumanni	Bulbul de Baumann	DD
Criniger olivaceus	Bulbul à barbe jaune	VU
Bathmocercus cerviniventris	Bathmocerque à capuchon	nt
Schistolais leontica	Prinia de Sierra Leone	VU
Illadopsis rufescens	Akalat à ailes rousses	nt
Lamprotornis iris	Choucador iris	DD

Statut de conservation (BirdLife International 2000):

VU = Vulnérable: espèce confrontée, à moyen terme, à un risque élevé d'extinction à l'état suavage.

DD = Insuffisamment Documenté: espèce pour laquelle on ne dispose pas d'assez d'informations pour évaluer son risque d'extinction.

nt = Quasi-menacé: espèce qui se rapproche de celles de la catégorie Vulnérable

paléarctiques européennes sont apparues aussi communes dans la forêt que dans les zones herbeuses.

Zone de basse altitude

Dans cette zone, 98 espèces ont été recensées (voir Annexe 6), parmi lesquelles sept dont la protection est d'intérêt mondial (BirdLife International 2000; Tableau 5.1). Deux d'entre elles sont classées comme « Vulnérables » (l'Échenilleur à barbillons *Lobotos lobatus* et le Bulbul à barbe jaune *Criniger olivaceus*), trois comme « Quasi-Menacées » (le Calao à casque *Ceratogymna elata*, le Bathmocerque à capuchon *Bathmocercus cerviniventris* et l'Akalat à ailes rousses *Illadopsis rufescens*), tandis que deux sont considérées comme relevant « Insuffisamment Documenté » (le Bulbul de Baumann *Phyllastrephus baumanni* et le Choucador iris *Lamprotornis iris*). Des 153 espèces du biome des forêts guinéo-congolaises recensées dans le pays (Robertson 2001a; cette étude), 98 (soit 64 %) ont été observées dans la zone de basse altitude.

Par ailleurs, quelques espèces rares et mal connues ont été observées. Parmi elles, un ibis forestier *Bostrychia rara/olivacea*, le Guêpier à tête bleue *Merops muelleri*, l'Indicateur à queue en lyre *Melichneutes robustus*, le Pririt à ventre doré *Dyaphorophyia concreta*, la Mésange enfumée *Parus funereus* et le Sénégali vert *Mandingoa nitidula*.

Sur chaque zone, l'objectif de détecter la présence de certaines espèces discrètes par l'utilisation de filets japonais a été atteint (voir Annexe 7). Au total, 181 individus de

56 espèces différentes furent capturés, parmi lesquels deux espèces dont la protection est d'intérêt mondial : le Bathmocerque à capuchon *Bathmocercus cerviniventris* et la Prinia de Sierra Leone *Schistolais leontica*. Parmi les autres espèces intéressantes, citons le Pigeon à masque blanc *Aplopelia larvata* et l'Agrobate du Ghana *Cercotrichas leucosticta* (voir ci-dessous). Le taux de capture de 5,5 oiseaux par 100 mètres de filet/heure est beaucoup plus élevé que ceux obtenus précédemment dans d'autres forêts ouest-africaines (voir, par exemple, Allport et al. 1989; Gartshore et al. 1995).

Six des 15 espèces à répartition restreinte, c'est-à-dire des espèces d'oiseaux terrestres dont l'aire de reproduction est inférieure à 50 000 km², qui composent la Zone d'Endémisme d'Oiseaux des forêts de Haute Guinée (Fishpool et Evans 2001, Stattersfield et al. 1998) ont été notées dans la forêt classée : l'Échenilleur à barbillons *Lobotos lobatus*, le Bulbul à barbe jaune *Criniger olivaceus*, le Bathmocerque à capuchon *Bathmocercus cerviniventris*, la Prinia de Sierra Leone *Schistolais leontica*, l'Apalis de Sharpe *Apalis sharpii* et l'Akalat à ailes rousses *Illadopsis rufescens*. Sept espèces ont été recensées pour la première fois en Guinée (voir Tableau 5.2). Seize espèces de rapaces ont été observées, parmi lesquels quelques uns des plus grands vautours et aigles d'Afrique, comme le Vautour africain *Gyps africanus*, l'Aigle couronné *Stephanoaetus coronatus* et l'Aigle martial *Polemaetus bellicosus*.

NOTES SUR DES ESPÈCES SPÉCIFIQUES

Voir Tableau 5.1 pour la définition du statut de conservation.

Espèces globalement menacées ou dont la protection est d'intérêt mondial

Ceratogymna elata Calao à casque jaune (NT). Cette espèce a été vue pendant quatre jours dans la zone de basse altitude. Un groupe comptait 14 oiseaux ; toutes les autres observations n'ont concerné que deux individus. Cette espèce était auparavant connue de trois autres sites en Guinée (Robertson 2001a). Il est à remarquer que peu de calaos, toutes espèces confondues, ont été observés pendant l'étude.

Tableau 5.2. Espèces nouvelles pour la Guinée

Bubo poensis	Grand-duc à aigrettes
Neafrapus cassini	Martinet de Cassin
Indicator willcocksi	Indicateur de Willcocks
Smithornis capensis	Eurylaime du Cap
Phyllastrephus baumanni	Bulbul de Baumann
Cercotrichas leucosticta	Agrobate du Ghana
Vidua camerunensis	Combassou du Cameroun

Lobotos lobatus Échenilleur à barbillons (VU). Un male a été observé à environ 750 mètres d'altitude, à l'est du camp installé sur les pentes inférieures. Il se nourrissait dans la voûte d'une forêt primaire, à une hauteur de 15-25 mètres, à proximité d'une clairière. Cette espèce était auparavant seulement connue en Guinée de la Forêt Classée de Ziama (Halleux 1994, Robertson 2001a).

Phyllastrephus baumanni Bulbul de Baumann (DD). Nous avons observé un couple en lisière de forêt, à environ 570 mètres. Le lendemain, à 100 mètres de cet endroit, nous avons rencontré deux oiseaux qui semblaient être le même couple dans un groupe plurispécifique. Ceci constitue la première observation de cette espèce en Guinée. Jusqu'à une date récente, il y avait très peu d'observations fiables de cette espèce dans l'ensemble de son aire de distribution (Fishpool 2000).

Criniger olivaceus Bulbul à barbe jaune (VU). Un couple a été observé dans un groupe plurispécifique en forêt primaire, à environ 570 mètres d'altitude. Il se nourrissait dans la strate moyenne, à 10-15 mètres de hauteur. Auparavant, cette espèce était seulement connue des Forêts Classées de Ziama et de Diécké (Robertson 2001a).

Bathmocercus cerviniventris Bathmocerque à capuchon (NT). Nous avons noté quatre mâles en chant et un couple chantant en duo à 550-580 mètres d'altitude, dans la végétation dense près de petits cours d'eau. Nous avons aussi capturé une femelle à proximité du camp de basse altitude, dans le territoire de l'un des quatre mâles. Auparavant, cette espèce locale n'était connue que de Ziama (Halleux 1994).

Schistolais leontica Prinia de Sierra Leone (VU). Au moins deux et probablement trois couples ont été notés à 1300-1350 mètres d'altitude. Un couple se trouvait sur des buissons dans une galerie forestière et un second a été régulièrement vu à la lisière d'une autre petite forêt galerie. Deux individus se sont pris dans les filets, de l'autre côté de cette galerie forestière. La tête et les parties inférieures d'un d'eux (un jeune ?) étaient légèrement plus pâles que celles de l'autre oiseau. Cette espèce est maintenant connue de quatre sites sur l'ensemble de son aire de répartition restreinte, qui ne comprend qu'un seul autre site en Guinée: les monts Nimba (Fishpool 2001; Okoni-Williams et al. 2001; Robertson 2001a,b; L.D.C. Fishpool comm. pers.).

Illadopsis rufescens Akalat à ailes rousses (NT). Cette espèce chantait en forêt primaire à environ 570, 650 et 1200 mètres d'altitude.

Lamprotornis iris Choucador iris (DD). Nous avons observé un groupe de dix individus dans la savane arborée au niveau du camp de Rio Tinto, et un individu dans un habitat semblable en zone de basse altitude.

Autres espèces rares ou mal connues
(Statuts en Afrique de l'Ouest selon Borrow et Demey 2001)

Bostrychia rara/olivacea Ibis vermiculé ou olive. Un individu a été vu par I. Herbinger (comm. pers.) à côté d'un cours d'eau forestier à 900 mètres d'altitude, près du camp. Il n'a pas été spécifiquement identifié mais chacune de ces deux espèces, qui sont rares et très localisées en Afrique de l'Ouest, serait nouvelle pour la Guinée.

Aplopelia larvata Pigeon à masque blanc. Un individu a été capturé à 570 mètres d'altitude, dans une forêt dont la canopée se limitait à une hauteur d'environ 15 mètres. Cette espèce est connue de peu de sites en Afrique de l'Ouest, de la Sierra Leone à l'ouest de la Côte d'Ivoire, où elle est rare ou peu commune (Demey et Rainey en préparation).

Merops muelleri Guêpier à tête bleue. Un couple a été vu en lisière de forêt à 570 mètres d'altitude et l'un des oiseaux a été capturé par la suite sur le même site. Cette espèce est rare et très localisée en Afrique de l'Ouest.

Melichneutes robustus Indicateur à queue en lyre. La parade d'un individu a été entendue chaque jour au dessus d'une galerie forestière à 560 mètres d'altitude. Cette espèce n'était auparavant connue que de la Forêt Classée de Ziama (Halleux 1994).

Mirafra africana Alouette à nuque rousse. Fréquemment vue dans les savanes et sur les pistes minières au-dessus de 1300 mètres. La sous-espèce en question, *henrici*, est seulement connue de quelques zones d'altitude de Haute Guinée. Un mâle en parade a été vu à plusieurs reprises, sautillant jusqu'à environ 80 centimètres du sol en battant bruyamment des ailes. Apparemment, ce comportement n'avait pas encore été décrit (Colston et Curry-Lindahl 1986, Keith et al. 1992, R. Safford comm. pers.).

Sheppardia cyornithopsis Rougegorge merle. Trois individus ont été capturés en forêt, à 570 et 1350 mètres d'altitude, tandis qu'un autre a été vu sur le premier site dans un groupe plurispécifique. Ce taux de rencontre est relativement important pour cette espèce rarement recensée en Afrique de l'Ouest.

Elminia nigromitrata Tchitrec à tête noire. Recensé presque quotidiennement, du haut en bas des pentes. En plus, trois individus ont été capturés dans la zone de basse altitude. Cette espèce est généralement peu commune en Afrique de l'Ouest mais semble fréquente dans le sud-est de la Guinée (Halleux 1994, cette étude).

Dyaphorophyia concreta Pririt à ventre doré. Rencontré aussi bien vers les sommets qu'au bas des pentes où deux individus ont été capturés. Cette espèce est généralement rare ou assez rare en Afrique de l'Ouest.

Parus funereus Mésange enfumée. Trois individus ont été observés, se nourrissant dans la canopée de grands arbres, à la lisière d'une clairière, à environ 570 mètres d'altitude. Cette espèce, généralement rare ou assez rare en Afrique de l'Ouest, était seulement connue, en Guinée, de la Forêt Classée de Ziama (Halleux 1994).

Ploceus preussi Tisserin de Preuss. Un individu a été vu en forêt, à 1350 mètres d'altitude. Cette espèce, précédemment recensé dans les Forêts Classées de Ziama et Diécké (Wilson 1990, Halleux 1994), est généralement peu répandue et très localisée en Afrique de l'Ouest.

Mandingoa nitidula Sénégali vert. Un individu a été capturé à 1350 mètres et un autre à 580 mètres d'altitude à la lisière de la forêt. Cette espèce est peu commune ou rare en Afrique de l'Ouest.

Euschistospiza dybowskii Sénégali à ventre noir. Trois mâles ont été capturés à 1350 mètres d'altitude, en lisière de la forêt. Cette espèce, peu commune, est très localisée en Afrique de l'Ouest.

Espèces nouvelles pour la Guinée

Bubo poensis Grand-duc à aigrettes. Un adulte a été identifié par son cri roulé enregistré le 6 décembre en forêt primaire, sur le site où *Criniger olivaceus* a été observé. Cette identification a été confirmée en comparant l'enregistrement avec ceux de Chappuis (2000). Cette espèce est peu à assez commune sur l'ensemble du bloc forestier guinéen.

Neafrapus cassini Martinet de Cassin. Un à trois individus ont été vus presque chaque jour dans la zone de basse altitude. Cette espèce résidente est localement commune avec une répartition irrégulière dans la zone de forêt ombrophile ouest-africaine.

Indicator willcocksi Indicateur de Willcocks. Le chant d'un individu a été entendu dans une galerie forestière et un autre oiseau vu à la lisière d'une clairière, respectivement à 560 et 570 mètres d'altitude. Cette espèce forestière est rare ou peu commune en Afrique de l'Ouest.

Smithornis capensis Eurylaime du Cap. Cinq individus ont été vus et entendus, paradant en forêt dans la zone de basse altitude. Cette espèce est généralement assez rare à rare, avec une distribution fragmentée en Afrique de l'Ouest.

Phyllastrephus baumanni Bulbul de Baumann. Voir ci-dessus. *Cercotrichas leucosticta* Agrobate du Ghana. Un couple a été capturé, à 570 mètres d'altitude, dans une forêt à la canopée d'une hauteur d'environ 15 mètres. Cette espèce très discrète, connue en Afrique de l'Ouest de la Sierra Leone au Ghana, est assez rare en milieu forestier.

Vidua camerunensis Combassou du Cameroun. Quatre combassous mâles vus à 1500 mètres d'altitude, en lisière de forêt, et deux autres en savane à une altitude de 560 mètres, ont été identifiés par leur bec blanc, leurs pattes mauve pâle et leurs rémiges brunes. Deux de ses hôtes potentiels ont été observés dans la forêt classée : le Sénégali à ventre noir et l'Amarante flambé *Lagonosticta rubricata*. Le statut et la distribution de cette espèce sont mal connus en raison de sa ressemblance avec d'autres combassous.

DISCUSSION

Le nombre total de 233 espèces recensées sur les deux sites est élevé, compte tenu de la courte période d'étude et en comparaison avec le nombre total d'environ 625 espèces répertoriées pour l'ensemble de la Guinée (Robertson 2001a). Ceci est un indice de la richesse de la Forêt Classée du Pic de Fon. En comparaison, 287 et 141 espèces ont été respectivement recensées dans les Forêts Classées de Ziama et de Diecké, les deux seuls autres sites du sud-est de la Guinée ayant fait l'objet d'études (Robertson 2001a). Après plusieurs années d'études intensives, 383 espèces ont été recensées dans la partie libérienne du Mont Nimba et les forêts avoisinantes (Colston et Curry-Lindahl 1986).

Le Mont Nimba présente de nombreuses similitudes avec le Pic de Fon de par son relief et ses différents habitats, donnant ainsi une indication de la richesse potentielle du Pic de Fon. Les 104 espèces inféodées au biome des forêts guinéo-congolaises que nous avons recensées dans la forêt classée représentent 68% des espèces de ce biome connues en Guinée, soit une forte proportion.

La présence de la Prinia de Sierra Leone dans la zone de haute altitude a été la découverte la plus importante de cette étude. Cette espèce n'est actuellement connue que de trois autres sites dans le monde. L'un d'entre eux, le Mont Nimba guinéen, a également fait l'objet de prospections minières, tandis qu'une partie de son habitat sur le versant libérien a déjà été détruit par les activités minières. Il semble que l'on ne trouve la Prinia de Sierra Leone que dans la végétation dense en lisière de forêt et le long de cours d'eau situés au-dessus de 700 mètres (Borrow et Demey 2001, cette étude). Elle pourrait être particulièrement sensible au changement des habitats d'altitude du Pic de Fon. Bien que cette espèce soit actuellement classée comme « Vulnérable »en raison de sa population adulte en diminution et fragmentée, estimée à moins de 10 000 individus, elle pourrait être reclassée comme « Menacée d'extinction » (un statut de conservation reflétant une menace plus grave) car, connue de moins de six localités, elle a probablement une aire de répartition inférieure à 500 km² (BirdLife International 2000). Même si certaines montagnes de la Haute Guinée, où la Prinia de Sierra Leone pourrait être présente, n'ont pas encore fait l'objet d'études ornithologiques, une révision du statut de conservation de cette espèce semble souhaitable, étant donnée la menace qui pèse sur l'avenir des sites où sa présence est connue.

Très peu de calaos, aussi bien en nombre qu'en espèces, ont été observés dans la forêt classée. La plupart des calaos forestiers ont été recensés dans les autres forêts protégées du sud-est guinéen (Robertson 2001a). Les calaos étant connus pour leur capacité à se déplacer sur de longues distances pour se nourrir (Kemp 1995, HJR obs. pers.), cette absence est peut-être liée à la phénologie locale des arbres fruitiers. En cette période de l'année, en Côte d'Ivoire, la plupart des calaos se trouvent dans les forêts du sud. Ils pourraient donc être absents du Pic de Fon parce que se trouvant plus au sud, peut-être dans les forêts libériennes. La chasse aux grands mammifères s'est avérée relativement intensive dans la forêt classée (voir Chapitre 8, Barrie et Kante 2004). Des discussions avec notre guide A. Camara ont révélé que les oiseaux étaient également la cible des chasseurs et ceci pourrait expliquer en partie l'absence ou la faible densité de grandes espèces telles que les pintades et calaos. Ni le Picatharte de Guinée *Picathartes gymnocephalus*, ni le Gobemouche du Libéria *Melaenornis annamarulae*, deux espèces dont la protection est d'intérêt mondial, n'ont été recensés au cours de l'étude. Vu les habitats du Pic de Fon et la présence de ces espèces sur des sites semblables aux alentours (Robertson 2001a), l'on peut cependant raisonnablement supposer qu'elles puissent s'y trouver.

Le grand nombre d'espèces dont la protection est d'intérêt mondial recensées au cours d'une étude si courte est une indication de la richesse potentielle de la forêt (voir Tableau 5.1). Ce site du Pic de Fon remplit les conditions pour être retenu comme Zone d'Importance pour la Conservation des Oiseaux, en raison du nombre d'espèces menacées (catégorie A1) et de la présence d'un grand nombre d'espèces à répartition restreinte (A2) ou à biome restreint (A3) (Fishpool et Evans 2001).

RECOMMANDATIONS POUR LA CONSERVATION

Compte tenu de la haute valeur de la forêt classée pour la conservation, les recommandations suivantes sont faites :

1. Il faudrait envisager de garder intacte une portion substantielle de la forêt classée sur l'ensemble des pentes, des pics jusqu'au piedmont, car certaines espèces peuvent, selon les saisons, avoir besoin de forêts situées à différents niveaux d'altitude. Peu de sites en Afrique de l'Ouest possèdent des forêts intactes qui s'étendent sur un gradient altitudinal aussi important. Les calaos sont connus pour avoir besoin de vastes étendues de couvert forestier pour satisfaire leurs besoins nutritionnels. Comme la phénologie des arbres peut varier suivant l'altitude, la survie des calaos dans la forêt classée dépend probablement de la présence de forêts sur la totalité du gradient altidudinal.

2. Les effets potentiels d'une modification de l'habitat devraient être pris en compte avant le lancement d'une exploitation minière. Si l'exploration procède à une étude de faisabilité, des études plus détaillées sur les effets sur les oiseaux devraient être entreprises en tant qu'élément d'une Evaluation de l'Impact Environnemental et Social (ESIA). Les effets potentiels sur la population de la Prinia de Sierra Leone sont particulièrement préoccupantes, ainsi que ceux sur les espèces confinées à la prairie d'altitude telles que l'Aigle martial et l'Alouette à nuque rousse.

3. Le défrichement de la forêt pour l'agriculture et les incendies volontaires sont largement répandus. Des tentatives de restauration des zones déjà défrichées devraient être mises en œuvre. Presque toutes les espèces aviennes forestières sont affectées par le défrichement ou la dégradation de la forêt et ce phénomène concerne la plupart des espèces menacées observées au cours de cette étude.

4. La chasse dans la forêt classée est importante et devrait être réduite. Nous avons constaté que les pintades et les calaos, qui font partie des plus grands oiseaux d'Afrique occidentale, étaient rares ; leur nombre ne pourrait qu'encore plus se réduire si la chasse s'intensifiait.

5. Des études complémentaires devraient être menées sur différents sites de la forêt classée. Ces inventaires devraient être effectuées à différentes saisons (par exemple au début de la saison des pluies, en avril-mai, quand davantage d'espèces, en particulier les coucous, les hiboux et certaines fauvettes, se reproduisent et sont vocalement les plus actives) et dans des habitats que nous n'avons pas pu couvrir, tels que les cours d'eau des pentes inférieures. Ce dernier habitat pourrait être important pour deux espèces menacées, la Chouette pêcheuse rousse *Scotopelia ussheri* et l'Onoré à crête blanche *Tigriornis leucolopha*. La première est très peu connue et devrait faire l'objet d'études spécifiques. Différentes espèces d'oiseaux migrateurs d'Europe pourraient être rencontrées et une migration altitudinale d'oiseaux à l'intérieur de la forêt pourrait être mise en évidence.

6. Des programmes de suivi devraient être mis en place pour évaluer l'utilité de certaines méthodes de gestion pour la faune sauvage. Si les financements sont disponibles, les chasseurs locaux, qui connaissent le mieux les forêts et leur faune, devraient être employés pendant les programmes de suivi de la faune. Leur fournir un emploi ne pourrait que contribuer à réduire d'autant l'importance des prélèvements cynégétiques.

RÉFÉRENCES

Allport, G.A., M. Ausden, P.V. Hayman, P. Robertson and P. Wood. 1989. The conservation of the birds of Gola Forest, Sierra Leone. Study Report No. 38 International Council for Bird Preservation. Cambridge, UK.

Barrie, A. and S. Kanté. 2004. Résultats de l'étude des grands mammiféres de la Forêt Classée du Pic de Fon, Guinée. In: McCullough, J. (ed.) 2004. A Rapid Biological Assessment of the Forêt Classée du Pic de Fon, Simandou Range, South-eastern Republic of Guinea. RAP Bulletin of Biological Assessment 35. Conservation International, Washington, DC.

BirdLife International. 2000. Threatened Birds of the World. Lynx Edicions and BirdLife International. Barcelona, Spain and Cambridge, UK.

Borrow, N. and R. Demey. 2001. Birds of Western Africa. Christopher Helm. London.

Brosset, A. 1984. Oiseaux migrateurs européens hivernant dans la partie guinéenne du Mont Nimba. Alauda 52: 81–101.

Chappuis, C. 2000. Oiseaux d'Afrique - African Bird Sounds, Vol. 2. 11 CDs with booklet. Société d'Etudes Ornithologiques de France, Paris, France and The British Library National Sound Archive, London, UK.

Colston, P.R. and K. Curry-Lindahl. 1986. The birds of Mount Nimba, Liberia. British Museum (Natural History). London.

Demey, R. and H.J. Rainey. *In press.* The birds of Haute Dodo and Cavally Forest Reserves. RAP report. Washington, DC: Conservation International.

Demey, R. and H.J. Rainey. In prep. An annotated checklist of the birds of Mont Sangbé National Park, Ivory Coast.

Fishpool, L.D.C. 2000. A review of the status, distribution and habitat of Baumann's Greenbul *Phyllastrephus baumanni*. Bulletin of the British Ornithologists' Club 120: 213-229.

Fishpool, L.D.C. 2001. Côte d'Ivoire. *In*: Fishpool, L.D.C. and M.I. Evans (eds.). Important Bird Areas in Africa and Associated Islands: Priority sites for conservation. Newbury and Cambridge, UK: Pisces Publications and BirdLife International. Pp. 219-232.

Fishpool, L.D.C. and M.I. Evans (eds.). 2001. Important Bird Areas in Africa and Associated Islands: Priority sites for conservation. BirdLife Conservation Series No. 11. Pisces Publications and BirdLife International. Newbury and Cambridge, UK.

Gartshore, M.E., P.D. Taylor and I.S. Francis. 1995. Forest Birds in Côte d'Ivoire. BirdLife International Study Report No. 58. BirdLife International. Cambridge, UK.

Halleux, D. 1994. Annotated bird list of Macenta Prefecture, Guinea. Malimbus 17: 85-90.

ICBP. 1992. Putting biodiversity on the map: priority areas for global conservation. International Council for Bird Preservation. Cambridge, UK.

Keith, S., E.K. Urban and C.H. Fry (eds.). 1992. The Birds of Africa, Vol. 4. London, UK: Academic Press.

Kemp, A. 1995. The Hornbills. Oxford University Press. Oxford, UK.

Okoni-Williams, A.D., H.S. Thompson, P. Wood, A.P. Koroma and P. Robertson. 2001. Sierra Leone. *In:* Fishpool, L.D.C. and M.I. Evans (eds.). Important Bird Areas in Africa and Associated Islands: Priority sites for conservation. Newbury and Cambridge, UK: Pisces Publications and BirdLife International. Pp. 769-778.

Robertson, P. 2001a. Guinea. *In:* Fishpool, L.D.C. and M.I. Evans (eds.). Important Bird Areas in Africa and Associated Islands: Priority sites for conservation. Newbury and Cambridge, UK: Pisces Publications and BirdLife International. Pp. 391-402.

Robertson, P. 2001b. Liberia. *In:* Fishpool, L.D.C. and M.I. Evans (eds.). Important Bird Areas in Africa and Associated Islands: Priority sites for conservation. Newbury and Cambridge, UK: Pisces Publications and BirdLife International. Pp. 473-480.

Stattersfield, A.J., M.J. Crosby, A.J. Long and D.C. Wege. 1998. Endemic Bird Areas of the World: Priorities for Biodiversity Conservation. BirdLife International. Cambridge, UK.

Wilson, R. 1990. Annotated bird lists for the Forêts Classées de Diécké and Ziama and their immediate environs. Unpublished report commissioned by IUCN.

Chapitre 6

Évaluation rapide des chiroptères dans la Forêt Classée du Pic de Fon, Guinée

Jakob Fahr et Njikoha Ebigbo

RÉSUMÉ

Nous rapportons ici les résultats d'une évaluation des populations de chauves-souris du Pic de Fon, dans la chaîne du Simandou, au sud-est de la Guinée. Nous présentons un ensemble de chauve-souris particulier, caractérisé par des espèces forestières telles que *Epomops buettikoferi*, *Rhinolophus guineensis*, *guineensis* et *Hipposideros jonesi* qui sont endémiques à la Haute Guinée ou à l'Afrique de l'Ouest. La présence de trois espèces sympatriques de *Kerivoula* est notable avec *K. K cuprosa* et *K. phalaena*, recensées pour la première fois en Guinée. De plus, trois individus Murin de Welwitsch, *Myotis welwitschii* ont été capturés pendant l'évaluation. C'est le premier recensement pour l'Afrique de l'Ouest et il représente une extension de 4 400 km au nord-ouest des plus proches localités connues. Nous avons révisé la distribution de cette espèce en Afrique et conclu que cette espèce présente un modèle de répartition para montagneux. Sept espèces, soit 33,3% du total des 21 espèces, sont enregistrées dans le dernier volume de la liste rouge de l'UICN : une espèce en tant que « Vulnérable » (*Epomops buettikoferi*) et six espèces « Presque Menacées » (*Rhinolophus alcyone*, *R. guineensis*, *Hipposideros jonesi*, *H. fuliginosus*, *Kerivoula cuprosa*, *Miniopterus schreibersii*). Nos résultats de l'évaluation RAP ainsi que la présence d'espèces de chiroptères endémiques aux zones d'altitude de la Haute Guinée, met en évidence l'importance régionale des habitats de montagne d'Afrique de l'Ouest en général, et de la chaîne du Simandou en particulier, pour la conservation des chiroptères d'Afrique.

INTRODUCTION

En Afrique de l'Ouest, le couvert forestier a été réduit à 14,4% de son étendue originelle et les forêts résiduelles continuent à être dégradées ou détruites. Les forêts de la région biogéographique de la Haute Guinée sont classées en tant que sites de première importance continentale et mondiale pour la biodiversité (Myers et al. 2000, Brooks et al. 2001). En conséquence, Conservation International a organisé un Atelier pour la Sélection des Priorités en 1999 afin de sélectionner les zones clefs pour la conservation de cette région, sur la base d'un consensus d'experts participants (Bakarr et al. 2001). L'atelier a abouti à la délimitation de régions prioritaires pour la conservation de la biodiversité (classées de haute à exceptionnellement haute priorité), incluant des zones qui ne sont pas protégées actuellement. De plus, l'atelier a mis en évidence la nécessité de conduire des évaluations biologiques imminentes, afin de produire les informations scientifiques de base qui font défaut pour nombre de ces régions. Il a été décidé que les zones prioritaires les moins connues devraient être explorées le plus tôt possible par le biais d'une évaluation biologique rapide (RAP) afin de rassembler les bases scientifiques indispensables à leur protection et au développement des capacités dans la région. La distribution de la plupart des espèces chiroptères en Afrique est encore insuffisamment connue. Ceci est particulièrement vrai en ce qui concerne l'Afrique de l'Ouest et l'Afrique Centrale où peu d'études ont encore été menées. De plus, de nombreux habitats disparaissent à un rythme alarmant, entraînant la perte de connaissances basiques mais inestimables sur les espèces et leur distribution. Ces données de base sont essentielles pour la conservation sensée et les plans de gestion des zones cibles, et par conséquent pour la protection effective des chiroptères.

Les chauves-souris (Chiroptera) sont l'un des groupes de vertébrés les plus écologiquement diversifiés et sont considérés comme des espèces clefs pour le maintien des fonctions des écosystèmes. Ceci est dû à leur grande richesse spécifique, leur importante biomasse et leur diversité trophique, particulièrement sous les tropiques où elles jouent un rôle important en tant que disséminatrices de semences, pollinisatrices ainsi que prédatrices d'insectes. Elles forment le seul groupe mammifère qui ait développé le vol actif, s'ouvrant ainsi à de nombreuses niches écologiques. Les chauves-souris constituent l'ordre mammifère le plus nombreux avec plus de 1100 espèces au monde, seulement surpassé (en terme de richesse spécifique) par les rongeurs. De plus, à l'échelle locale, elles sont souvent le groupe mammifère le plus riche en espèces parmi les communautés animales tropicales. Bien que la taille des chauves-souris soit comparativement petite, elles sont caractérisées par une durée de vie remarquablement longue et un faible taux de reproduction. Ces caractéristiques biologiques évolutives, de même que des besoins spécifiques précis et spécialisés en terme d'habitats, exposent les chauves-souris à de multiples menaces réelles et potentielles telles que : la dégradation ou la perte d'habitat, l'altération ou la destruction de leurs perchoirs diurnes, voire leur exploitation directe, et enfin l'empoisonnement à travers l'accumulation de substances toxiques dans la chaîne alimentaire. Environ un quart des chauves-souris (23,8%) sont mondialement menacées (« Gravement Menacées d'Extinction », « Menacées » ou « Vulnérables ») et 12 espèces sont éteintes (Hutson et al. 2001, IUCN 2002). Leurs réactions sensibles aux modifications de l'environnement font des chauves-souris un groupe indicateur idéal pour l'évaluation et le suivi de l'évolution et de l'ampleur des dégradations des habitats naturels.

Les forêts de Haute Guinée, qui abritent une faune chiroptère unique et très diversifiée, disparaissent rapidement et l'inclusion des chauves-souris dans le présent RAP du Pic de Fon, dans la région du Simandou, est justifiée par leur importance écologique reconnue en tant qu'espèces clefs et par leur forte valeur en tant qu'espèces indicatrices très sensibles.

MATÉRIEL ET MÉTHODES

Site d'étude
La Forêt Classée du Pic de Fon, couvrant une surface de 25 600 ha, a été créée en 1953 et entoure le Pic de Fon. Les zones d'altitude sont dominées par des prairies montagneuses. Les versants occidentaux présentent de nombreuses reliques de forêts primaires le long de ravins, reliées aux forêts qui se trouvent à la base de la chaîne montagneuse. Sur la partie orientale seuls de petits blocs forestiers sont présents le long des torrents jaillissant tout au long de la chaîne du Simandou. Les plaines adjacentes sont dominées par des savanes arborées avec de petites inclusions de galeries forestières. Les précipitations annuelles varient de 1 700 à 2 000 mm, avec une longue saison sèche de cinq à six mois (de novembre à avril). Pour une description détaillée de la végétation, du climat et de la géologie, se reporter au chapitre 2 (botanique) et au chapitre 1 (introduction) de ce rapport. Deux sites ont été explorés : le premier est proche du sommet du Pic de Fon (zone d'altitude ; 8°32'N, 8°54'W, 1 350 m), le second, sur le versant occidental, proche du village de Banko (piedmont ; 8°31'N, 8°56'W, 600 m). Les deux sites sont distants d'environ 3,6 km. Les positions géographiques des deux sites ont été déterminées à l'aide d'un GPS portatif (GPS Garmin 12).

Récolte et analyse des données
Les travaux de terrain ont été effectués par N.E. Des filets de capture ont été installés durant sept nuits du 28 novembre au 6 décembre 2002 (pas de filets les 1er et 2 décembre). Des filets de deux tailles différentes ont été employés : 12 x 2,8 m et 6 x 2,8 m (mailles de 16 mm ; filins 2 x 70 d) avec respectivement 5 et 4 niveaux. La plupart des filets ont été fixés à des perches proches du sol ou légèrement surélevés par rapport à la végétation environnante (couche herbeuse) dans la forêt, la savane et l'habitat sommital. Sur le site de piedmont, les filets ont été posés à la fois en forêt relativement intacte et au sein de bosquets de bananiers. En plus des filets placés au niveau du sol, un système de filet aérien a été installé sur le site de piedmont. Un système de corde et poulie a été utilisé pour monter trois filets de 12 m de long dont le sommet atteignait 12 m de hauteur. Ce système a été installé à la lisière d'une forêt intacte de *Parinari excelsa*. Enfin, un piège en harpe à trois rangées (3,4 m2 de surface de capture) a été installé sur chacun des sites pendant 10 nuits du 27 novembre au 6 décembre. Les filets de captures étaient déployés de 18:00 à 0:00 heures, alors que le piège en harpe était ouvert de 18:00 à 6:00 heures. L'effort total de capture était de 194,8 filets heures ou 16,2 filets nuit (calculé pour un équivalent de 12 mètre de filet) et 111,5 piège en harpe. heure soit 9,3 piège en harpe. Nuit (Tab. 6.1).

Une courbe d'accumulation spécifique aléatoire a été calculée avec EstimateS 6.0b1 (Colwell 2000). Bien qu'il existe plusieurs méthodes statistiques pour estimer le nombre total d'espèces échantillonnées (Colwell 2000), nous n'avons pas été en mesure d'utiliser ces méthodes car l'effort d'échantillonnage et les méthodes variaient trop d'un site à l'autre.

Les appels par écholocation de chauves-souris captives ont été enregistrés avec un détecteur D 240x de Pettersson (à la fois en mode d'expansion x10 et x20) et transférés sur un Walkman Sony professionnel WM-D6C. Ces appels ont ensuite été analysés sur un PC standard avec le

Tableau 6.1: L'effort de capture pour chaque méthode utilisée.

Méthode	L'effort de capture	
	Filets Heures[1]	Filets Nuits[2]
12 m filets de capture, sol	144,1	12,0
12 m filets de capture, auvent	28,4	2,4
6 m filets de capture, sol	44,7	3,7
Piège en harpe	111,5	9,3

[1]une filet-heure = un filet de capture/ Piège en harpe installé pour 1 heure

[2]une filet-nuit = un filet de capture/ Piège en harpe installé pour 12 heures

logiciel Avisoft-SASLab Pro 4.2 et utilisés pour identifier les taxons difficiles. Dans les familles *Rhinolophidae* et *Hipposideridae*, les composants de la fréquence constante (CF) des appels par écholocation sont particuliers à chaque espèce. Nous avons mesuré la fréquence CF (amplitude maximum, harmonie seconde) avec des spectrogrammes (logiciel Hanning window, transformation de Fourrier - fenêtre d'échantillonnage 512 unités). Vingt quatre spécimens échantillons ont été sacrifiés pour vérifier leurs identifications et pour documenter la faune chiroptère de la région.

Ces spécimens sont actuellement dans la collection de JF (Département d'Ecologie Expérimentale, Université d'Ulm). Des notes taxinomiques sur des espèces particulières capturées lors de cette étude sont rapportées ailleurs (Fahr and Ebigbo 2003). Nous n'avons pas comparé nos données avec celles de Konstantinov et al. (2000) car leurs résultats indiquent des erreurs d'identifications importantes.

RÉSULTATS

Les chauves-souris de la chaîne du Simandou

Nous avons capturé un total de 276 chauves-souris, comprenant 21 espèces, 14 genres et 6 familles (Tableaux 6.2 et 6.3). Pour les filets proches du niveau du sol, le taux de capture global était de 0,38 chauve-souris capturée pour 12 m de filet.heure (c/fh), dont 0,30 c/fh pour les Mégachiroptères et 0,08 c/fh pour les Microchiroptères. Pour les filets surélevés le taux de capture était de 2,57 c/fh, dont 2,50 c/fh pour les Mégachiroptères et 0,07 c/fh pour les Microchiroptères. Le taux de capture global pour les pièges en harpe était de 1,26 chauves-souris par piège.heure (c/ph), dont 0,03 c/ph pour les Mégachiroptères et 1,23 c/ph pour les Microchiroptères. Les méthodes d'échantillonnage ont produit des taux de capture respectifs de 63 individus et 12 espèces dans les filets proches du niveau du sol, 73 individus et 7 espèces dans les filets de canopée, et 140 individus et 12 espèces dans les pièges en harpe. Sur le plan de l'efficacité d'échantillonnage,

Tableau 6.2. Effort de capture (nh :filet/piège heure), succès de capture (nombre d'individus ; chauves souris par filet/piège heure), couverture d'espèces (Total : toutes les espèces ; Excl. : exclusivement capturée avec une méthode) décomposés par méthode de prélèvement

	Effort [nh]	N° d'indiv.	Chauves souris / nh	Espèces Total	Espèces Excl.
Filets niveau du sol	166,4	63	0,38	12	5
Filets surélevés	28,4	73	2,57	7	1
Piège en harpe	111,5	140	1,26	12	7
Total	—	276	—	21	13

nous avons capturé sept espèces, soit 33,3 % du total de 21 espèces, exclusivement au piège en harpe. Nous avons échantillonné cinq espèces, soit 23,8 % du total des espèces, uniquement avec les filets proches du niveau du sol, et avec les filets de canopée une espèce soit 4,8 % du total des espèces. Dans l'ensemble, 13 espèces soit 61,9 % des 21 espèces ont été échantillonnées par le biais d'une seule méthode (Tableau 6.2). Au cours de l'étude une seule grotte habitée par des chauves-souris (8°32'N, 8°53'W) a été trouvée, abritant une seule espèce : *Rousettus aegyptiacus* (Roussette d'Egypte). Nous avons estimé que la colonie comprenait au moins quelques milliers d'individus au moment de la visite.

Sur le site du piedmont nous avons enregistré un total de 17 espèces, dont 11 espèces qui n'ont été trouvées que sur ce site. Le site d'altitude présentait un total moins élevé de 10 espèces, dont quatre espèces exclusivement enregistrées sur ce site. Parmi les 21 espèces qui ont été

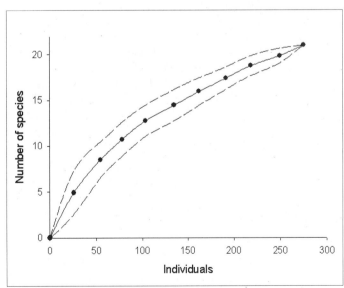

Figure 6.1. Courbe aléatoire d'accumulation spécifique pour les chiroptères échantillonnés durant l'expertise RAP (ligne pleine et points : randomisation, 100 tests ; lignes de tirets + 1 sd.).

récoltées sur la chaîne du Simandou, seules six d'entre elles (28,6 %) ont été enregistrées à la fois sur le site d'altitude et sur celui du piedmont, démontrant une rotation des espèces comparativement élevée entre les deux zones d'échantillonnage. Il faut noter que l'effort de capture pour les pièges en harpe sur le site d'altitude et sur celui du piedmont était très similaire, avec respectivement 59,2 et 52,3 pièges heures. L'effort de capture pour les filets proches du niveau du sol était aussi relativement identique, avec respectivement 89,4 filets heures en montagne et 77,0 filets heures sur le site du piedmont (calculés pour un équivalent de 12 m. filet). Par contre, le système de canopée, qui a présenté un taux de capture élevé sur le site du piedmont, n'a pas été utilisé sur le site d'altitude.

L'ensemble des populations de chiroptères comprend des espèces à la fois forestières et savanicoles, même si les espèces essentiellement associées aux habitats forestiers (catégorie « F » ou « F(S) » dans les Tableau 6.3) prédominent dans l'échantillon (13 espèces soit 61,9%). Les espèces trouvées principalement soit dans des habitats savanicoles (« S » ou « (F)S »), soit à la fois en forêt et en savane (« FS »), constituent de plus faibles fractions (4 espèces soit 19,0 % pour chaque catégorie, Tableau 6.3). Cet ensemble est également caractérisé par un grand nombre d'espèces qui dépendent – partiellement- de grottes en tant que perchoirs diurnes (*Rousettus aegyptiacus*, *Rhinolophus* spp., *Hipposideros* spp., *Miniopterus schreibersii*). La courbe d'accumulation des espèces s'élève fortement et aucun plateau asymptotique n'est discernable (Figure 6.1), démontrant que l'inventaire de la faune chiroptère de la chaîne du Simandou est loin d'être complet. *Myotis welwitschii*, *Kerivoula phalaena* et *K. cuprosa* sont recensées pour la première fois en Guinée.

Myotis welwitschii

Etonnamment, nous avons capturé trois individus de *M. welwitschii*, un chiroptère jusqu'ici seulement connu en Afrique de l'Est et en Afrique australe. Cette chauve-souris se distingue de toutes les autres espèces de *Myotis* présentes en Afrique par la coloration distincte de ses membranes et de son pelage. Les trois individus (2 mâles, 1 femelle) ont été capturés aux niveaux supérieurs de filets de 12 m (2-3 m au-dessus du niveau du sol) en habitat de lisière pendant la nuit du 29 au 30 novembre à 19h09, 21h27 et 0h10 respectivement. Les sites de capture au filet en altitude étaient installés dans une parcelle de prairie montagneuse située entre deux étendues de forêt dévalant le versant occidental. Les deux zones de forêt se rejoignent plus bas pour former une seule grande forêt. Les mâles ont été pris en bordure d'une rangée d'arbres de la prairie s'étirant en parallèle à la forêt plus étendue. La femelle a été capturée à peine 150 m en dessous du lieu de capture des mâles, directement à la lisière de la forêt, à 200 m environ du point de convergence des forêts.

Ce résultat intéressant nous a incités à réviser la distribution documentée de *M. welwitschii* en nous basant sur la littérature, les spécimens de musée et nos communications personnelles (Fahr et Ebigbo 2003). *Myotis welwitschii* est désormais documentée pour les quinze pays suivants : Angola, Burundi, Congo (K.), Ethiopie, Guinée, Kenya, Malawi, Mozambique, Rwanda, Afrique du Sud, Soudan, Tanzanie, Ouganda, Zambie, et Zimbabwe. La plupart des sites se trouvent en zone d'altitude ou en montagne (voir carte, p. 20). La répartition altitudinale s'étend de 55 à 2 311 m au-dessus du niveau de la mer avec une moyenne de 1 209 m (n=61). L'analyse des écorégions occupées révèle une préférence prononcée pour les zones boisées et les mosaïques de forêts et savanes (Fahr et Ebigbo 2003). Jusqu'à présent, il n'existe aucun rapport sur la zone forestière fermée du Bassin central du Congo.

DISCUSSION

Myotis welwitschii

L'inventaire pour la Guinée correspond à une extension de la zone de répartition de 4 400 km depuis les plus proches localités connues le long du Rift Albertine (Fahr et Ebigbo 2003). Cette exceptionnelle augmentation de la zone de répartition complète notre connaissance fragmentaire de la distribution des chauves-souris en Afrique et met en évidence le besoin d'études supplémentaires, particulièrement dans les régions présentant un intérêt important pour la conservation, telles que soulignées par l'Atelier pour la Sélection des Priorités (Bakarr et al. 2001).

Le modèle de distribution de *M. welwitschii* est manifestement lié aux zones d'altitudes ou aux montagnes (voir carte, p. 20; 500 m alt.: n=9; 1000 m alt.: n=12; 1500 m alt.: n=22; 2000 m alt.: n=11; > 2000 m alt.: n=7). Cela correspond à un schéma de distribution « paramontagneux », c.a.d. que les espèces sont liées aux régions montagneuses bien que leur répartition altitudinale couvre à la fois les basses et hautes altitudes. Revus en détail, de nombreux inventaires sont situés dans des zones topographiquement complexes, c'est-à-dire sur les flancs de montagne ou sur les plateaux d'altitude. La raison de cette répartition para montagneuse de *M. welwitschii* est difficile à expliquer. Chez d'autres espèces de chauves souris, cette caractéristique est probablement liée à une forte dépendance aux cavités rocheuses en tant que perchoirs diurnes (Fahr et al. 2002b). La densité des cavités est généralement beaucoup plus importante dans les régions montagneuses, ce qui explique probablement le schéma de répartition para montagneux de certaines espèces. Toutefois, il a été rapporté que *M. welwitschii* se perche de façon non dissimulée dans la végétation, par exemple sur des buissons ou des arbres, et a parfois été trouvée dans des feuilles de bananier enroulées (voir Fahr et Ebigbo 2003).

Bien que cette espèce soit largement distribuée en Afrique, elle est rarement recensée dans les inventaires de chauves-souris. Le nombre de localités connues est presque équivalent au nombre de spécimens inventoriés. Ceci s'explique peut-être par le fait que l'abondance de

M. welwitschii est généralement très faible, ou parce qu'il est difficile de la capturer avec les méthodes actuelles d'échantillonnage. Notre inventaire pour la Guinée est remarquable dans ce sens car nous avons capturé trois individus au cours de la même nuit.

LA POPULATION DE CHAUVE-SOURIS DE LA CHAÎNE DU SIMANDOU

Efficacité de l'échantillonnage

Tenant compte des résultats de notre étude, 63 espèces sont documentées pour la Guinée (Tableau 6.4; Barnett et Prangley 1997, Ziegler et al. 2002, Fahr et al. 2002b, Fahr et Ebigbo 2003). Nous avons recensé 21 espèces pour la chaîne du Simandou, soit un tiers des chauves-souris connues pour le pays, pendant la très brève période de notre étude RAP. Toutefois, le nombre réel d'espèces pour le pays et la chaîne du Simandou est certainement beaucoup plus important que celui connu actuellement. Au niveau local, c'est-à-dire pour la chaîne du Simandou, cela est indiqué par l'accroissement brutal de la courbe d'accumulation des espèces, qui montre que beaucoup plus d'espèces pourraient être trouvées durant des études continues. Deuxièmement, une comparaison avec les listes d'espèces provenant de sites alentours – tels que le Mont Nimba, la chaîne du Ziama et la forêt de Diécké – nous permet d'identifier plusieurs espèces manquantes, qui ont été trouvées dans seulement un ou plusieurs sites de cette région, bien que les types d'habitats soient très similaires (voir ci-dessous, Tableau 6.5). Dans ce contexte, il est nécessaire de souligner que les listes d'espèces, même provenant des sites les mieux échantillonnés tels que le Mont Nimba sont incomplètes car : a) aucun de ces sites n'a fait l'objet d'études sur le long terme et couvrant toutes les saisons, b) ni les pièges en harpe ni les filets de capture n'ont été utilisés dans des études antérieures, bien que ces méthodes soient indispensables pour parvenir à des inventaires complets (voir Fahr 2001 et les résultats de cette étude), et c) les échantillonnages précédents ont été surtout faits de façon opportuniste, c'est-à-dire que la majorité des types d'habitats n'ont pas été échantillonnés de façon exhaustive et constante.

Sept espèces de chauves-souris frugivores (Pteropodidae) ont été prises, représentant la plus large proportion des espèces capturées au niveau familial (33,3 %, Table 6.4). Les Vespertilionidae, avec 23,8% du total des espèces, étaient la seconde plus importante famille dans notre étude. Aucun membre des familles Rhinopomatidae (Chauve-souris à queue longue), Emballonuridae (Chauve-souris à queue engainée) et Megadermatidae (Chauve-souris à ailes jaunes) n'a été recensé durant cette étude. En Afrique de l'Ouest, 120 espèces sont actuellement recensées entre la Mauritanie et le Nigeria (Fahr et al. 2001), les Vespertilionidae et les Molossidae en constituant la majeure partie.

Compte tenu de la distribution des chauves-souris en Afrique de l'Ouest et en Haute Guinée, et de leurs habitats de prédilection respectifs, nous estimons que le nombre total d'espèces sera au minimum de 50 espèces, et avoisinera probablement les 65 espèces. Nous prévoyons en particulier des espèces supplémentaires dans les familles suivantes : Pteropodidae – 4 espèces, Emballonuridae – 2 espèces, Nycteridae – 6 espèces, Rhinolophidae – 3 espèces, Hipposideridae – 3 espèces, Vespertilionidae – 17 espèces, et Molossidae – 9 espèces.

La Guinée est potentiellement un pays de «mégadiversité» au sein de l'Afrique de l'Ouest en général et de la région de Haute Guinée en particulier. Toutefois, la faune chiroptère est l'une des moins connues et plusieurs espèces supplémentaires sont susceptibles de se trouver dans ce pays. Ceci est illustré par nos captures de *Myotis welwitschii*, *Kerivoula phalaena* et *K. cuprosa*, qui sont toutes recensées pour la première fois en Guinée. On estime actuellement que la Côte d'Ivoire et le Ghana, des pays qui ont davantage été explorés, abritent respectivement 87 et 82 espèces, même si ces chiffres ne sont absolument pas complets (J. Fahr en préparation). La position biogéographique particulière de la Guinée inclut la plupart des biomes d'Afrique de l'Ouest, tels que mangrove, savane côtière, forêts de plaine, savane guinéenne et savane soudanaise. En outre, ainsi que nous le détaillons plus bas, les forêts de montagne et les prairies de Haute Guinée abritent des espèces restreintes ou endémiques de ces habitats. Aussi nous attendons-nous à un nombre total d'espèces pour la Guinée proche de 100 espèces.

D'autre part, notre combinaison de méthodes d'échantillonnage (filets de capture proches du niveau du sol et de la canopée, pièges en harpe, investigations de perchoirs) montre qu'une part substantielle de la population chiroptère locale peut être suivie au cours d'une période d'étude relativement courte. Même si le total actuel de 21 espèces est certainement très inférieur au nombre réel d'espèces, il ne faut pas oublier que nous avons réussi à recenser plusieurs espèces rares, bien que la période d'échantillonnage n'ait duré que 10 jours. De nombreuses études insistent sur le fait que des périodes d'échantillonnage beaucoup plus longues sont nécessaires pour parvenir à un inventaire presque complet. Cette étude a été réalisée au début de la saison sèche. Les résultats d'études à long terme menées dans les Parcs Nationaux de Comoé et de Taï, en Côte d'Ivoire, montrent que le plus grand nombre d'espèces est capturé en début de saison des pluies, alors que les prises sont nettement réduites en saison sèche (J. Fahr en préparation). De plus, la présente étude n'a porté que sur deux zones, comparativement petites, et s'est concentrée sur les habitats forestiers et les prairies d'altitude proches, tandis que les savanes arborées extensives n'ont pas été échantillonnées. Il est recommandé que les prochaines études soient réalisées au début de la saison sèche (avril-juin), et qu'elles incluent des types d'habitats et des zones que la présente étude n'a pas couverts. En outre, davantage de temps et d'efforts doivent être investis dans la recherche et l'échantillonnage des nombreuses grottes qui sont susceptibles de se trouver sur le site de l'étude.

Tableau 6.3. Espèces enregistrées pendant le RAP dans la chaîne du Simandou. GN : filet niveau du sol ; CN : filet surélevé ; HT :piège en harpe (en gras : espèce attrapée avec une seule méthode). Piedmont et Altitude se refèrent à des relevés aux deux différentes localités de prélèvement (Voir Matériel et Méthodes) ; Habitat : attribution grossière aux types d'habitat préférés (F :forêt, S : savanes et zones boisées ; entre parenthèses : représentation marginale du type respectif d'habitat).

Espèces	N° Ind.	GN	CN	HT	Piedmont	Altitude	Habitat	
Pteropodidae								
Epomophorus g. gambianus (OGILBY, 1835)	1	x			+			S
Micropteropus pusillus (PETERS, 1868)	19	x	x	x	+	+		S
Epomops buettikoferi (MATSCHIE, 1899)	3	x	x		+		F	(S)
Megaloglossus woermanni PAGENSTECHER, 1885	8	x	x		+		F	
Myonycteris torquata (DOBSON, 1878)	3		x	x	+		F	(S)
Rousettus aegyptiacus unicolor (GRAY, 1870)	87	x	x		+	+	F	S
Eidolon h. helvum (KERR, 1792)	2	x				+	F	S
Nycteridae								
Nycteris grandis PETERS, 1865	1	x			+		F	
Rhinolophidae								
Rhinolophus alcyone TEMMINCK, 1853	1			x	+		F	(S)
Rhinolophus guineensis EISENTRAUT, 1960	7	x		x	+	+	F	
Rhinolophus l. landeri MARTIN, 1838	1			x	+			S
Hipposideridae								
Hipposideros jonesi HAYMAN, 1947	1	x				+	F	S
Hipposideros cf. *caffer* (SUNDEVALL, 1846)	5			x	+		F	S
Hipposideros cf. *ruber* (NOACK, 1893)	5			x	+	+	F	(S)
Hipposideros fuliginosus (TEMMINCK, 1853)	116	x	x	x	+	+	F	
Vespertilionidae								
Myotis welwitschii (GRAY, 1866)	3	x				+	(F)	S
Kerivoula lanosa muscilla THOMAS, 1906	1			x	+		F	(S)
Kerivoula phalaena THOMAS, 1912	1			x		+	F	
Kerivoula cuprosa THOMAS, 1912	5			x	+		F	
Miniopterus schreibersii villiersi AELLEN, 1956	4	x		x	+	+	F	(S)
Molossidae								
Mops spurrelli (DOLLMAN, 1911)	1		x		+		F	

Nous concluons, pour les raisons énumérées ci-dessus, que la faune chiroptère de la chaîne du Simandou est sous-échantillonnée et nous recommandons des études supplémentaires utilisant une combinaison de méthodes incluant les filets de capture proches du niveau du sol et de la canopée, des pièges en harpe, des investigations de perchoirs (en particulier pour les études de grottes) et le suivi acoustique (enregistrements des appels par écholocation).

Composition spécifique et habitats préférentiels
La faune chiroptère de la chaîne du Simandou est largement dominée par les espèces forestières. Plusieurs d'entre elles atteignent la limite septentrionale de leur distribution dans la région, telles que *Megaloglossus woermanni*, *Nycteris grandis*, *Hipposideros fuliginosus*, *Kerivoula* spp., et *Mops spurrelli*. La présence sympatrique de trois espèces de *Kerivoula*, un genre de chauve-souris sylvicole très spécialisé, est notable dans la mesure où ces chauves-souris ont rarement été recensées au cours d'études antérieures. *K. cuprosa* n'est connue que dans sept localités à travers le monde. Plusieurs espèces préférant les habitats ouverts ont également été recensées, reflétant la structure en mosaïque du site d'étude, qui comprend des forêts de plaine, de ravins et des forêts galeries, des savanes arborées et des prairies d'altitude. La diversité des types

d'habitats de la zone est due au gradient d'élévation et à la position géographique de la chaîne du Simandou, situé dans l'écotone de la zone guinéenne où les savanes du nord et les formations forestières du sud forment une mosaïque complexe. Sachant que de nombreuses espèces de chiroptères sont relativement spécialisées pour des types d'habitats spécifiques, nous nous attendons à ce que la diversité d'habitats de la chaîne du Simandou recèle un assemblage de chauves-souris d'une richesse spécifique exceptionnelle, soulignant le caractère unique de cet habitat.

La faune chiroptère est aussi caractérisée par la forte proportion d'espèces qui dépendent strictement (cinq espèces soit 23,8 % du total des espèces : *Rousettus aegyptiacus, Rhinolophus landeri, R. guineensis, Hipposideros jonesi, Miniopterus schreibersii*) ou partiellement (également *schreibersii* cinq espèces : *Nycteris grandis, Rhinolophus alcyone, Hipposideros fuliginosus, H.* cf. *caffer, H.* cf. *ruber*) des grottes. La présence de ces espèces sur un site donné dépend dans une large mesure de la disponibilité de grottes susceptibles d'offrir des perchoirs, ce qui entraîne en général des zones de distribution clairsemées.

Espèces endémiques et recommandations pour la conservation

La chaîne du Simandou est bordée de trois zones auxquelles a été attribué un rang global de priorité « élevé » durant l'Atelier pour la Sélection des Priorités, la zone Fon-Tibé (B7), la chaîne Wonegizi-Ziama (B8) et la zone du Mont Bero-Tetini (B9). Dans le cas des mammifères seulement, la chaîne du Simandou est incluse dans une zone qui a été classée à un rang de priorité « extrêmement élevé » pour la conservation (Bakarr et al. 2001). Ces régions font partie du massif montagneux de Haute Guinée, qui comprend aussi le Mont Nimba, la forêt de Diecké, les montagnes de Loma-Tingi et le Fouta Djallon. L'une des caractéristiques remarquables de la faune chiroptère des montagnes de Haute Guinée est le nombre démesurément élevé d'espèces qui sont soit endémiques à la Haute Guinée soit représentées

dans cette région par des populations isolées (Tableau 6.5). Plusieurs de ces espèces sont aussi strictement ou partiellement dépendantes des grottes comme perchoirs diurnes (*Coleura afra, Rhinolophus* spp., spp., *Hipposideros* spp., *Myotis tricolor, Miniopterus* spp.), ce que leur permet la grande diversité topographique de cette région.

Plusieurs espèces sont non seulement endémiques à la Haute Guinée, mais s'avèrent également occuper des aires de répartition étonnamment faibles, comme *Rhinolophus maclaudi, R. ziama, Hipposideros marisae* et *H. lamottei*. Parmi elles, *H. marisae* est classée comme « Vulnérable » et *H. lamottei* comme « Insuffisamment Documenté » selon la dernière Liste Rouge de l'UICN (Hutson et al. 2001, IUCN 2002). Fahr et al. (2002b) ont proposé de classer *R. maclaudi* comme « Menacé d'Extinction » et le récemment décrit *R. ziama* comme « Insuffisamment Documenté ». *R. ziama* et *H. lamottei* sont tous deux connus de deux localités, et presque rien n'est connu de leur biologie. Ces espèces mettent en relief l'importance des zones montagneuses de Haute Guinée pour la conservation des chauves-souris à aires de répartition isolées ou restreintes. De plus, nous soulignons que ceci concerne aussi les chauves-souris qui ont une aire de répartition plus étendue, puisque sept espèces, soit 33,3 % du total de 21 espèces recensées au cours de ce RAP, sont enregistrées dans la dernière Liste Rouge des Espèces Menacées de l'UICN (Hutson et al. 2001, IUCN 2002) : une espèce comme « Vulnérable » (*Epomops buettikoferi*) et six espèces comme «Faible Risque: Presque Menacé « (*Rhinolophus alcyone, R. guineensis, Hipposideros jonesi, H. fuliginosus, Kerivoula cuprosa, Miniopterus schreibersii*).

Il est aussi remarquable que sur neuf espèces décrites du continent africain ces vingt dernières années, trois (notées avec un astérisque) aient été découvertes dans les montagnes de Haute Guinée : *Epomophorus minimus* Claessen & De Claessen & De Vree, 1991, *Rhinolophus hillorum**, *R.*

Tableau 6.4. Comparaison entre le nombre d'espèces de chauves souris par famille de la Forêt Classée du Pic de Fon et celui de l'Afrique de l'Ouest (du Sénégal au Nigeria)

| | No. d'espèces -- % du no.total d'espèces | | | | | |
	Pic de Fon		Guinée		Afrique de l'Ouest	
Pteropodidae	7	33,3 %	11	17,5 %	14	11,7 %
Rhinopomatidae	0	—	0	—	2	1,7 %
Emballonuridae	0	—	2	3,2 %	5	4,2 %
Nycteridae	1	4,8 %	7	11,1 %	9	7,5 %
Megadermatidae	0	—	1	1,6 %	1	0,8 %
Rhinolophidae	3	14,3 %	9	14,3 %	11	9,2 %
Hipposideridae	4	19,0 %	9	14,3 %	11	9,2 %
Vespertilionidae	5	23,8 %	18	28,6 %	46	38,3 %
Molossidae	1	4.8 %	6	9,5 %	21	17,5 %
Total	21	100,0 %	63	100,0 %	120	100,0 %

Tableau 6.5. Espèces de chauves souris avec des occurences isolées au sein de la région de Haute Guinée (HG) ou endémiques à la région. #. Espèces dépendantes des grottes

Espèces	PdF	Nimba	Ziama	Wonegizi	Distribution
Emballonuridae					
Coleura afra (PETERS, 1852)					occurrence isolée HG, #
Rhinolophidae					
Rhinolophus guineensis	+	+	+	+	esp. endémique à HG
Rhinolophus maclaudi POUSARGUES, 1897					esp. endémique à HG
Rhinolophus ziama FAHR et al. 2002			+	+	esp. endémique à HG
Rhinolophus denti knorri EISENTRAUT, 1960					esp. endémique à HG
Rhinolophus hillorum KOOPMAN, 1989		+		+	occurrence isolée HG,
Rhinolophus simulator alticolus SANBORN, 1936		+		+	occurrence isolée HG,
Hipposideridae					
Hipposideros jonesi	+		+	+	esp. endémique à HG
Hipposideros marisae AELLEN, 1954		+	+	+	esp. endémique à HG
Hipposideros lamottei BROSSET, 1985		+			esp. endémique à HG
Vespertilionidae					
Myotis welwitschii	+				occurrence isolée HG,
Myotis tricolor (TEMMINCK, 1832)					occurrence isolée HG,
Hypsugo crassulus bellieri DE VREE, 1972				+	esp. endémique à HG
Miniopterus schreibersii villiersi	+		+	+	occurrence isolée HG,
Miniopterus i. inflatus THOMAS, 1903		+		+	occurrence isolée HG,
Molossidae					
Chaerephon b. bemmeleni (JENTINK, 1879)		+			occurrence isolée HG,

Pdf: Pic de Fon, cette étude; Nimba: Chaîne de Mont Nimba, Guinée et Libéria (Aellen 1963, Hill 1982, Brosset 1985, J. Fahr unpubl. data); Ziama: Chaîne de Ziama, Guinée (Roche 1972, Fahr et al. 2002b, J. Fahr unpubl. data); Wonegizi: Chaîne de Wonegizi, Libéria (Koopman 1989, Koopman et al. 1995, Fahr et al. 2002b).

*ziama**, *R. maendeleo* Kock, Csorba & Howell, 2000, *R. sakejiensis* Cotterill, Cotterill, 2002, *Hipposideros lamottei**, *Plecotus balensis* Kruskop & Kruskop & Lavrenchenko, 2000, *Glauconycteris curryae* Eger & Schlitter, Eger & Schlitter, 2001, *Scotophilus nucella* Robbins, 1984. La préservation des habitats de montagne non dégradés dans la région de Haute Guinée est d'une importance considérable, non seulement au niveau régional mais aussi pour la conservation des chauves-souris à l'échelle continentale et mondiale. Tous les efforts sont requis pour protéger les quelques zones non dégradées telle que la chaîne du Simandou, en raison de l'extrême limitation des habitats de montagne et du large pourcentage qui en a déjà été perdu par l'utilisation anthropique des terres.

En plus de la dégradation des habitats à plus ou moins grande échelle, le dérangement ou l'exploitation des chauves-souris dans leurs aires de repos diurnes représente une menace sérieuse (Hutson et al. 2001, Fahr et al. 2002a). Alors que de nombreuses espèces de chauves-souris sont connues pour nicher dans les arbres creux, sous l'écorce ou entre les feuilles de palmiers qui sont relativement à l'abri d'une exploitation directe, ce sont en particulier les espèces cavernicoles qui sont menacées par les dérangements et les prélèvements pendant la journée. La population biologique de la plupart des espèces chiroptères supporte mal une exploitation même faible, car au regard de leur petite taille, les chauves-souris sont des mammifères dont la longue durée de vie et le faible taux de reproduction sont inhabituels. La grande majorité des espèces de chauves-souris africaines se reproduisent seulement une à deux fois par an et les femelles ne donnent naissance qu'à un seul petit. L'exploitation des chauves-souris sur leurs sites de repos diurnes est très aisée et productive, entraînant une chute rapide des populations locales. Toute utilisation de chiroptères cavernicoles comme viande de brousse est donc hautement insoutenable et peut mener à l'extinction des populations locales ou même des espèces qui ont des aires de répartition restreintes. Les dérangements fréquents des nichoirs diurnes peuvent aussi en provoquer l'abandon.

En conséquence, une gestion intégrée de la Forêt Classée du Pic de Fon devrait inclure comme composante importante un programme spécifique sur les chauves-souris cavernicoles, avec une attention particulière pour les espèces chiroptères endémiques et menacées des montagnes de Haute Guinée (Tableau 6.5). Nous suggérons que ce programme cible les priorités suivantes :

- Etudes continues pour évaluer l'intégralité de la faune chiroptère de la chaîne du Simandou.

- Etude des grottes : cartographie des sites, identification des espèces chiroptères et estimation de la taille des colonies.

- Recherches spécifiques sur les espèces de chauve-souris endémiques aux montagnes de Haute Guinée (*Rhinolophus ziama*, *R. maclaudi*, *Hipposideros lamottei*, *H. marisae*) et études approfondies de leur biologie.

- Programme de suivi à long terme de grottes sélectionnées ayant d'importantes colonies de chauves-souris (à la fois en terme de taille des colonies et de rareté/endémisme des espèces).

- Enquêtes auprès des communautés locales pour évaluer les niveaux d'exploitation potentiels.

- Programmes de sensibilisation et d'éducation pour contrer l'utilisation insoutenable des chauves-souris comme viande de brousse.

RECOMMANDATIONS POUR LA CONSERVATION

En résumé, nous avons trouvé que 1) la région du Simandou héberge une population particulière de chauves-souris caractérisée par des espèces sylvicoles et cavernicoles, 2) parmi celles-ci, une espèce est considérée comme « Vulnérable » et six espèces sont « Presque Menacées » selon la Liste Rouge de l'UICN, 3) les montagnes de Haute Guinée hébergent nombre d'espèces endémiques à l'Afrique de l'Ouest et dans certains cas endémiques à ces montagnes (espèces à zone de répartition restreinte), 4) plusieurs de ces espèces sont mondialement menacées ou proches de l'être, et 5) plusieurs nouvelles espèces ont récemment été trouvées dans la région, indiquant que de nombreuses découvertes sont encore à faire dans cette zone.

La chaîne du Simandou, ainsi que les autres zones forestières protégées de Guinée (Ziama, Diécké, Mont Nimba, Mont Béro, Mont Tétini) forment le dernier bastion d'habitats d'altitude non dégradés et protégés de la zone forestière de Guinée. Considérant l'extension très limitée des habitats de montagne en Afrique de l'Ouest, nous soulignons le caractère unique de ces habitats ainsi que de leur faune et flore respectives (Lamotte 1998). Nous prévoyons des conséquences désastreuses pour les populations de chiroptères de la région si ces zones devaient être détruites par l'exploitation minière à ciel ouvert, la déforestation et l'empiétement villageois. Le recensement inespéré de *M. welwitschii*, ainsi que celui d'espèces endémiques aux montagnes de Haute Guinée ou d'Afrique de l'Ouest soulignent l'importance de la région pour la conservation en général et plus particulièrement en ce qui concerne la chaîne du Simandou. Nos découvertes confirment les conclusions de l'Atelier pour la Sélection des Priorités qui a classé les habitats de montagne de Guinée comme étant d'extrêmement haute priorité pour les mammifères et confère à la région de Haute Guinée le statut de site d'importance mondiale pour la conservation de la biodiversité.

RÉFÉRENCES

Aellen, V. 1956. Speologica africana. Chiroptères des grottes de Guinée. Bull. Inst. Fr. Afr. Noire Ser. A Sci. Nat. 18(3): 884-894.

Aellen, V. 1963. La Réserve Naturelle Intégrale du Mont Nimba. XXIX. Chiroptères. Mém. Inst. Fr. Afr. Noire 66: 629-638.

Bakarr, M., B. Bailey, D. Byler, R. Ham, S. Olivieri and M. Omland (eds.). 2001. From the Forest to the Sea: Biodiversity Connections from Guinea to Togo. Conservation International, Washington, DC. 78 pp. www.biodiversityscience.org/priority_outcomes/west_africa

Barnett, A.A. and M.L. Prangley. 1997. Mammalogy in the Republic of Guinea: An overview of research from 1946 to 1996, a preliminary check-list and a summary of research recommendations for the future. Mammal Rev. 27(3): 115-164.

Brooks, T., A. Balmford, N. Burgess, J. Fjeldså, L.A. Hansen, J. Moore, C. Rahbek and P.H. Williams. 2001. Toward a blueprint for conservation in Africa. BioScience 51(8): 613-624.

Brosset, A. 1985 [for 1984]. Chiroptères d'altitude du Mont Nimba (Guinée). Description d'une espèce nouvelle, *Hipposideros lamottei*. Mammalia 48(4): 545-555.

Coe, M. 1975. Mammalian ecological studies on Mount Nimba, Liberia. Mammalia 39(4): 523-588.

Colston, P.R. and K. Curry-Lindahl. 1986. The Birds of Mount Nimba, Liberia. British Museum (Natural History). London. 129 pp.

Colwell, R.K. 2000. EstimateS: Statistical estimation of species richness and shared species from samples. Version 6.0b1. Application and user's guide. http://viceroy.eeb.uconn.edu/estimates

Eisentraut, M. 1960. Zwei neue Rhinolophiden aus Guinea. Stuttgarter Beitr. Naturk. (39): 1-7.

Eisentraut, M. H. and Knorr. 1957. Les chauves-souris cavernicoles de la Guinée française. Mammalia 21(4): 321-340.

Fahr, J. 2001. A fresh look at Afrotropical bat assemblages: Combining different sampling techniques and spatial scales. Bat Research News 42(3): 98.

Fahr, J. and N.M.Ebigbo. 2003. A conservation assessment of the bats of the Simandou Range, Guinea, with the first record of Myotis welwitschii (GRAY, 1866) from West Africa. Acta Chiropterologica 5(1): 125-141.

Fahr, J., N.M. Ebigbo and P. Formenty. 2002a. Final Report on the Bats (Chiroptera) of Mount Sangbé-National Park, Côte d'Ivoire. Afrique Nature, Abidjan. 32 pp.

Fahr, J., N.M. Ebigbo and E.K.V. Kalko. 2001. The influence of local and regional factors on the diversity, structure, and function of West African bat communities (Chiroptera). In: BIOLOG – German Programme on Biodiversity and Global Change, p. 144-145. Status Report, BMBF & DLR, Bonn. 247 pp.

Fahr, J., H. Vierhaus, R. Hutterer and D. Kock. 2002b. A revision of the Rhinolophus maclaudi species group with the description of a new species from West Africa (Chiroptera: Rhinolophidae). Myotis 40: 95-126.

Gatter, W. 1997. Birds of Liberia. Pica Press, The Banks, Mountfield. 320 pp.

Hill, J.E. 1982. Records of bats from Mount Nimba, Liberia. Mammalia 46(1): 116-120.

Hutson, A.M., S.P. Mickleburgh and P.A. Racey (comp.). 2001. Microchiropteran Bats: Global Status Survey and Conservation Action Plan. IUCN/SSC Chiroptera Specialist Group. IUCN, Gland, Switzerland. x + 258 pp.

IUCN. 2002. 2002 IUCN Red List of Threatened Species. www.redlist.org.

Konstantinov, O.K., A.I. Pema, V.V. Labzin and G.V. Farafonova. 2000. Records of bats from Middle Guinea, with remarks on their natural history. Plecotus et al. 3: 129-148.

Koopman, K.F. 1989. Systematic notes on Liberian bats. Am. Mus. Novitates (2946): 1-11.

Koopman, K.F., C.P. Kofron and A. Chapman. 1995. The bats of Liberia: Systematics, ecology, and distribution. Am. Mus. Novitates (3148): 1-24.

Lamotte, M. 1942. La faune mammalogique du Mont Nimba (Haute Guinée). Mammalia 6: 114-119.

Lamotte, M. (ed.). 1998. Le Mont Nimba. Réserve de Biosphère et Site du Patrimoine Mondial (Guinée et Côte d'Ivoire). Initiation à la Géomorphologie et à la Biogéographie. UNESCO Publishing, Paris. 153 pp.

Myers, N., R.A. Mittermeier, C.G. Mittermeier, G.A.B. da Fonseca and J. Kent. 2000. Biodiversity hotspots for conservation priorities. Nature 403: 853-858.

Roche, J. 1972 [for 1971]. Recherches mammalogiques en Guinée forestière. Bull. Mus. natn. Hist. nat. (3)16: 737-781.

Toure, M. and J. Suter. 2001. Workshop report of the 1st trinational meeting (Côte d'Ivoire, Guinea, Liberia), 12-14 September 2001, Man, Côte d'Ivoire. Initiating a Tri-national Programme for the Integrated Conservation of the Mount Nimba Massif. Fauna & Flora Int., Conservation International & BirdLife Int., Abidjan. 56 pp. http://www.fauna-flora.org/around_the_world/africa/mount_nimba.htm

Toure, M. and J. Suter. 2002. Workshop report of the 2nd trinational meeting (Côte d'Ivoire, Guinea, Liberia), 12-15 February 2002, N'Zérékoré, Guinea. Initiating a Tri-national Programme for the Integrated Conservation of the Nimba Mountains. Fauna & Flora Int., Conservation International & BirdLife Int., Abidjan. 82 pp.

Verschuren, J. 1977 [for 1976]. Les cheiroptères du Mont Nimba (Liberia). Mammalia 40(4): 615-632.

Wolton, R.J., P.A. Arak, H.C.J. Godfray and R.P. Wilson. 1982. Ecological and behavioural studies of the Megachiroptera at Mount Nimba, Liberia, with notes on Microchiroptera. Mammalia 46(4): 419-448.

Ziegler, S., G. Nikolaus and R. Hutterer. 2002. High mammalian diversity in the newly established National Park of Upper Niger, Republic of Guinea. Oryx 36(1): 73-80.

Chapitre 7

Une évaluation rapide des micro mammifères terrestres (musaraignes et rongeurs) de la Forêt Classée du Pic de Fon, Guinée

Jan Decher

RÉSUMÉ

Une étude des petits mammifères non volants, durant l'évaluation biologique rapide (RAP) de deux sites, sur le versant ouest de la montagne du Pic de Fon (chaîne du Simandou, Guinée) a permis le recensement de trois espèces de musaraignes et huit espèces de rongeurs. Six espèces supplémentaires d'écureuils ont été observées. Aucun petit mammifère particulièrement rare ni localement endémique n'a été observé. Le taux de succès du piégeage était beaucoup moins important (4-7 %) sur le site d'altitude (env. 1330 m) que sur le site moins élevé (22-32 %; env. 620 m). Les résultats étaient caractéristiques des forêts semi-sempervirentes pluviales de montagne, avec un haut niveau de biomasse de petits mammifères à basse altitude, ce qui suggère un accroissement de la productivité forestière due à l'abondance d'eau, de plantes hautes et d'une diversité de micro habitats, et dans les zones principales, un niveau de destruction relativement faible. Les chasseurs locaux et les guides nous ont confirmé la présence de l'endémique micropotamogale de Lamotte (*Micropotamogale lamottei*), sur la base de dessins et de descriptions de l'animal, mais cette espèce n'a été ni observée ni capturée pendant le RAP.

INTRODUCTION

En 1999, durant l'Atelier pour la Sélection des Priorités de Conservation International au Ghana (Bakarr et al. 2001), les montagnes de Guinée et la région Fon-Tibé ont été respectivement désignées comme étant d'une « extrêmement haute » et « très haute » priorité de conservation sur la carte des mammifères et les cartes du consensus intégré. La montagne du Pic de Fon (chaîne du Simandou) en Guinée Orientale, est géographiquement proche et présente des similarités édaphiques et topographiques avec la chaîne des Monts Nimba, juste au sud à la frontière de la Côte d'Ivoire et du Liberia. Les Monts Nimba et la chaîne des Monts Ziama près de Sérédou à la frontière du Liberia, sont connus pour héberger un faible nombre de petits mammifères endémiques mais une importante diversité de mammifères en général (Coe 1975, Gautun et al. 1986, Heim de Balsac 1958, Heim de Balsac et Lamotte 1958, Roche 1971, Verschuren et Meester 1977).

Les petits mammifères terrestres constituent une part importante de l'écologie des forêts tropicales. Il a été montré que les rongeurs, par exemple, de même que les chauves-souris, les primates et les oiseaux, sont d'importants disséminateurs de semences à la fois des basses et hautes strates arboricoles (Gautier-Hion et al. 1985; Longman and Jeník 1987), contribuant à la recolonisation des trouées forestières.

MATÉRIELS ET MÉTHODES

Les méthodes d'étude des micro-mammifères sont celles décrites par Voss et Emmons (1996) ainsi que Martin et al. (2001), et sont conformes aux instructions et méthodes standards recommandées pour les travaux de terrain sur les mammifères (Animal Care and Use Committee 1998, Wilson et al. 1996).

Sur les deux sites du Pic de Fon, trois et quatre lignes de pièges ont été installées, utilisant un total de 56 pièges vif Sherman, 12 pièges à rat de type Victor et 8 grands pièges vif Tomahawk. De plus, à une altitude de 620 m, deux lignes de pitfalls ont été posées avec des clôtures et neuf seaux en plastique. L'effort de piégeage était d'un total de 606 nuit.pièges (nombre de piège par nombre de nuits de piégeage). A 1 350 m d'altitude l'état rocheux du sol n'a pas permis de creuser de pitfall. Des pièges Sherman et des tapettes ont été appâtés avec de la drupe de noix de palme fraîche. Les pièges Tomahawk ont été appâtés avec du poisson séché et du manioc (*Manihot esculenta*).

Les lignes de pièges étaient relevées chaque matin et les animaux capturés étaient mesurés, identifiés et relâchés sur le site de capture, ou conservés comme spécimens de collection pour identification ultérieure. Les spécimens de collection ont été déposés au Museum Alexander Koenig de Bonn, Allemagne (musaraignes) et au United States National Museum, Washington, D.C. (rongeurs). Pour chaque animal capturé des informations sur les micro-habitats ont été relevées, comprenant le pourcentage de couverture de la canopée, la distance de l'arbre le plus proche et du tronc tombé le plus proche, la hauteur du piège, la pente, et le pourcentage de cinq types de couverture du sol. Les coordonnées approximatives de la distribution des sites de piégeage, obtenues avec un GPS Garmin 12 sont décrites ci-dessous (uniquement l'extrémité des lignes de piège) :

Site 1 (1 350 m):

Camp 1:	8° 31' 52.0" N	8° 54' 21.3" W
Ligne A:	8° 31' 52.1" N	8° 54' 23.8" W
Ligne B:	8° 31' 53.5" N	8° 54' 27.1" W
Ligne C:	8° 31' 54.6" N	8° 54' 28.2" W

Site 2 (620 m):

Camp 2:	8° 31' 29.2" N	8° 56' 12.2" W
Ligne A:	8° 31' 32.8" N	8° 56' 06.0" W
Ligne B:	8° 31' 35.6" N	8° 56' 04.4" W
Ligne C:	8° 31' 19.6" N	8° 56' 11.0" W
Ligne D:	8° 31' 37.5" N	8° 56' 09.1" W

Les noms scientifiques et la taxinomie utilisés dans ce rapport proviennent de Wilson et Reeder (1993), les noms communs sont tirés de Wilson et Cole (2000).

RÉSULTATS

Le Tableau 7.1 présente une récapitulation de la distribution des résultats obtenus par un piégeage standardisé durant la période d'échantillonnage. Le site 1 a été échantillonné pendant cinq nuits, le site 2 durant quatre nuits. Quatre

Tableau 7.1. Revue des captures des petits mammifères non-volants par piégeage sur deux sites du Pic de Fon (chaîne du Simandou) en Guinée du 28 novembre au 7 décembre 2002. Deux espèces additionnelles, *Hybomis trivirgatus* et *Mus musculoides* ne sont pas dans ce tableau car elles ont été identifiées par d'autres méthodes

Sites:	Site 1 (Élévation ca. 1350 m)					Site 2 (Élévation ca. 620 m)				
Date:	28 Nov.	29 Nov.	30 Nov.	1 Dec.	2 Dec.	4 Dec.	5 Dec.	6 Dec.	7 Dec.	**Totals**
ORDER/Genre Espèces										
INSECTIVORA										
Crocidura foxi	1									1
Crocidura cf. *denti*					1					1
Crocidura grandiceps						1		1		2
RODENTIA										
Hybomys planifrons	1	2		1		2	2	2		10
Hylomyscus alleni			1			2	2	5	2	12
Malacomys edwardsi	1			1						2
Mus setulosus			1							1
Praomys rostratus	1	1	3	3		16	22	10	13	69
Atherurus africanus								1		1
Totals Captures:	4	3	5	5	1	21	26	19	15	99
No. d'Espèces	4	2	3	3	1	4	4	5	2	9
No. Pièges:	68	72	72	72	14	81	81	77	69	-
Nuit. Pièges :	68	72	72	72	14	81	81	77	69	606
Pièges succès (%):	5,9	4,2	6,9	6,9	7,1	25,9	32,1	24,7	21,7	16,3

Une évaluation rapide des micro mammifères
terrestres (musaraignes et rongeurs) de la Forêt
Classée du Pic de Fon, Guinée

vingt dix neuf (99) individus ont été capturés pendant 606 nuit.pièges, soit 9 espèces recensées sur les deux sites; 3 espèces de musaraignes (*Soricidae*) et six espèces de rongeurs (*Muridae* et *Hystricidae*). De plus, un individu de *Mus musculoides* a été trouvé mort sur une piste proche du Camp 2 et un individu de *Hybomys trivirgatus* a été observé le jour du départ le long de la piste entre le Site 2 et le village de Banko. Les espèces d'écureuils, principalement observées par les participants au RAP travaillant sur d'autres groupes taxinomiques, étaient des Ecureuil de Fernando Po (*Paraxerus poensis*), Ecureuil d'Aubinn (*Protoxerus aubinii*), Heliosciure de Gambie (*Heliosciurus gambianus*) et d'autres (voir chapitre 8). Seul l'Ecureuil d'Aubinn (*Protoxerus aubinii*) est une espèce endémique de Haute Guinée (voir aussi les résultats sur les grands mammifères, Chapitre 8).

Représentant 69,7% des captures, *Praomys rostratus* était l'espèce la plus souvent capturée, suivie de *Hylomyscus alleni* (12,1 %) et *Hybomys planifrons* (10,1 %). Aucune de ces trois espèces de musaraignes ne semble être endémique de Haute Guinée. Les musaraignes identifiées comme *Crocidura* cf. *denti* demandant des examens plus détaillés (R. Hutterer com. pers.). *Crocidura grandiceps* n'était jusqu'à présent connue que des forêts de basse altitude en Côte d'Ivoire, au Ghana et au Nigeria (Hutterer et Happold 1983). La Figure 7.1 montre la courbe d'accumulation des espèces capturées pendant la période d'échantillonnage, y compris *H. trivirgatus*.

Dans la Figure 7.2, les moyennes des variables des micro-habitats sélectionnés, enregistrées sur chaque site de piégeage, sont comparées pour les trois espèces de musaraignes et les cinq espèces de rongeurs capturées, et les moyennes de tous les sites sont respectivement combinées à la haute et basse altitudes. La Figure 7.2a montre que la couverture de la canopée est relativement élevée sur tous les sites de piégeage à l'exception d'une unique capture de la musaraigne *Crocidura* cf. *denti* et de deux captures du rongeur *Praomys rostratus* dans la Ligne 1C, placée en prairie à 1 330 m. La couverture moyenne de la canopée à 1 330 m était de 79,5 %, et de 93,6 % à 630 m d'altitude. La Figure 7.2b montre clairement que le niveau de roche exposée était beaucoup plus élevé à 1 330 m (23,4 %) qu'à 630 m (6,6 %). A l'inverse, la couverture de litière était deux fois plus importante au sol à 630 m (65,6 %) qu'à 1 330 m (35,6 %).

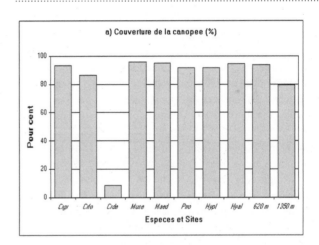

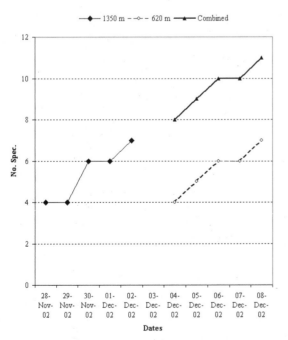

Figure 7.1. Courbe d'accumulation pour onze espèces de micro-mammifères échantillonnées entre le 28 novembre et le 8 décembre 2002 à 1 350 et 620 m d'altitude sur le Pic de Fon, Guinée (y compris *Hybomys trivirgatus* observé le dernier jour).

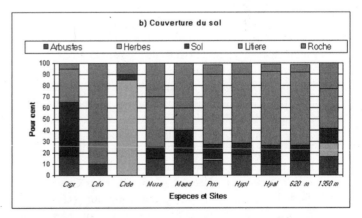

Figure 7.2. Données des micro-habitats sélectionnés pour les individus échantillonnés des espèces de micro-mammifères et pour chaque niveau d'altitude (620 m et 1 350 m) sur le Pic de Fon, Guinée. Les abréviations et tailles d'échantillons sont : Crgr = *Crocidura grandiceps* (n=2); Crfo = *Crocidura foxi* (n=1); Crde = *Crocidura* cf. *denti* (n=1); Muse = *Mus setulosus* (n=1); Maed = *Malacomys edwardsi* (n=1); Prro = *Praomys rostratus* (n=67); Hyal = *Hylomyscus alleni* (12); Hypl = *Hybomys planifrons* (n=10); 620 m d'altitude (n=79); 1 350 m d'altitude (n=16).

DISCUSSION

En nous basant sur la courbe d'accumulation des espèces, sur les 45 petits mammifères recensés du Mont Nimba (Heim de Balsac 1958, Heim de Balsac et Lamotte 1958, Coe 1975, Verschuren et Meester 1977, Gautun et al. 1986) et sur les 33 espèces recensées de la chaîne des Monts Ziama (Roche 1971; voir Appendice 6) nous pouvons être raisonnablement certains que le nombre d'espèces de petits mammifères présentes sur le Pic de Fon excède le nombre recensé durant cette brève étude. En outre, pendant ce RAP, les prairies montagneuses n'ont été échantillonnées que pendant deux nuits et sur un seul site.

Les genres *Malacomys* et *Hylomyscus* sont souvent plus communs que *Praomys* dans les zones de forêts tropicales plus étendues et plus humides, telles que les forêts du Haute Dodo et du Haute Cavally récemment étudiées en Côte d'Ivoire (Decher et al. 2004). Dans les forêts semisempervirentes plus sèches et plus montagneuses telles que sur le Pic de Fon, *Praomys* est dominant et les autres genres sont moins communs ou absents, ce qui a également été vu dans des forêts insulaires de la plaine d'Accra du Ghana (Decher et Bahian 1999) et dans les restes de forêts des Monts Togo du Ghana (Decher et Abedi-Lartey 2002). *Malacomys, Hybomys,* et *Lophuromys* semblent beaucoup plus sensibles aux destructions forestières ou à leur insolation que *Praomys* et le sylvicole *Hylomyscus*. La présence de *Mus setulosus* et *Mus musculoides* a aussi été rapportée au Mont Nimba (Heim de Balsac et Lamotte 1958, Coe 1975) et dans la région de Sérédou (Roche 1971). La capture de l'athérure africain (*Atherurus africanus*) a été faite à proximité du village en forêt secondaire. Coe (1975 : 563) a déjà insisté sur le fait que cette espèce « est en mauvaise posture au Nimba en raison du prix élevé qu'elle rapporte en tant que nourriture ». Les villageois de Banko ont aussi déclaré que le rat géant africain (*Cricetomys* sp.) était communément chassé dans la zone. Dans une étude de 1985 concernant les espèces commercialisées pour la viande de brousse sur un marché d'Accra, au Ghana, le rat géant africain est classé quatrième, l'aulacode (*Thryonomys swinderianus)* étant classé premier et l'athérure classé en onzième position des 18 espèces recensées (Ntiamoa-Baidu 1987). *Atherurus* et *Cricetomys* sont tous deux d'importants disséminateurs de semences, entassant ou cachant de grandes quantités de fruits et noix (Gautier-Hion et al. 1985). La présence continue de ces espèces de gros rongeurs sera seulement assurée si la Forêt Classée du Pic de Fon reste intacte et si les restrictions sur la chasse sont renforcées, à la fois en vue de constituer une aire protégée source de ces espèces pour les zones alentour où la pression de chasse est plus importante, et pour que ces espèces puissent contribuer à la régénération de la forêt.

Les espèces les plus rares qui devraient servir d'indicateurs potentiels d'une haute forêt relativement intacte sont la musaraigne *Crocidura nimbae*, endémique de la haute Guinée et trouvée précédemment en Guinée à Iti, Mont Nimba (Verschuren et Meester 1977), et les rongeurs tels que *Dendromus melanotis, Grammomys buntingi*, endémique de Haute Guinée, *Lophuromys sikapusi, Dephomys defua (D. eburnea inclus),* trois espèces de loirs africains (*Graphiurus,* Myoxidae) et jusqu'à trois espèces d'écureuils volants (*Anomaluridae*), tous rapportés du Mont Nimba ou de la chaîne du Ziama (Coe 1975, Gautun et al. 1986, Heim de Balsac et Lamotte 1958, Roche 1971).

La composition des espèces devrait présenter une différence significative entre la prairie et la forêt, si chacun des écosystèmes était échantillonné de façon plus adéquate. Les espèces de prairie incluent l'aulacode (*Thryonomys swinderianus*) mentionné ci-dessus, un certain nombre d'espèces de musaraignes du genre *Crocidura* et des rongeurs de savane tels que *Lemniscomys striatus, Tatera kempi,* et *Uranomys ruddi*. Aucun d'entre eux n'a été trouvé au cours de cette étude.

L'endémique et aquatique micropotamogale (*Micropotamogale lamottei*) a seulement été déclaré présent par nos informateurs locaux à la vue de la photo de l'animal. Cette espèce existe très probablement au Pic de Fon puisque Guth et al. (1959) ont rapporté des spécimens de Ziéla et Sérédou en Guinée, à moins de 100 km de distance du présent site d'étude. Roche (1971) a capturé un spécimen immature dans la région de Sérédou au cours de deux mois de piégeage intensif. Vogel (1983) a eu plus de succès en utilisant des pièges Sherman et des pièges aquatiques pour obtenir des spécimens vivants en Côte d'Ivoire.

Cette brève étude au Pic de Fon a révélé une communauté de petits mammifères typiquement forestiers qui semble quelque peu appauvrie. Elle est plus caractéristique des forêts semi-sempervirentes de montagne que des grandes forêts humides continues. Pourtant, compte tenu des informations disponibles pour le Mont Nimba et Ziama (Annexe 8) qui sont basées sur des études beaucoup plus importantes, nous pouvons être raisonnablement confiants dans le fait que de nouvelles espèces seront trouvées au Pic de Fon avec un effort d'échantillonnage plus important.

RECOMMANDATIONS POUR LA CONSERVATION

Cette étude rapide de la faune des petits mammifères du Pic de Fon est incomplètement échantillonnée et les informations éparses obtenues devraient être considérées comme une esquisse préliminaire de la diversité des petits mammifères de cette zone, en comparaison des autres sites de montagne de Guinée Occidentale (Annexe 8). Une étude continue est fortement recommandée en particulier pour vérifier la présence d'autres espèces de musaraignes y compris celle de l'endémique micropotamogale qui a été capturé au Mont Nimba et dans la Forêt Classée de Ziama, en employant des méthodes de capture variées pendant de longues périodes et sur d'autres sites du Pic de Fon.

La disparition à grande échelle de végétation de surface et de terre au sommet de la montagne est tout aussi préoccupante pour les populations de petits

Une évaluation rapide des micro mammifères
terrestres (musaraignes et rongeurs) de la Forêt
Classée du Pic de Fon, Guinée

mammifères que l'avance rapide des plantations, et l'utilisation par les populations locales de feux pour nettoyer la forêt, à proximité du Site 2. L'herbe exotique et invasive *Chromolaena odorata* pousse déjà à proximité de la lisière forestière, ce qui constitue un risque supplémentaire d'incendie en saison sèche (Gautier 1988-89). Certaines espèces arboricoles telles que les écureuils volants (Anomaluridae) et le loir (Myoxidae) présentent probablement une faible tolérance aux trouées dans la canopée résultant de l'empiètement par l'agriculture sur les lisières forestières, ou à l'exploitation excessive de la canopée. Leur densité au Pic de Fon est sans doute déjà trop faible pour être détectée durant une étude rapide d'une semaine.

Des études menées en zone tempérée ont prouvé que certaines activités minières pouvaient affecter négativement la diversité et l'abondance des petits mammifères (Kirkland 1976, 1982).

La pollution des eaux ou des réductions du niveau d'eau risquent aussi d'affecter les invertébrés aquatiques et de ce fait affecter indirectement les petits prédateurs tels que l'endémique micropotamogale et deux autres espèces de loutres, *Aonyx capensis* et *Lutra maculicollis*, qui sont probablement toujours présentes en faible nombre dans la région. Les petits mammifères ont aussi été utilisés comme bio-indicateurs pour confirmer la pollution par les métaux lourds, dans des zones minières dont la terre n'était auparavant pas exposée à la surface (McBee et Bickham 1990). Ces études et d'autres devraient être prises en considération lors de la planification des activités minières potentielles, pour réduire les menaces potentielles ou les nuisances pour les populations résidentes de petits mammifères.

Une récente critique des pratiques scientifiques et/ou politiques concernant la biodiversité en Guinée, a mis en avant le fait que la conservation présente, a priori, un biais relatif à « la richesse et la conservation de la biodiversité dans les réserves, et la destruction de la biodiversité dans les zones inhabitées » (Fairhead and Leach 2002:105). Notre expérience de l'emploi d'experts locaux pour la capture d'*Atherurus africanus* à proximité du village de Banko, et notre confiance dans les informateurs locaux au sujet de la présence de l'endémique micropotamogale, soulignent la nécessité d'une coopération plus étroite avec les populations locales et leur savoir traditionnel, ainsi que d'une représentation de la Forêt Classée du Pic de Fon et des villages alentours comme formant partie d'un tout et devant coexister en interdépendance.

En conclusion, afin de mieux évaluer la diversité spécifique réelle des petits mammifères du Pic de Fon et l'impact des destructions anthropiques résultant de l'exploration minière et de la déforestation, et dans le but de suivre les effets des futures activités dans la zone, je recommande une étude plus approfondie des petits mammifères impliquant d'autres sites d'échantillonnage dans des habitats perturbés et non perturbés ainsi qu'à différentes altitudes, dans les forêts comme dans les prairies du Pic de Fon.

RÉFÉRENCES

Animal Care and Use Committee. 1998. Guidelines for the capture, handling, and care of mammals as approved by The American Society of Mammalogists. Journal of Mammalogy, 79:1416-1431.

Bakarr, M., B. Bailey, D. Byler, R. Ham, S. Olivieri, and M. Omland. 2001. From the forest to the sea: biodiversity connections from Guinea to Togo. Conservation International, Washington, D. C., 78 pp.

Coe, M. 1975. Mammalian ecological studies on Mount Nimba, Liberia. Mammalia, 39(4): 523-581.

Decher, J. and L. K. Bahian. 1999. Diversity and structure of terrestrial small mammal communities in different vegetation types on the Accra Plains of Ghana. Journal of Zoology, London, 247:395-407.

Decher, and Abedi-Lartey. 2002. Small mammal zoogeography and diversity in forest remnants in the Togo Highlands of Ghana. Final Report to the National Geographic Society for Research and Exploration, the Ghana Wildlife Division and Conservation International. 32 pp.

Decher, J., B. Kadjo, M. Abedi-Lartey, E. O. Tounkara, and S. Kante. 2004. Small mammals (shrews, rodents, and bats) from the Haute Dodo and Cavally Forests, Côte d'Ivoire. In F. Lauginie, G. Rondeau, and L.E. Alonso (editors). A biological assessment of two classified forests in South-Western Cote d'Ivoire. RAP Bulletin of Biological Assessment 34. Conservation International. Washington, DC.

Fairhead, J. and M. Leach. 2002. Practising 'Biodiversity': The articulation of international, national and local science/policy in Guinea. IDS Bulletin, 33:102-110.

Gautier, L. 1988-89. Contact forêt-savane en Côte-d'Ivoire centrale: rôle de *Chromolaena odorata* (L.) King & Robinson dans la dynamique de la végétation. Centre Suisse de Recherches Scientifiques en Côte-d'Ivoire (Adiopodoumé), 1988-89:100-108.

Gautier-Hion, A., J.M. Duplantier, R. Quris, F. Feer, C. Sourd, J.-P. Decoux, G. Dubost, L. Emmons, C. Erard, P. Hecketsweiler, A. Moungazi, C. Roussilhon and J.-M. Thiollay. 1985. Fruit characters as a basis of fruit choice and seed dispersal in a tropical forest vertebrate community. Oecologia, 65:324-337.

Gautun, J.C., I. Sankhon and M. Tranier. 1986. Nouvelle contribution à la connaissance des rongeurs du massif guinéen des Monts Nimba (Afrique occidentale). Systématique et aperçu quantitatif. Mammalia, 50:205-217.

Guth, C., H. Heim de Balsac and M. Lamotte. 1959. Recherches sur la morphologie de *Micropotamogale lamottei* et l'évolution des Potamogalinae. I. Écologie, denture, anatomie crânienne. Mammalia, 23:423-447.

Heim de Balsac, H. 1958. XIV. Mammifères Insectivores. Memoires de l'Institut Français d'Afrique Noire, 53 (Fasc. IV): 301-357.

Heim de Balsac, H. and M. Lamotte. 1958. La réserve naturelle intégrale du Mont Nimba: 15. Mammifères

Rongeurs. Memoires de l'Institut Fondamental d'Afrique Noire, 53:339-357.

Hutterer, R. 1983 *Crocidura grandiceps*, eine neve Spitzmaous and Westafrika. Revue Suisse de Zoologie, 90: 699-707.

Hutterer, R. 1993, Order Insectivora. In: Wilson, D.E. and D.M. Reeder (eds.). Mammal species of the world: a taxonomic and geographic reference. 2nd edition. Smithonian Institution Press, Washington, DC. 1,206pp.

Hutterer, R. and D.C.D. Happold. 1983. The shrews of Nigeria (Mammalia: Soricidae). Bonner Zoologische Monographien, 18:1-79.

Kirkland, G.L., Jr. 1976. Small mammals of a mine waste situation in the Central Adirondacks, New York: A case of opportunism by Peromyscus maniculatus. The American Midland Naturalist, 95:103-110.

Kirkland, G.L., Jr. 1982. Ecology of small mammals in iron and titanium open-pit mine wastes in the central Adirondacks mountains, New York. National Geographic Society Research Reports, 14:371-380.

Lim, B.K. and P.J. van de Groot Coeverden. 1997. Taxonomic report of small mammals from Côte d'Ivoire. Journal of African Zoology, 111:261-279.

Longman, K.A. and J. Jeník. 1987. Tropical forest and its environment. 2nd ed. Longman Singapore Publishers Ltd., Singapore, 347 pp.

Martin, R.E., R.H. Pine and A.F. Deblase. 2001. A manual of mammology with keys to families of the world. 3rd ed. Mc Graw Hill, Bostos, xi+333 pp.

McBee, K. and J.W. Bickham. 1990. Mammals as bioindicators of environmental toxicity. Pp. 37-88. *In:* Genoways, H.H. (ed.). Current Mammalogy. Vol. 2. Plenum Publishing Corporation.

Ntiamoa-Baidu, Y. 1987. West African wildlife: a resource in jeopardy. Unasylva, 39:27-35.

Roche, J. 1971. Recherches mammalogiques en Guinée forestière. Bulletin du Muséum Nationale d'Histoire Naturelle 3eme serie(16) :737-781.

Verschuren, J. and J. Meester. 1977. Note sur les Soricidae (Insectivora) du Nimba libérien. Mammalia. 41(3): 291-299.

Vogel, P. 1983. Contribution a l'écologie et a la zoogéographie de *Micropotamogale lamottei* (Mammalia, Tenrecidae). Revue de Ecologie (La Terre et la Vie), 38:37-49.

Voss, R.S. and L. H. Emmons. 1996. Mammalian diversity in Neotropical lowland rainforests: a preliminary assessment." Bulletin of the American Museum of Natural History, 230:1-115.

Wilson, D.E. and F. R. Cole. 2000. Common names of mammals of the world. Smithsonian Institution Press, Washington, xiv+204 pp.

Wilson, D.E., F.R. Cole, J.D. Nichols, R. Rudran and M.S. Foster. 1996. Measuring and monitoring biological diversity. Standard methods for mammals. Smithsonian Institution Press, Washington, xxvii + 409 pp.

Ziegler, S., G. Nikolaus and R. Hutterer. 2002. High mammalian diversity in the newly established National Park of Upper Niger, Republic of Guinea. Oryx 36(1): 73-80.

Chapitre 8

Résultats de l'étude des grands mammifères de la Forêt Classée du Pic de Fon, Guinée

Abdulai Barrie et Soumaoro Kanté

RÉSUMÉ

Cette publication présente les résultats d'une étude RAP menée dans la Forêt Classée du Pic de Fon du 27 novembre au 12 décembre 2002. L'objectif de cette étude était d'évaluer la diversité biologique des grands mammifères dans la région. Nous avons utilisé les observations visuelles et auditives ainsi que celles des traces et des pièges photographiques pour étudier la présence des grands mammifères. Nous avons confirmé la présence de 31 et 34 grands mammifères sur les sites 1 et 2 de la Forêt Classée du Pic de Fon. Au total, nous avons confirmé la présence de 38 mammifères dans ces forêts. Le Site 1 de la Forêt Classée du Pic de Fon est actuellement prospecté pour son gisement de fer, tandis que le Site 2 est soumis à une surexploitation du gibier et à des empiètements agricoles. En dépit des lois guinéennes restreignant la chasse, nous avons trouvé des preuves évidentes d'activités cynégétiques dans les deux forêts. Les grands mammifères tels que les primates et les céphalophes ont rarement été directement observés.

INTRODUCTION

La Forêt Classée du Pic de Fon se trouve dans la chaîne du Simandou. Il existe très peu d'informations pour cette région, située au sein du *hotspot* des forêts guinéennes tel que désigné par CI, et qui consiste en forêts courant perpendiculairement à la côte d'Afrique de l'Ouest. Parmi les *hotspots* désignés par CI, le *hotspot* des forêts guinéennes héberge la plus grande diversité de mammifères avec 551 espèces estimées présentes (Myers 1998; Bakarr et al. 2001). Bien que le nombre d'espèces endémiques soit relativement bas, la forêt reste importante pour la conservation des mammifères (Sayer et al. 1992 ; Kingdon 1997; Mittermeier et al. 1999), et elle est l'une des deux régions du monde qui sont de la plus haute priorité pour la conservation des primates. La Forêt Classée du Pic de Fon est le centre d'un immense bloc forestier qui n'a pas été autant dégradé ou fragmenté que la majorité des écosystèmes des forêts de Haute Guinée, et offre de ce fait de fantastiques opportunités pour la conservation.

Barnett et Prangley (1997) ont listé 190 espèces de mammifères pour la Guinée, d'après la révision de 26 publications. Aucune de ces publications ne concernait les espèces du Pic de Fon, car il existe très peu d'informations écologiques pour la région du Simandou. La majorité des études de l'écologie des mammifères sur le terrain ont été menées au Mont Nimba, et complétées par des études dans les zones adjacentes du Liberia et proches de la Sierra Leone (Barnett and Prangley 1997). L'UICN (1988) donne une liste partielle des mammifères recensés au Mont Nimba et a classé 10 espèces comme « Menacé » pour la Guinée. Les études primatologiques ont aussi été concentrées sur le Mont Nimba. Bien que certains mammifères tels que le bongo (*Tragelaphus euryceros*) et les carnivores endémiques telles que la genette de Johnston (*Genetta johnstoni*) et la mangue du Liberia (*Liberiictis kuhnii*) soient connues par de petites populations en Guinée (Rosevear 1974 ; et Coe 1975) au Liberia et en Côte d'Ivoire, les grands mammifères n'ont pas encore fait l'objet de suivis systématiques. Dans l'ordre des Artiodactyles, trois espèces de céphalophes « Vulnérable » du genre *Cephalophus* (*C. jentinki*, *C. niger* et *C. zebra*) et la « Presque Menacé » petite antilope royale, *Neotragus pygmaeus*, sont endémiques (Kingdon 1997), ce qui renforce l'importance du *hotspot* de Haute Guinée (Tableau 8.1).

D'autres grands mammifères importants sont le léopard (*Panthera pardus*) et l'éléphant (*Loxodonta africana*). La composition des espèces du Pic de Fon est similaire à celle du Mont Nimba. Pourtant, aucun suivi systématique des grands mammifères terrestres n'a été effectué dans la zone du Pic de Fon.

MATÉRIEL ET MÉTHODES

Zone d'étude

Nous avons mené notre étude au début de la saison sèche sur deux sites du Pic de Fon : le Site 1 qui se trouvait sur la chaîne de montagnes en altitude (08°31'52.0"N, 08°54'21.3"W) du 27 au 30 novembre 2002, et le Site 2, dans la plaine (08°31'29.2"N, 08°56'12.2"W) du 2 au 7 décembre 2002. Les Sites 1 et 2 étaient respectivement à 1350 m et 600 m environ. Les forêts classées en Guinée sont des forêts de production. Le Site 1 est actuellement prospecté pour des gisements de fer et les forêts autour du Site 2 ont été défrichées pour des cultures (café, cacao, plantains et riz). Les cours d'eau à l'intérieur des forêts s'écoulaient normalement pour cette période de l'année.

Méthodes

Nous avons employé des méthodes actives et passives pour déterminer la présence des grands mammifères. La méthode active incluait l'observation directe des espèces, l'identification des traces et des sons, les nids, les excréments et d'autres informations indirectes pour déterminer la présence d'espèces de grands mammifères non volants dans les deux zones d'études. Les observations directes et les identifications des empreintes et des sons ont été réalisées au cours d'excursions quotidiennes menées à partir du camp de base. Les suivis ont eu lieu la nuit à l'aide de projecteurs. Comme nos collègues ont aussi rassemblé des données sur les grands mammifères de façon opportuniste, et que certaines observations se sont sans doute répétées, nous n'avons utilisé ces informations que pour attester la présence des espèces.

La méthode d'observation passive comprenait l'utilisation de neuf pièges photographiques de type CamTrakker (CamTrakker Atlanta, Georgia) employés sur chaque site d'étude. Les pièges photographiques CamTrakker étaient activés par des détecteurs de chaleur. Chaque CamTrakker était équipé d'un objectif 35mm Samsung Vega 77i en mode autofocus, chargé de film argentique 200 ISO. Le délai entre la réception du détecteur et la prise de vue était de 0.6 secondes. Les appareils photos étaient réglés pour fonctionner de façon continue (bouton de contrôle 1 activé) et attendre 20 secondes au maximum entre les prises de vue (boutons de contrôle 6 et 8 activés). Les appareils photos ont été placés dans des lieux

Tableau 8.1 Espèces de grands mammifères endémiques et presque endémiques à l'Afrique de l'Ouest

Ordre	Famille	Espèces	Nom commun
Primates	Cercopithecinae	*Cercopithecus (mona) campbelli*	Singe des Palétuviers
		Cercopithecus diana diana	Cercopithèque Diane
		Cercopithecus (c.) p. buettikoferi	Hocheur Blanc-nez
	Colobidae	*Procolobus verus*	Colobe de Van Beneden
		Colobus p. polykomos	Colobe Blanc et Noir d'Afrique Occidentale
		Colobus vellerosus	Colobe de Geoffroy
		Piliocolobus badius badius	Colobe Bai
Carnivora	Herpestidae	*Liberiictis kuhni*	Mangue du Libéria
	Viverridae	*Genetta johnstoni*	Genette de Johnston
		Genetta bourloni	Genette de Bourlon
		Poiana leightoni	Poiane d'Afrique occidentale
Artiodactyla	Hippopotamidae	*Hexaprotodon liberiensis*	Hippopotame Nain
	Notragini	*Neotragus pygmaeus*	Antilope Royale
	Cephalophini	*Cephalophus jentinki*	Céphalophe de Jentink
		Cephalophus zebra	Céphalophe Rayé
		Cephalophus niger	Céphalophe Noir

susceptibles d'être fréquentés par des espèces mammifères variées. Les sites de repos, les sentes et les sites de nourrissage tels que les arbres fruitiers étaient les endroits types choisis pour placer les appareils photos. Ceux-ci étaient situés à environ 500 m les uns des autres et au moins à 500 m du camp de base. Nous avons utilisé cette méthode pour calculer les taux d'observation pour chaque site de la même façon que les transects standards sont utilisés. Au lieu que les observateurs fassent des observations en suivant un itinéraire, les « observations » se déplaçaient le long d'itinéraires face aux appareils photos fixes (les observateurs). Pour les mammifères discrets subissant une forte pression de chasse, la méthode des pièges photographiques est peut-être plus efficace que celle qui consiste à marcher le long de transects, particulièrement quand les observateurs ont des capacités d'expertise différentes et variées.

RÉSULTATS

Nous avons observé, identifié à l'ouïe ou photographié 31 espèces de grands mammifères sur le Site 1 de la Forêt Classée du Pic de Fon (Tableau 8.2) et 34 espèces sur le Site 2 (Tableau 8.3) pour un total combiné de 39 espèces de grands mammifères. Les pièges photographiques ont fourni deux photos de chimpanzés sur le Site 1 pour un taux de réussite photographique d'une photo pour trois jours de piégeage. Pour le Site 2, nous avons obtenu un total de trois photos d'une espèce pour un taux de réussite photographique d'une photo pour 1,7 jours de piégeage. Les grands mammifères les plus communément photographiés ont été *Atherurus africanus*, l'athérure africain, (3 photos sur 4) et *Pan troglodytes verus* le chimpanzé d'Afrique Occidentale (1 photo sur 4). Aucun léopard, oryctérope, ou bongo n'a été observé, mais les chasseurs locaux ont signalé que ces espèces étaient toujours présentes dans la Forêt Classée du Pic de Fon. D'autres espèces, signalées par les chasseurs, mais dont les descriptions et identifications n'étaient pas claires, ont été ignorées. D'après des empreintes, des fèces et les informations des chasseurs, nous avons noté la présence du buffle africain dans la Forêt Classée du Pic de Fon.

Nous avons observé des primates en deux occasions et nos collègues en ont observé à six reprises durant notre étude. Les signes de présence de sept chimpanzés, les appels et les nids confirment la présence continue des chimpanzés sur les Sites 1 et 2 de la Forêt Classée du Pic de Fon. Les mammifères rencontrés sur un seul des deux sites étaient l'écureuil fouisseur, la genette commune, la civette africaine et le daman de roche pour le Site 1 ; et pour le site 2, le hocheur, le galago du Sénégal, le potto, l'héliosciure de Gambie, l'écureuil géant de Stanger, le pangolin géant et le phacochère. Nous pensons que ces différences sont liées à la courte durée de notre étude et non à des différences fondamentales des faunes de mammifères entre les deux zones d'études.

DISCUSSION

Les grands mammifères sont extrêmement rares en Guinée et dans une grande partie de l'Afrique de l'Ouest, du fait d'une exploitation non contrôlée, de la perte des habitats et de la demande croissante en viande de brousse (Lowes 1970, Davies 1987, Starin 1989, Martin 1991, McGraw 1998). Dans cette région, une large part de la forêt a subi des changements importants en terme de superficie et de composition, en raison de la fragmentation des milieux. Ce qu'il subsiste de la grande forêt est un mélange d'espèces sempervirentes et semi-décidues, particulièrement en forêt secondaire. L'augmentation de la population humaine accélère la conversion des habitats forestiers restants en zones où domine l'activité humaine et en terres cultivées. La demande locale en viande de brousse, le développement des cultures et des plantations réduisent l'étendue et le potentiel futur des forêts restantes en Afrique de l'Ouest, et les cultures agraires (spécialement les plantations) ont accéléré la fragmentation des forêts et la perte de grands mammifères. Les forêts primaires et secondaires en dehors des zones protégées sont les cibles de l'exploitation des ressources (Lebbie 2002) et les demandes globales en ressources risquent d'avoir un impact aussi grave que ceux de la chasse et de l'agriculture.

La Forêt Classée du Pic de Fon est aussi située à proximité des frontières du Liberia et de la Côte d'Ivoire et la population a augmenté du fait que les réfugiés et les déplacés fuient les guerres civiles de ces pays. Même si la forêt est classée et sensée être une réserve forestière interdite à la chasse, la pression de chasse y est très forte. D'après des entretiens avec des chasseurs et des guides locaux, les primates et les céphalophes sont les espèces les plus fréquemment chassées et piégées, un phénomène signalé comme courant dans d'autres pays d'Afrique de l'Ouest (Ausden et Wood 1990, Lebbie 1998, Eves et Bakarr 2001). La plupart des chasseurs exploitent ces espèces pour la filière commerciale plutôt que dans un but alimentaire, et la grande masse corporelle de ces espèces en fait des cibles évidentes. La viande de brousse est une source de protéines importante en Afrique de l'Ouest et la demande en est également élevée (Asibey 1976, Jeffrey 1977, Ajayi 1979, Martin 1983, Falconer et Koppell 1990, Njiforti 1996, Bowen-Jones et Pendry 1999). Les populations urbaines ont contribué à la demande en gibier dans les régions où les sources de protéines alternatives ont un coût élevé (Wilkie et al. 1992). La plupart des ménages dans les zones rurales et urbaines de pays d'Afrique de l'Ouest comme le Ghana, le Liberia, le Sénégal, la Guinée Equatoriale et la Sierra Leone, consomment régulièrement de la viande de brousse, et la chasse et le braconnage sont signalés comme des activités lucratives (Cremoux 1963, Ajayi 1979, Martin 1983, Addo et al. 1994, Barrie 2002).

La chasse au gibier, parallèlement à la perte de l'habitat, est la principale menace à la survie des mammifères en Afrique de l'Ouest (Bakarr et al. 2001). Récemment,

Tableau 8.2. Grands mammifères dont la présence a été confirmée sur le Site 1 du Pic de Fon, 2002 (H = entendu, S = vu, T = traces, P = photographié, M = nombreux, ou O = autre évidence). Les espèces en gras ont été documentées seulement sur ce site. Les noms scientifiques sont cités d'après Kingdon (1997).

Ordre	Famille	Espèces	Nom commun	H	S	T	P	O
Primates	Hominidae	*Pan troglodytes verus*	Chimpanzé					nids
	Cercopithecidae	*Cercocebus atys atys*	Cercocèbe Enfumé	1				
		Cercopithecus diana diana	Cercopithèque Diane	1				
		Cercopithecus campbelli campbelli	Singe de Palétuviers	2				
		Cercopithecus petaurista buettikoferi	Hocheur Blanc-nez					
	Galagonidae	*Galagoides demidoff*	Galago de Demidoff					
Rodentia	Sciuridae	***Euxerus erythropus***	Ecureuil fouisseur		5			
		Paraxerus poensis	Ecureuil de Fernando Po		3			
		Heliosciurus rufobrachium	Héliosciure à pattes rousses		2			
	Hystricidae	*Hystrix cristata*	Porc épic					épines
		Atherurus africanus	Athérure			4		
	Muridae	*Cricetomys emini*	Rat d'emin					excréments
	Thryonomyidae	*Thryonomys swinderianus*	Aulacode					emprisonné
Carnivora	Herpestidae	*Herpestes ichneumon*	Mangouste ichneumon		1			
		Herpestes sanguinea	Mangouste rouge		2			
		Crossarchus obscurus	Mangouste brune		1			
	Viverridae	*Civettictis civetta*	Civette			2		
		Genetta geneta	Genette commune					Equipes de Rio Tinto
		Nandinia binotata	Nandinie		2			
Pholidota	Manidae	*Uromanis tetradactyla*	Pangolin à Longue Queue			2		
		Phataginus tricuspis	Pangolin Commun					Se nourrissant
Hyracoidea	Procaviidae	***Procavia capensis***	Daman des rochers	1				
Artiodactyla	Suidae	*Potamochoerus porcus*	Potamochère		M			
	Bovidae	*Syncerus caffer*	Buffle d'Afrique			2		
		Tragelaphus scriptus	Guib harnaché			3		
	Antelopinae	*Sylvicapra grimmia*	Céphalophe Couronné			3		
		Cephalophus maxwelli	Céphalophe de Maxwell			M		excréments
		Cephalophus niger	Céphalophe Noir			4		
		Cephalophus dorsalis	Céphalophe Bai			3		excréments
		Cephalophus silvicultor	Céphalophe à Dos Jaune			5		
	Neotragini	*Neotragus pygmaeus*	Antilope Royale					excréments

Tableau 8.3. Grands mammifères dont la présence a été confirmée sur le Site 2 du Pic de Fon, 2002 (H = entendu, S = vu, T = traces, P = photographié, M = nombreux, ou O = autre évidence). Les espèces en gras ont été documentées seulement sur ce site. Les noms scientifiques sont cités d'après Kingdon (1997).

Ordre	Famille	Espèces	Nom commun	H	S	T	P	O
Primates	Hominidae	*Pan troglodytes verus*	Chimpanzé	2	7		2	
	Cercopithecidae	*Cercocebus atys atys*	Cercocèbe Enfumé		6			
		Cercopithecus diana diana	Cercopithèque Diane	2				
		Cercopithecus campbelli campbelli	Singe des Palétuviers		3			
		Cercopithecus nictitans	Hocheur	2				
		Cercopithecus petaurista buettikoferi	Hocheur Blanc-nez	2				
	Galagonidae	*Galagoides demidoff*	Galago de Demidoff	4	7			
		Galagoides senegalensis	Galago du Senegal		1			
	Loridae	***Perodicticus potto***	Potto		1			
Rodentia	Sciuridae	*Paraxerus poensis*	Ecureuil de Fernando Po		4			
		Protoxerus aubinnii	Ecureuil d'Aubinn		1			
		Heliosciurus rufobrachium	Héliosciure à pattes rousses		3			
		Protoxerus stangeri	Ecureuil géant de Stanger	1				
		Heliosciurus gambianus	Héliosciure de Gambie		1	2		
	Hystricidae	*Hystrix cristata*	Porc épic			2		épines
		Atherurus africanus	Athérure			2	3	décomposé
	Muridae	*Cricetomys emini*	Rat d'emin					excréments
	Thryonomyidae	*Thryonomys swinderianus*	Aulacode					attrapé
Carnivora	Herpestidae	*Herpestes ichneumon*	Mangouste ichneumon		1			
		Herpestes sanguinea	Mangouste rouge		2			
	Viverridae	*Civettictis civetta*	Civette		1			
Pholidota	Manidae	*Uromanis tetradactyla*	Pangolin à Longue Queue			1		
		Phataginus tricuspis	Pangolin Commun					Se nourrissant
		Smutsia gigantica	Pangolin Géant			3		
Artiodactyla	Suidae	*Potamochoerus porcus*	Potamochère		7	M		
		Phacochoerus africanus	Phacochère					fouillant
	Bovidae	*Syncerus caffer*	Buffle d'Afrique			2		
		Tragelaphus scriptus	Guib harnaché			5		
	Antelopinae	*Sylvicapra grimmia*	Céphalophe Couronné			3		
		Cephalophus silvicultor	Céphalophe à Dos Jaune			7		
		Cephalophus maxwelli	Céphalophe de Maxwell			M		excréments
		Cephalophus niger	Céphalophe Noir			1		
		Cephalophus dorsalis	Céphalophe Bai			2		
	Neotragini	*Neotragus pygmaeus*	Antilope Royale					excréments

l'apparente extinction de *Procolobus badius waldroni* a été a été attribuée à la chasse et à la demande en gibier de cette région (Oates et al. 2000). La viande de brousse est une source de protéines particulièrement importante, ainsi qu'une source de revenus pour de nombreuses personnes dans la région, et une grande variété d'espèces sont chassées. Les antilopes, les potamochères et les primates dominent le marché de la viande de brousse alors que le grand aulacode (*Thryonomys swinderianus*) et le rat géant (*Cricetomys* spp.) sont préférés par les populations rurales. L'étendue de ces pratiques a poussé les gouvernements à promulguer des interdictions de la chasse bien que la législation soit difficilement applicable et ne puisse être renforcée (Sayer et al. 1992).

Pour mener notre étude sur le Site 1, nous avons utilisé le système de pistes créé pour les activités minières par Rio Tinto. Les routes taillées perpendiculairement aux pentes de la montagne ont causé des phénomènes d'érosion et dans certains cas exposé le soubassement rocheux. Nous avons aussi utilisé le réseau extensif de pistes familières à nos guides dans les forêts des Sites 1 et 2.

Bien qu'en Guinée la chasse soit interdite dans tout le pays, nous avons trouvé 11 cartouches, 2 pièges et un certain nombre de campements de chasse sur le Site 1 et nous avons trouvé sur le Site 2, 36 cartouches, 38 pièges ou lignes de pièges et entendu des tirs de fusils chaque jour durant nos travaux diurnes et à quatre reprises durant la nuit. Le contrôle des activités cynégétiques est sous la responsabilité de l'agence gouvernementale responsable de l'exploitation du bois en Guinée, la Direction Nationale des Eaux et Forêts. Plusieurs villages étaient localisés autour ou à proximité de ces forêts. La combinaison de la chasse illégale et de l'empiètement humain nuit fortement à la perpétuation des grands mammifères dans la Forêt Classée du Pic de Fon. Nos résultats indiquent et sont représentatifs du syndrome de la forêt vide dans laquelle les populations de grands mammifères sont, une à une, réduites en densité et finalement exterminées de vastes zones (Sanderson et al. En press).

Nos résultats indiquent que le riche assortiment biologique de mammifères présent au Mont Nimba, et dans les Parcs Nationaux de Taï et de Sapo, se trouve aussi dans la Forêt Classée du Pic de Fon. Durant notre bref séjour, nous avons relevé la présence de nombreuses espèces listées comme « Menacé » par l'UICN : *Syncerus caffer* (buffle africain) listé comme « Dépendant des mesures de conservation » et *Cephalophus maxwelli* (céphalophe de Maxwell), *Cephalophus niger* (céphalophe noir), *Cephalophus dorsalis* (céphalophe à bande dorsale noire), *Cephalophus silvicultor* (céphalophe à dos jaune), et *Neotragus pygmaeus* (antilope royale) listés par l'UICN comme « Presque Menacé » (Tableau 8.2 et 8.3). Ces informations montrent que l'inclusion de la forêt du Pic de Fon dans un réseau d'aires protégées pourrait permettre d'accroître les populations régionales de ces espèces.

RECOMMANDATIONS POUR LA CONSERVATION

Le manque de mise en application des lois concernant les activités illégales dans les forêts classées, telles que l'empiètement humain et la chasse au gibier, n'est pas différent de la situation rencontrée au sein du système de parc nationaux en Guinée, où de nombreux parcs ne bénéficient pas de la protection adéquate et sont surnommés « parcs de papier ». La structure politique de la conservation des forêts en Guinée a été établie. Les enjeux concernent maintenant la mise en œuvre de cette politique sous forme d'action concrète. Les activités minières, la chasse au gibier et les activités d'empiètement nécessitent une évaluation et un suivi attentifs pour que la Forêt Classée du Pic de Fon ne se transforme pas rapidement en bois ouverts n'entretenant plus rien de la riche biodiversité trouvée en forêt primaire.

RÉFÉRENCES

Addo, F., E. O. A. Asibey, K. B. Quist and M. B. Dyson. 1994. The economic contribution of women and protected areas: Ghana and the bushmeat trade. Pages 99-115 *In:* M. Munasinghe and J. McNeely, (eds.) Protected area economics and policy: linking conservation and sustainable development. World Bank and World Conservation Union, Washington, D.C.

Ajayi, S. S. 1979. Food and animal production from tropical forest: utilization of wildlife and by-products in West Africa. FAO, Rome, Italy.

Asibey, E. O. A. 1976. The effects of land use patterns on future supply of bushmeeat in Africa south of the Sahara. Working Paper on Wildlife Management and National Parks, 5th Session.

Ausden, M. and P. Wood. 1990. The wildlife of the Western Area Forest Reserve, Sierra Leone. February 22nd - April 23rd 1990. RSPB.

Bakarr, M.I. 1992. Sierra Leone: Conservation of Biological Diversity. An assessment report prepared for the Biodiversity Support Program, Washington, DC.

Bakarr, M.I., G.A.B. da Fonseca, R. Mittermeier, A.B. Rylands and K.W. Painemilla (ed.). 2001. Hunting and Bushmeat Utilization in the African Rain Forest. Advances in Applied Biodiversity Science Number 2. Conservation International. Washington, DC.

Barnett, A. A. and Prangley, M. L. 1997. Mammalogy in the Republic of Guinea: An overview of research from 1946 to 1996, a preliminary check-list and a summary of research recommendations for the future. Mammal Rev. 1997; 27(3): 115-164.

Barrie, A. 2002. Post conflict conservation status of large mammals in the Western Area Forest Reserve (WAFR), Sierra Leone. M. Sc Dissertation, Njala University College, Freetown.

Bowen-Jones, E. and S. Pendry. 1999. The threat to primates and other mammals from the bushmeat trade in

Africa, and how this threat could be diminished. Oryx 33(3):233-246.

Coe, M.J. 1975. Mammalian ecological studies on Mount Nimba, Liberia. Mammalia, 39: 523-587.

Cremoux, P. 1963. The importance of game meat consumption in the diet of sedentary and nomadic peoples of the Senegal River Valley. Pp. 127-129 in Conservation of Nature and Natural Resources in Modern African States (G. G. Watterson, editor). IUCN publications new series No. 1.

Davies, A.G. 1987. The Gola Forest Reserves, Sierra Leone: Wildlife Conservation and Forest Management. IUCN, Gland, Switzerland. 126p.

Eves, H.E. and M I. Bakarr. 2001. Impacts of bushmeat hunting on wildlife populations in West Africa's Upper Guinea Forest Ecosystem. In Bakarr, M.I., G.A.B. da Fonseca, R. Mittermeier, A.B. Rylands and K.W. Painemilla (eds). Hunting and Bushmeat Utilization in the African Rain Forest: Perspectives Toward a Blueprint for Conservation Action. Advances in Applied Biodiversity Science Number 2:39-57. Conservation International, Washington, DC.

Falconer, J. and C. Koppell. 1990. The Major Significance of Minor Forest Products: The Local Use and Value of Forests in the West African Humid Forest Zone. FAO, Community Forests Note 6. Rome.

IUCN. 1988. Guinea: Conservation of Biological Diversity. World Conservation Monitoring Centre, Cambridge.

IUCN. 1990. 1990 IUCN Red List of Threatened Animals. IUCN, Gland, Switzerland and Cambridge, UK. 228 pp.

Jeffrey, S. 1977. How Liberia uses wildlife. Oryx 14:168-173.

Kingdon, J. 1997. The Kingdon Field Guide to African Mammals. Harcourt Brace & Company, New York.

Lebbie, A. 1998. The No. 2 River Forest River, Sierra Leone: Managing for Biodiversity and the Promotion of Ecotourism. Report Prepared for the United Nations (UN); Project No. SIL/93/002.

Lebbie, A. 2001. Distribution, Exploitation and Valuation of Non-Timber Forest Products from a Forest Reserve in Sierra Leone. PhD Dissertation, University of Wisconsin-Madison, USA.

Lebbie, A. 2002. Western Guinea Lowland Forest. In: New Map of the World. National Geographic Society - World Wildlife Fund Publication. Washington, DC.

Lee, P.J., J. Thornback and E.L. Bennett. 1988. Threatened Primates of Africa. The IUCN Red Data Book. IUCN, Gland, Switzerland and Cambridge, UK.

Lowes, R.H.G. 1970. Destruction in Sierra Leone. Oryx 10(5):309-310.

Martin, C. 1991. The rainforests of West Africa: Ecology, Threats and Conservation. Birkhauser Verlag, Boston.

Martin, L.G.H. 1983. Bushmeat in Nigeria as a natural resource with environmental implications. Environmental Conservation 10(2):125-132.

McGraw, W.S. 1998. Three Monkeys nearing extinction in the forest reserves of eastern Cote d'Ivoire. Oryx 32(3): 233-236.

Mittermeier. R.A., N. Myers, and C.G. Mitteermeier (eds.). 1999. Hotspots: Earth's Biologically Richest and most Endangered Terrestrial Ecoregions. Cemex, Conservation International. 430p.

Myers, N. 1998. Threatened biotas: 'hotspots' in tropical forests. Environmentalist 8:187-208.

Njiforti, H.L. 1996. Preferences and present demand for bushmeat in north Cameroon: some implications for wildlife conservation. Environmental Conservation 23(2):149-155.

Oates, J.F. 1986. Action Plan for African Primate Conservation 1986-1990. IUCN/SSC Primate Specialist Group. Stony Brook, New York, USA.

Oates, J.F., M. Abedi-Lartey, S. McGraw, T.T. Struhsacker, and G.H. Whitesides. 2000. Extinction of a Western African Red Colobus Monkey. Conservation Biology. 14(5):1526-1533.

Rosevear, D.R., 1974. Carnivores of West Africa: British Museum (Natural History), London.

Sayer, J.A., C.S. Harcourt and N.M. Collins. 1992. The Conservation Atlas of Tropical Forests: Africa. IUCN and Simon & Schuster, Cambridge.

Starin, E.D. 1989. Threats to the monkeys of The Gambia. Oryx 23(4): 208-214 24.

Wilkie, D.S., J.G. Sidle and G.C. Boundzanga. 1992. Mechanised logging, market hunting, and a bank loan in Congo. Conservation Biology 6(4):570-580.

Chapitre 9

Une évaluation rapide des primates de la Forêt Classée du Pic de Fon, Guinée

Ilka Herbinger et Elhadj Ousmane Tounkara

RÉSUMÉ

Une évaluation rapide de la faune des primates a été menée entre le 27 novembre et le 7 décembre 2002 dans la Forêt Classée semi-sempervirente du Pic de Fon, dans la chaîne du Simandou, au sud-est de la Guinée. Sur deux sites, la présence et l'abondance des espèces de primates ont été estimées selon une méthode de transects linéaires. Au total, huit espèces de primates ont été recensées, y compris deux prosimiens *(Perodicticus potto* et *Galagoides demidoff)*, cinq singes anthropoïdes *(Cercocebus atys atys, Cercopithecus campbelli campbelli, Cercopithecus petaurista buettikoferi, Cercopithecus nictitans,* et *Cercopithecus diana diana)* ainsi qu'un grand singe hominidé, le chimpanzé d'Afrique Occidentale *(Pan troglodytes verus)*. La présence de cinq autres espèces a été rapportée par les villageois et chasseurs locaux et semble très probable *(Papio anubis, Erythrocebus patas, Cercopithecus aethiops sabaeus, Colobus polykomos polykomos* et peut-être *Procolobus verus)*. Quatre des 13 taxons sont classés parmi les espèces de primates « Presque Menacées » ou « Menacées d'Extinction ». Bien que les densités pour la plupart des espèces de singes paraissent relativement faibles (taux de rencontre < 0,25 par heure), la densité estimée pour la population de chimpanzés (0,64 chimpanzés/km2) était supérieure à celle observée dans des forêts dégradées (0,4 chimpanzés/km2). Ceci indique que la Forêt Classée du Pic de Fon héberge probablement toujours un nombre important d'individus de plusieurs espèces de primates. Ainsi, la population de primates du Simandou, avec ses nombreuses espèces, est une représentation importante de la diversité régionale spécifique des primates pour la Région de Haute Guinée, et leur conservation devrait être de la plus haute priorité.

INTRODUCTION

Les primates sont l'un des composants majeurs des écosystèmes forestiers tropicaux et jouent un rôle important, par exemple en tant que disséminateurs de semences, dans la structuration des habitats forestiers (Chapman 1995, Chapman et Onderdonk 1998, Lambert et Garber 1998, Chatelain et al. 2001). Ils sont prédateurs d'autres mammifères et sont aussi la proie de grands carnivores, de rapaces et de serpents. Leur présence ou leur absence a des répercussions sur la pérennité d'une grande variété de plantes, d'espèces d'invertébrés et de vertébrés. Dans certaines forêts d'Afrique de l'Ouest et de l'Est, les primates forment ce que l'on appelle des associations poly-spécifiques dans lesquelles jusqu'à neuf espèces coexistent, interagissent, et vivent à de fortes densités (Galat-Luong et Galat 1978, Whitesides et al. 1988, Bshary 1995). Les primates sont les principaux participants à la diversité de mammifères de telles forêts.

La Forêt Classée du Pic de Fon dans la chaîne du Simandou en Guinée, est supposée abriter un total de 15 espèces de primates. La chaîne du Simandou fait partie de l'écosystème de la Région de Haute Guinée qui inclut les forêts de la Sierra Leone jusqu'à l'est du Togo. Elle est considérée comme l'une des 25 zones au monde prioritaires pour la conservation, en raison de son haut degré de biodiversité et d'endémisme (Mittermeier et al. 1999). Malheureusement, la Région de Haute Guinée est aussi fortement menacée et a récemment souffert de taux de déforestation dramatiques avec une estimation d'une perte de 80% du couvert forestier dans les années 80 (Martin 1989). De nombreuses populations de primates ont ainsi décliné de façon drastique et certaines espèces ont complètement disparu de certaines zones dans certains pays (plusieurs espèces de primates au Ghana (comme le colobe bai de Miss Waldron), en Sierra Leone, au Liberia ; Lee et al. 1988). Dus aux effets combinés de la destruction des habitats et de la forte pression de chasse, la Région de Haute Guinée abrite aujourd'hui la majorité des communautés de primates les plus menacées en Afrique (Lee et al. 1988). Le chimpanzé d'Afrique Occidentale est estimé disparu de quatre pays d'Afrique de l'Ouest (Togo, Bénin, Gambie et Burkina Faso). La Guinée est l'un des seuls pays, avec la Côte d'Ivoire, la Sierra Leone et le Liberia, qui conserve des populations qui pourraient être viables sur le long terme (Kormos et Boesch 2003).

L'objectif de cette étude était de collecter des informations sur les espèces de primates présentes dans la Forêt Classée du Pic de Fon en Guinée, et de proposer une estimation préliminaire de leur abondance relative. De plus, nous avons voulu évaluer les menaces qui pèsent actuellement sur cette population de primates et proposer les mesures nécessaires à leur protection.

MÉTHODES

L'inventaire a été mené sur deux sites de la partie sud-ouest de la Forêt Classée du Pic de Fon, entre le 27 novembre et le 7 décembre 2002. L'équipe, composée des deux auteurs et de guides locaux, a visité différentes vallées forestières autour des deux sites pendant respectivement trois et cinq jours.

Site 1: Pic de Fon (08°31'52"N, 08°54'21"W; 1350 m; 28-30 novembre) y compris trois vallées forestières de plus haute altitude.

Site 2: 'Banko' (08°31'29"N, 08°56'12"W; 600 m; 2-7 décembre) y compris six vallées forestières de relativement plus basse altitude.

Tous les suivis ont été effectués à pied en marchant lentement (approximativement 0,5 km/h) le long de transects linéaires de longueurs variées (500-4000 m) en faisant des pauses régulières pour observer et écouter les primates. Nous avons répertorié les signes directs de primates, telles que les observations, et les signes indirects, telles que les vocalisations et dans le cas des chimpanzés, les nids. Chaque transect était parcouru une seule fois et nous avons soit utilisé les pistes pré-existantes (principalement celles des chasseurs) soit choisi des directions relevées au compas. En raison de l'escarpement et de la végétation parfois très dense, nous n'avons pas toujours été en mesure de suivre une ligne droite mais nous avons essayé de conserver un transect aussi rectiligne que possible. Nous

avons mesuré les distances parcourues pour chaque transect à l'aide d'un topo fil et recensé les types d'habitats. Nous avons concentré tous les suivis sur les zones de forêts et négligé les savanes, car la végétation herbacée y atteignait une hauteur de 2 mètres et empêchait toute observation des primates. Nous avons aussi recherché les primates dans la savane lors de nos déplacements en véhicule entre les sites, mais n'en avons jamais observé. Tous les habitats forestiers étaient des forêts galeries ou des forêts de ravins le long de cours d'eau inégalement répartis dans différentes vallées.

Lorsque nous avons observé ou entendu des singes, nous avons tenté de déterminer les espèces, le nombre de groupes ou d'individus et leur sexe. Nous avons aussi noté l'heure et la position sur le transect et estimé la distance perpendiculaire à l'individu observé ou entendu. Dans la majorité des cas, nous avons identifié les singes par leurs cris spécifiques d'alarme à longue distance, lancés par les mâles. Lorsque nous avons détecté des nids de chimpanzés, les mesures suivantes ont été notées : position sur le transect, distance perpendiculaire au nid, hauteur estimée du nid, le diamètre de l'arbre de nidification à hauteur de poitrine (dbh), et l'âge des nids (Frais : il n'y a que des feuilles fraîches dans un nid intact, parfois de l'urine, des fèces; Récent : nid intact, mais commençant à se dessécher, présence de feuilles jaunes ; Vieux : nid presque intact, uniquement des feuilles jaunes ; Très vieux : trous dans le fond du nid ou feuilles tombées). Nous avons aussi noté les groupes de nids (définis comme un ensemble de nids qui ne sont pas éloignés de plus de 50 m et qui sont du même âge) et leur taille.

Les recensements diurnes ont eu lieu entre 6h30 le matin et 17h00 le soir. A plusieurs occasions nous avons écouté les vocalisations des primates depuis des collines surélevées, tôt le matin et dans la soirée. Nous avons aussi marché le long de pistes la nuit entre 20h00 et 23h00 pour recenser les espèces de prosimiens nocturnes en repérant le reflet de leurs yeux à l'aide d'une lampe frontale. Sur le premier site, nous avons passé un total de 26 heures et sur le second un total de 45 heures pour recenser les primates (cela inclut le temps passé en forêt mais pas sur le transect, ex : pendant le retour au camp). En plus des 71 heures totales de recensement, 7 heures (2 heures pour le premier et 5 heures pour le second site) ont été passées en recensement nocturne.

Le faible nombre de signes directs et indirects de singes nous empêche de calculer les estimations de densité pour les différentes espèces. Nous avons cependant déterminé la densité des chimpanzés en appliquant la méthode du comptage des nids perchés (« standing crop nest count » - Plumptre et Reynolds 1996). Cette méthode ne requiert qu'un seul recensement pour chaque zone et permet une estimation de la densité en tenant compte des distances perpendiculaires aux nids le long des transects et du taux de détérioration des nids, de telle façon que le comptage soit corrigé par la production journalière de nids. Nous avons utilisé un taux de détérioration des nids de 221 ± 22 jours (validé pour 21 nids dans la région du Fouta Djallon en Guinée par R. Kormos, communication personnelle) et un taux de production journalier de 1,15 ± 0,047 pour les corrections (les chimpanzés fabriquent plus d'un nid par jour en raison de la nidification journalière et du fait qu'ils réutilisent rarement les vieux nids). Nous avons utilisé le logiciel DISTANCE pour analyser les données conformément au standard d'analyse des transects linéaires, selon lequel la diminution du nombre d'observations d'un objet avec une augmentation de la distance perpendiculaire, est modélisée pour obtenir une estimation de la probabilité d'observations de l'objet (Buckland et al. 1993).

Nous avons aussi questionné les populations locales et les chasseurs au sujet de la présence des espèces de primates. Il a premièrement été demandé aux gens de décrire les primates (couleur, forme, espèce arboricole ou terrestre), d'imiter leurs vocalisations et de donner leurs noms locaux vernaculaires avant de leur montrer des images pour l'identification. Nous avons considéré que tel ou tel primate était présent sur la chaîne du Simandou seulement si la description, le nom local (vérifié sur une liste de toutes les

espèces de primates dans plusieurs langues) et l'image choisie correspondaient.

RESULTATS

Nous pouvons confirmer, dans la Forêt Classée du Pic de Fon sur la chaîne du Simandou, la présence de deux espèces de prosimiens (le potto de Bosman *Perodicticus potto* et le galago de Demidoff *Galagoides demidoff*), cinq espèces de primates anthropoïdes (le mangabey fuligineux *Cercocebus atys atys*, le cercopithèque de Campbell *Cercopithecus campbelli campbelli*, le hocheur blanc-nez *Cercopithecus petaurista buettikoferi* et le cercopithèque hocheur *Cercopithecus nictitans*, et enfin le cercopithèque Diane *Cercopithecus diana diana,*); ainsi qu'une espèce de grand singe hominidé, le chimpanzé d'Afrique Occidentale *Pan troglodytes verus*, (Tableau 9.1). De plus, suite aux enquêtes auprès des villageois locaux, nous supposons la présence quasi certaine de quatre ou cinq autres espèces de primates (le babouin doguera *Papio anubis,* le patas *Erythrocebus*

Tableau 9.1. Espèces de primates de la Forêt Classée du Pic de Fon dans la chaîne du Simandou en Guinée, enregistrées durant les parcours pour les inventaires et par un travail de recensement externe pour chaque site

Espèces	Nom commun	Pic de Fon	Banko Forêt	Confirmation (N)
Perodicticus potto	Potto	-	+	S(1)
Galagoides demidoff	Galago de Demidoff	+	+	S(17), H(37)
Cercocebus atys atys	Cercocèbe Enfumé	+	(+)	S(3), H(2)
Cercopithecus campbelli campbelli	Singe des Palétuviers	+	(+)	H(6)
Cercopithecus petaurista buettikoferi	Hocheur Blanc-nez	+	(+)	S(1), H(2)
Cercopithecus nictitans	Hocheur	-	(+)	S(1)
Cercopithecus diana diana	Cercopithèque Diane	-	+	H(1)
Pan troglodytes verus	Chimpanzé	+	+	S(2), H(28), N(117)
Total		**5**	**8**	**S(25), H(76), N(117)**

+	espèces présentes	S	aperçu
-	espèces absentes	H	entendu
(+)	espèces enregistrées par d'autres membres de l'équipe RAP en dehors du travail de recensement	N	nid

Tableau 9.2. Espèces de primates dont la présence est probable dans la Forêt Classée du Pic de Fon, non enregistrées au cours de l'étude mais identifiées par les villageois et chasseurs locaux.

Espèces	Nom commun	Pic de Fon	Banko forêt
Papio anubis	Babouin Doguera	+	dans le passé
Erythrocebus patas	Singe Rouge, Patas	+	+
Cercopithecus aethiops sabaeus	Singe Vert	+	+
Colobus polykomos polykomos	Colobe Blanc et Noire d'Afrique Occidentale	+	+
(?) *Procolobus verus*	Colobe de Van Beneden	-	+
(?) *Galago senegalensis*	Galago du Senegal	-	+

+	espèces présentes
-	espèces absentes
(?)	en attente de confirmation

patas, le singe vert *Cercopithecus aethiops sabaeus*, le colobe noir et blanc d'Afrique Occidentale *Colobus polykomos polykomos* et peut-être le colobe de Van Beneden *Procolobus verus* Tableau 9.2). Nous n'avons pas été en mesure de recenser ces espèces durant notre évaluation rapide, mais les populations villageoises et les chasseurs nous en ont donné des descriptions précises. Une autre espèce de prosimien, le galago du Sénégal, *Galago senegalensis*, n'a pas été clairement décrite ni recensée mais pourrait être présente dans les habitats de savane. La Forêt Classée du Pic de Fon héberge donc 8 et peut-être jusqu'à 14 espèces différentes de primates, parmi les 15 espèces potentiellement présentes, une représentation importante de la diversité régionale des primates. La seule espèce de primate qui pourrait être présente dans cette zone mais n'a pas été observée et n'est pas connue par les populations locales, est le colobe bai d'Afrique Occidentale *(Procolobus badius)*. De plus, parmi les 14 espèces qui sont (potentiellement) présentes sur le Pic de Fon, quatre sont « Presque Menacées » ou « Menacées d'extinction » *(Cercocebus atys atys, Cercopithecus diana diana, Pan troglodytes verus,* et *Procolobus verus)*.

Malgré un grand nombre d'espèces, l'abondance des différents primates semble faible. Durant 64 heures de recensement diurne en forêt, nous avons rencontré (vu ou entendu) toutes les espèces de singes diurnes de une à six fois au maximum (taux de rencontre par heure ≤ 0,25; Tableaux 9.3, 9.4). Les chimpanzés, au contraire, ont été rencontrés plus souvent (taux de rencontre par heure ≤ 0,63; Tableaux 9.3, 9.4). Les prosimiens nocturnes semblent être présents en plus grande densité que les espèces diurnes (taux de rencontre par heure ≤ 1,43; Tableaux 9.4, 9.5). Nous avons pu rencontrer huit espèces différentes sur le second site de « Banko » et seulement cinq espèces sur le premier site du Pic de Fon, et nous avons confirmé la présence de la plupart des espèces de primates en les entendant vocaliser ou en les observant à plusieurs reprises dans différentes vallées sur le site du Pic de Fon, mais ils n'ont été vus ou entendus qu'une seule fois sur le site de Banko (Tableau 9.3, 9.5). Seuls le galago de Demidoff et les chimpanzés ont été vus ou entendus à plusieurs reprises sur le second site (Tableau 9.3, 9.4, 9.5).

Les conclusions sur l'espèce la plus abondante sur la base du nombre de nos observations doivent être tirées avec précaution. Les espèces telles que le potto de Bosman ou le hocheur blanc-nez sont plus discrètes et/ou secrètes comparées au galago de Demidoff ou au mangabey fuligineux. Cependant, nous avons été en mesure de confirmer la présence du galago de Demidoff, du mangabey fuligineux, du cercopithèque de Campbell et du chimpanzé d'Afrique Occidentale plus de trois fois par signes directs ou indirects et ceci indique probablement que ces espèces sont plus abondantes que les autres.

En plus des vocalisations de chimpanzés, parfois à plusieurs reprises le même jour au second site de « Banko», nous avons pu également faire des observations directes à deux reprises : la première fois un mâle seul est passé près de

nous, et la seconde fois nous avons trouvé un groupe de sept chimpanzés (une femelle en oestrus (réceptive) et d'autres femelles avec des jeunes) se nourrissant dans un *Nauclea diderrichii*. Autour du *Nauclea* nous avons observé, pendant notre étude, d'autres fruits connus pour être consommés par les chimpanzés tels que *Parinari excelsa, Vitex doniana, Detarium microcarpum, Cola cordifolia, Irvingia gabonensis* et *Pseudospongias microcarpa*. Au total, durant six jours d'étude dans la zone de « Banko », nous avons entendu les chimpanzés vocaliser 25 fois, rencontré probablement 9 sous-groupes différents de chimpanzés et entendu tambouriner trois mâles différents (Tableau 9.3).

Sur 17,2 km de transects nous avons observé un total de 117 nids (10 nids supplémentaires n'ont été vus qu'après avoir quitté le transect pour en prendre les mesures). Bien que nous ayons entendu des chimpanzés à des altitudes inférieures à 600 m, les nids ont été trouvés seulement au-dessus de 900 m d'altitude et jusqu'à plus de 1 400 m. Nous avons identifié 33 groupes de nids différents, la taille moyenne d'un groupe étant de 3 nids et le groupe de plus grande taille comportant 12 nids (Tableau 9.6). Nous avons cependant probablement surestimé le nombre de groupes de nids et sous-estimé la taille moyenne d'un groupe car les nids groupés et rapprochés montraient souvent différents degrés de détérioration et étaient de ce fait comptés en tant que groupes séparés, bien qu'ils aient probablement appartenu au même groupe et que seul l'état de dégradation des espèces d'arbres ait différé. La grande majorité des nids (90 %) étaient vieux ou très vieux et nous n'avons observé que neuf nids récents et deux qui étaient frais (probablement des nids de jour). Malgré le nombre relativement faible de nids frais, les chimpanzés occupaient clairement la zone de recensement pendant notre courte période d'étude, révélés par des signes directs et des vocalisations quotidiennes sur le second site. En moyenne, les chimpanzés construisaient leurs nids à 18 m de haut sur des arbres de taille moyenne avec un diamètre de 37 cm. Généralement ils préféraient les pentes comme sites de nidification.

En utilisant le logiciel DISTANCE et en choisissant le modèle le plus approprié (« courbe exponentielle négative»), la densité des chimpanzés pour les zones forestières du Pic de Fon (10 000 ha sur un total de 27 000 ha, estimé par images satellites) est estimée à 0,64 chimpanzés/km2, avec une population totale moyenne de 64 chimpanzés (18-226, intervalle de confiance à 95 %). Comme seuls les chimpanzés adultes et les jeunes sevrés construisent des nids, nous avons dû effectuer une correction pour la part de la population qui ne nidifiait pas (17,5 %, Ghiglieri 1984, Plumptre et Reynolds 1996), ce qui donne une population totale moyenne de 75 chimpanzés (21-246, intervalle de confiance à 95 %). La taille des communautés de chimpanzés varie de 10 à plus de 100 individus (Goodall 1986, Nishida et al. 1990, Boesch et Boesch-Achermann 2000, Herbinger et al. 2001). Il est donc probable que la Forêt Classée du Pic de Fon héberge entre une et peut-être trois ou quatre communautés différentes de chimpanzés.

Tableau 9.3. Recensements diurnes : nombre d'observations (S), vocalisations (H), ou nids (N) par km de transect dans la Forêt Classée du Pic de Fon. Tous les transects ont été tracés dans l'habitat de forêt galerie.

Piste	Longueur de transect (m)	*Cercocebus atys atys*	*Cercopithecus campbelli campbelli*	*Cercopithecus petaurista buettikoferi*	*Cercopithecus nicitans*	*Cercopithecus diana diana*	*Pan troglodytes verus*
Pic de Fon							
I a (vallée SW du camp)	2100	2 H (1 S, 3 ind.)	(I H)	-	-	-	-
I b (4 vallées S du camp)	1500	-	Groupe non identifié*		-	-	-
I b (vallée contigu S de Ib)	point d'écoute	1 H	2 H	1 H	-	-	-
I c (Pic Dalbatini vallée NW)	550	-	2 H	1 H	-	-	25 N
Banko							
II a (vallée NE du camp)	4000	-	-	-	-	1 H	5 H (1 groupe)
II b (vallée NW de le camp, W de la riviére)	2650	-	-	-	-	-	2 H (1 groupe), 35 N
II c (vallée W du camp)	2500	-	-	-	-	-	9 H, (3 groupes), 10 N
II d (vallée NW du camp, East du riviére)	1400	-	-	-	-	-	1 H, 37 N
II e (2 vallées NW du camp)	2500	-	-	-	-	-	4 H & 1 S (1 mâle), 1 S (7 ind., femelles et jeunes), 10 N
Total (durant transect)	17200	3 H (1 S)	5-6 H	2-3 H	-	1 H	21 H, 2 S, 117 N
Total (hors transect)**		2 S	1 H	-	1 S	-	4 H (1 groupe), 1 N
Total	**17200**	**3 H, 3 S**	**6-7 H**	**2-3 H**	**1 S**	**1 H**	**25 H, 2 S, 118 N**

Ind. Individus

* Le guide a signalé bruyamment l'observation des singes et ceux-ci se sont enfuis (Probablement *campbelli* et *petaurista*)

(S), (H) Observations et vocalisations rapportées par d'autres membres de l'équipe du RAP sur le transect (ils ont observé le même groupe de *Cercocebus* que nous avons entendu).

** Observations et vocalisations additionnelles rapportées par d'autres membres du RAP

Un groupe de *Cercocebus* (4 ind.) observé à côté du camp du Pic du Fon (N0)

Un groupe d'espèces mélangées de *Cercocebus* (10 ind.), *Cercopithecus petaurista* et *nictitans* a été aperçu au NE du camp de Banko (zone IIb)

Un *Cercopithecus campbelli campbelli* mâle a été entendu à côté du camp de Banko (NO)

Un groupe de chimpanzés *Pan troglodytes verus* a été entendu près du camp de Banko (NO)

Un nid a été aperçu (vallée NE du camp, zone IIb)

Tableau 9.4. Nombre d'observations directes et taux de rencontre par heure pour toutes les espèces de primates examinées

	Aire totale	Pic de Fon	Banko	Aire totale
	été entendus/vus	taux de rencontre par heure		
Perodicticus potto	1	0	0,2	0,14
Galagoides demidoff	10	0,5	1,8	1.43
Cercocebus atys atys	3	0,13	0	0,05
Cercopithecus campbelli campbelli	5-6	0,21-0,25	0	0,08-0,09
Cercopithecus petaurista buettikoferi	2-3	0,08-0,13	0	0,03-0,05
Cercopithecus diana diana	1	0	0,03	0,02
Pan troglodytes verus	8 groupes	0	0,2	0,13
Pan troglodytes verus	25 individus	0	0,63	0,39

Les calculs sont basés sur un total de 64 heures d'observations pendant la journée et 7 heures d'observations pendant la nuit.

Cercopithecus nictitans a été aperçu une fois par l'équipe RAP de grands mammifères

DISCUSSION

Les primates sont éthologiquement des animaux complexes. Plusieurs vivent en groupes sociaux structurés dans lesquels ils reconnaissent leurs parents sur plusieurs générations, forment des liens durables avec les autres membres du groupe, et dans de nombreux cas requièrent des échanges entre groupes avant de se reproduire. Les changements environnementaux, tels que la destruction de leurs habitats ou la pression de chasse, diminuent le potentiel reproductif des primates, facteur qui allié à leur longue durée et la lenteur de leur reproduction, peut rapidement mener à leur extinction locale. Les populations de petite taille et qui sont génétiquement et socialement isolées sont incapables de résister à des problèmes de maladies, de consanguinité et de pression humaine et leur survie est fortement menacée. La plupart des primates ayant tendance à vivre à l'intérieur ou à proximité de leur zone de naissance, les possibilités de migrer pour échapper aux effets des bouleversements de l'habitat sont rares.

Dans la Forêt Classée du Pic de Fon, la population de primates est régulièrement menacée par la chasse, la destruction de son habitat due aux activités agricoles des populations locales, et probablement aussi la pollution sonore et l'accès croissant par les pistes minières à des forêts galeries auparavant peu perturbées, qui résultent des activités d'exploration minières. Actuellement, les effets des activités agricoles sont surtout visibles dans les basses altitudes, alors que les activités minières affectent les hautes altitudes.

La pression de chasse très élevée dans la zone affecte certainement les populations de primates, même si la grande majorité de la population est musulmane et ne consomme pas les primates. Les pratiques de chasse (pièges à collet) ne sont pas sélectives et les primates qui se déplacent au sol (spécialement les chimpanzés et les cercocèbes, mais aussi d'autres espèces de *Cercopithecinae)* en sont probablement souvent les victimes.

Les chimpanzés adultes sont connus pour être capables de se libérer seuls d'un piège, mais y perdent aussi souvent un membre ou meurent des infections bactériennes consécutives (Goodall 1986, Boesch et Boesch-Achermann 2000). Les nourrissons ou les chimpanzés juvéniles ne sont pas capables de se libérer seuls des pièges et sont connus pour refuser l'aide de leur mère et meurent le plus souvent

Tableau 9.5. Observations nocturnes des primates dans la Forêt Classée du Pic de Fon, comprenant les observations de tous les membres d'équipe RAP.

Espèces	Pic de Fon	Banko forêt	Habitat
Perodicticus potto	-	1 S	forêt galerie
Galagoides demidoff	2 H*	17 S, 35 H**	forêt galerie et secondaire
S	a été aperçu		
H	a été entendu		

*1 H par un autre membre d'équipe RAP

**11 S, 32 H par différents membres d'équipe RAP

Tableau 9.6. Paramètres de nid de chimpanzé pour les différents transects

Piste	Altitude (m)	# de Nids	Groupes de nids	Taille Moyenne de Nid Groupe	Max. Taille de Nid Groupe	Classes d'âge de nid				Hauteur moyenne (m)	Moyenne (cm)
						Frais	Récent	Vieux	Très vieux		
PIC DE FON I c (550 m)	1 100-1 417	25	9	3	6	-	2 (8%)	5 (20%)	18 (72%)	12	26
BANKO II b (2 650 m)	900-1 077	35	9	4	8	1 (3%)	-	17 (48,5%)	17 (48,5%)	20	41
II c (2 500 m)	900-1 050	10	3	3	4	-	-	5 (50%)	5 (50%)	17	36
II d (1 400 m)	900-1 100	37	9	4	12	-	-	20 (54%)	17 (46%)	20	43
II e (2 500 m)	900-1 100	10	3	3	7	1 (10%)	7 (70%)	2 (20%)	-	20	38
TOTAL	900-1 417	117*	33	3	12	2 (1,7%)	9 (7,7%)	49 (41,9%)	57 (48,7%)	17.8	36,8

* 10 nids ont été aperçus quand nous avons quitté le transect; ils n'ont pas été inclus dans l'évaluation de la densité.

des infections consécutives (Boesch et Boesch-Achermann 2000). De plus une minorité de gens dont les croyances religieuses n'interdisent pas la consommation de primates chasse toutes les espèces (communication personnelle de la population locale).

Nous avons observé une pression de chasse beaucoup plus importante dans les basses altitudes de la Forêt Classée du Pic de Fon. Alors que les équipes primates et mammifères du RAP ont trouvé deux collets, plusieurs campements et pistes de chasseurs, et un total de 11 cartouches autour des sites les plus élevés du Pic de Fon, les deux groupes ont collecté un total de 35 collets et 32 cartouches et observé plusieurs sites de feux et de campement ainsi que des pistes sur toute l'étendue du site de basse altitude proche de Banko. Nous avons aussi entendu plusieurs coups de fusil durant le jour et la nuit, bien qu'il ait été demandé aux villageois d'éviter toute activité de chasse pendant notre séjour. Le faible taux d'observation de toutes les espèces de primates anthropoïdes sur le second site de Banko pourrait être le reflet de la pression de chasse plus prononcée dans cette zone. Soit les singes restent plus discrets et secrets, soit ils sont effectivement présents en plus faible densité aux basses altitudes.

Le fait que nous n'ayons pas trouvé de nids en dessous de 900 m d'altitude indique que le coeur du domaine des chimpanzés, c'est-à-dire l'endroit où ils passent la majeure partie de leur temps, se trouve probablement en altitude. En raison de la forte pression de chasse, ils ne doivent pas se considérer en sécurité dans les basses altitudes (les chimpanzés ont été observés en basses et hautes altitudes depuis environ 150 m jusqu'à plus de 1500 m). En conséquence, les activités minières concentrées dans les zones de plus haute altitude pourraient avoir un impact significatif sur la population de chimpanzés, particulièrement si la chasse et la pression agraire au niveau des zones les moins élevées se poursuivent sans relâche et que les chimpanzés sont contraints à se déplacer sans cesse vers les zones supérieures pour échapper à de telles pressions.

La présence des chimpanzés a été confirmée seulement dans les vallées sans accès immédiat par la piste et de ce fait moins altérées, ce qui semble montrer que la détérioration de l'habitat et les dérangements sonores dus aux activités actuelles d'exploration minière ont déjà eu un impact sur la population de chimpanzés.

Malgré des menaces variées, la Forêt Classée du Pic de Fon, dans la chaîne des Monts Simandou héberge toujours une représentation importante de la biodiversité de l'ordre des primates pour la Région de Haute Guinée. Toutefois, bien que le nombre d'espèces soit relativement important (8 et probablement 14), leur faible abondance (taux de rencontre < 0,25 par heure pour toutes les espèces diurnes) indique que leur survie est déjà fortement menacée. Des taux de rencontre similaires ou légèrement supérieurs, issus d'une évaluation rapide dans le Parc National de la Marahoué en Côte d'Ivoire, sont connus pour deux espèces, *Cercopithecus campbelli* et *C. petaurista* (0,05-0,42 par heure;

Struhsaker and Bakarr 1999). Dans cette étude, comme dans notre recensement, la pression de la chasse était corrélée negativement avec les taux de rencontre. L'abondance des espèces de singes dans d'autres sites protégés ouest africains comme par exemple dans le Parc National de Taï en Côte d'ivoire, semble plus élevée (il est possible de voir ou d'entendre une espèce donnée plusieurs fois par jour, en raison d'une densité de 2 à plus de 100 individus au km², selon les espèces (Galat-Luong et Galat, rapport non publié)). Si une protection plus efficace est mise en oeuvre, les densités de population des diverses espèces devraient normalement augmenter au Pic de Fon.

D'après d'autres études, principalement menées dans des forêts pluviales de plaine (Marchesi et al. 1995 ; Plumptre et Reynolds 1996), des déterminations de densité de 1- 2 chimpanzés/km² sont connues pour les forêts primaires intactes, alors qu'elles sont estimées plus basses pour les forêts dégradées (0,4 chimpanzés/km²) ou les forêts anthropisées et les habitats fragmentés (0,09 chimpanzés/km²). Avec une densité qui se situe entre celle de la forêt primaire et celle de la forêt dégradée, la population de chimpanzés de la Forêt Classée du Pic de Fon est très certainement menacée et déclinante mais comporte encore un nombre important d'individus. Une densité moyenne similaire de population (0,45 chimpanzés/km², variant de 3,0 ind./km² dans les forêts galeries à 0,01 ind./km² en savane) a aussi été trouvée pour les chimpanzés de la forêt de la Mafou dans le Parc National du Haut Niger, plus au nord en Guinée dans un habitat de savane plus sèche (Fleury-Brugiere et Brugiere 2002). A partir d'une étude nationale, Ham (1998) a estimé la densité de population moyenne à 0,16-0,34 chimpanzés/km² pour les habitats potentiels de chimpanzés en Guinée, ce qui indique que la population de chimpanzés de la Forêt Classée du Pic de Fon, dont la densité est au-dessus de la moyenne, est une population importante pour le pays. Les chimpanzés sont classés comme «En danger» par l'UICN et cités en Annexe I de la CITES (Espèces les plus gravement menacées d'extinction). Parmi les trois sous-espèces, le chimpanzé d'Afrique occidentale est celle la plus menacée par la destruction de son habitat, la pression de la chasse et le commerce de la viande de brousse et celui des animaux de compagnie ainsi que par la transmission des maladies. On présume qu'il reste actuellement 25 000 à 58 000 chimpanzés dans toute l'Afrique de l'Ouest, dont la majorité vit dans des zones non protégées (Kormos et Boesch 2003). La protection des populations restantes et la création de nouvelles zones protégées sont des mesures cruciales. De plus, même dans les régions où les populations de chimpanzés sont les plus denses, comme par exemple à l'ouest de l'Afrique équatoriale, la récente propagation de la fièvre hémorragique Ebola est avec la chasse une menace importante pour les anthropoïdes et a mené leur population locale proche de l'extinction (ex: dans la forêt de Minkébé, dans le nord du Gabon, où la densité de grands singes a chuté de 99% durant les dix dernières années, Walsh et al. 2003). Pour cette raison, la conservation des plus petites populations de chimpanzés devient de plus en plus importante lorsque les grandes populations s'avèrent tout aussi menacées d'extinction.

Il a été montré que différentes populations de chimpanzés étaient culturellement distinctes. Elles ont des traits de caractères comportementaux uniques transmis d'une génération à l'autre, une caractéristique que ces animaux ne partagent qu'avec l'homme du point de vue de sa complexité (Whiten et al. 1999, Whiten et Boesch 2001). Protéger les populations de chimpanzés de la chaîne du Simandou nous permettra peut-être de sauvegarder des caractéristiques comportementales uniques qui n'ont pas encore été découvertes ou qui sont seulement connues des chimpanzés de cette région. Durant des observations à long terme d'une communauté de chimpanzés à Bossou, dans la région du Mont Nimba, à environ 100 km au sud et dans un habitat similaire à celui de la région du Simandou, certains comportements uniques, inconnus dans d'autres régions, ont été observés, (par exemple, broyage au pilon (broyage de la couronne du palmier avec le pétiole des feuilles de palme), broyage des insectes (baguette utilisée pour écraser les insectes), extraction de la résine par pilonnage, crochetage de branches (une branche est utilisée pour en crocheter une autre), creusage (bâton utilisé comme une bêche pour creuser les termitières), pêche aux termites à l'aide d'une nervure de feuille, récolte d'algues à l'aide d'une baguette) (Whiten et al. 1999).

RECOMMANDATIONS POUR LA CONSERVATION

Afin de garantir la survie des primates de la Forêt Classée du Pic de Fon, nous faisons les recommandations suivantes :

- La chasse devrait être totalement bannie pour toutes les espèces « Presque Menacées » ou « Menacées d'Extinction » et sévèrement restreinte pour les espèces à faible risque, non seulement sur le plan légal mais aussi en pratique.

- Les activités agricoles au sein des limites de la Forêt Classée doivent être identifiées et interrompues et des programmes permettant aux populations humaines locales de maintenir leur niveau de vie sans dégrader la forêt devraient être développés.

- Un nombre suffisamment élevé de gardes est nécessaire et urgent, pour faire respecter de manière efficace la législation actuelle protégeant la Forêt Classée, avec un équipement leur permettant de surveiller la totalité de la zone.

- L'éducation environnementale est essentielle pour modifier les pratiques de chasse ainsi que les comportements, de sorte que les primates ne soient plus perçus comme des espèces abondantes, source de viande ou nuisibles pour les cultures. Grâce à la similarité reconnue entre les chimpanzés et les hommes, les chimpanzés pourraient jouer un rôle important en tant qu'espèces phares dans les campagnes d'éducation.

- Il est essentiel que des études hydrologiques soient menées avant toute activité qui pourrait causer des modifications du régime et de la qualité des eaux ou des bouleversements climatiques régionaux, y compris l'abaissement de la crête de la chaîne du Simandou de plusieurs centaines de mètres. Des changements climatiques et la perturbation des ressources en eaux souterraines de la chaîne montagneuse risquent d'entraîner une réduction du couvert forestier. Les primates étant très dépendants des habitats forestiers, ceci aurait un impact majeur sur l'ensemble de la population.

- Les effets du bruit et des perturbations sonores sur les populations de primates requièrent des analyses supplémentaires, tout comme les conséquences potentielles sur les populations de primates de l'abaissement du niveau de la crête de la chaîne du Simandou.

- Une protection immédiate et totale est nécessaire, en particulier pour les quatre espèces de primates « Presque Menacées » ou « Menacées d'Extinction » du Pic de Fon : le mangabey fuligineux, le cercopithèque Diane, le colobe de Van Beneden et le chimpanzé d'Afrique Occidentale, afin de garantir leur survie, compte tenu des nombreuses menaces pesant sur ces espèces.

- Si aucune mesure immédiate de conservation n'est soutenue et mise en œuvre, nous craignons que la plupart des espèces de primates disparaissent de cette Forêt Classée dans un futur proche.

Il est fortement recommandé que des études supplémentaires soient menées pendant des saisons différentes et couvrant un laps de temps plus long, pour confirmer la distribution, les schémas de répartition, l'abondance et les statuts de la population de primates le long de la chaîne de montagne du Simandou, afin d'assurer une protection adéquate aux primates de la Forêt Classée du Pic de Fon.

RÉFÉRENCES

Boesch, C. and H. Boesch-Achermann. 2000. The Chimpanzees of the Taï Forest: Behavioural Ecology and Evolution. Oxford University Press, Oxford.

Bshary, R. 1995. Rote Stummelaffen, Colobus Badius und Diana Meerkatzen, Cercopithecus Diana, im Taï-Nationalpark, Elfenbeinküste: wozu assoziieren sie? Dissertation, Universität München.

Buckland, S.T., D.R. Anderson, K.P. Burnham and J.L. Laake. 1993. Distance Sampling: Estimating Abundance of Biological Populations. Chapman & Hall, London.

Chapman, C.A. 1995. Primate seed dispersal: Coevolution and conservation implications. Evol. Anthrop. 4 (3): 74-82.

Chapman, C.A. and D.A. Onderdonk. 1998. Forests without primates: Primate/plant dependancy. Am. J. Primatol 45 (1): 127-141.

Chatelain, C., B. Kadjo, I. Kone and J. Refisch. 2001. Relations Faune-Flore dans le Parc National de Taï: une étude bibliographique. Tropenbos-Côte d'Ivoire Série 3.

Fleury-Brugiere, M.-C. and D. Brugiere. 2002. Estimation de la population et analyse du comportement nidificateur des chimpanzés dans la zone intégralement protégée Mafou du Parc national du Haut-Niger. Report to the Parc National du Haut-Niger/AGIR project, Faranah.

Galat-Luong, A. and G. Galat. 1978. Abondance relative et associations plurispécifiques des primates diurnes du parc national de Taï (Côte d'Ivoire). ORSTOM.

Galat-Luong, A. and G. Galat. unpublished report. Les Primates des Monts Nimba. Operation Pertubations et grande faune sauvage, Institut de Recherche pour le Developpement (IRD), Senegal, 1999.

Ghiglieri, M.P. 1984. The Chimpanzees of Kibale Forest: A Field Study of Ecology and Social Structure. Columbia University Press, New York.

Goodall, J. 1986. The Chimpanzees of Gombe. Belknap Press, Harvard University, Cambridge, MA.

Ham, R. 1998. Nationwide chimpanzee survey and large mammal survey, Republic of Guinea. Unpublished report for the European Communion, Guinea-Conakry.

Herbinger, I., C. Boesch and H. Rothe. 2001. Territory characteristics among three neighboring chimpanzee communities in the Taï National Park, Côte d'Ivoire. Int. J. Primatol. 22: 143-167.

Kormos, R. and C. Boesch. 2003. Regional Action Plan for the Conservation of Chimpanzees in West Africa. IUCN/SSC Action Plan. Washington, DC: Conservation International.

Lambert, J.E. and P.A. Garber. 1998. Evolutionary and ecological implications of primate seed dispersal. Am. J. Primatol. 45 (1): 9-28.

Lee, P.C., J. Thornback and E.L. Bennett. 1988. Threatened Primates of Africa, The IUCN Red Data Book. IUCN Gland, Switzerland and Cambridge, U.K.

Marchesi, P., M. Marchesi, B. Fruth and C. Boesch. 1995. Research Report: Census and distribution of chimpanzees in Côte d'Ivoire. Primates 36: 591-607.

Martin, C. 1989. Die Regenwälder Westafrikas: Ökologie – Bedrohung - Schutz. Birkhäuser Verlag, Basel.

Mittermeier, R.A., N. Myers, C.G. Mittermeier and P.R. Gil. 1999. Hotspots: Earth's Biologically Richest and Most Endangered Terrestrial Ecoregions. CEMEX.

Nishida, T., H. Takasaki and Y. Takahata. 1990. Demography and reproductive profiles. *In*: Nishida, T. (ed.). The Chimpanzees of the Mahale Mountains. Tokyo: Tokyo Univ. Press, Pp. 63-97.

Plumptre, A. J. and V. Reynolds. 1996. Censusing Chimpanzees in the Budongo Forest, Uganda. Int. J. Primatol. 17: 85-99.

Struhsaker, T.T. and M.I. Bakarr. 1999. A Rapid Survey of Primates and Other Large Mammals in Parc National de la Marahoue, Cote d'Ivoire. RAP Working Papers 10.

Walsh, P.D., K.A. Abernethy, M. Bermejo, R. Beyers, P. De Wachter, M.E. Akou, B. Huijbregts, D.I. Mambounga, A.K. Toham, A.M. Kilbourn, S.A. Lahm, S. Latour, F. Maisels, C. Mbina, Y. Mihindou, S.N. Obiang, E.N. Effa, M.P. Starkey, P. Telfer, M. Thibault, C.E.G. Tutin, L.J.T. White and D.S. Wilkie. 2003. Catastrophic ape decline in western equatorial Africa. Nature advance online publication, 6 April 2003 (doi:10.1038/nature01566).

Whiten, A., J. Goodall, W.C. McGrew, T. Nishida, V. Reynolds, Y. Sugiyama, C.E.G. Tutin, R.W. Wrangham and C. Boesch. 1999. Cultures in chimpanzees. Nature 399: 682-685.

Whiten, A. and C. Boesch. 2001. The cultures of chimpanzees. Sci. Am. 284 (1): 60-67.

Whitesides, G.H., J.F. Oates, S.M. Green and R.P. Kluberdanz. 1988. Estimating Primate Densities from Transects in a West African Rain Forest: A Comparison of Techniques. J. Anim. Ecol. 57 (2): 345-367.

Zuberbühler, K. and D. Jenny. 2002. Leopard predation and primate evolution. J. Hum. Evol. 43: 873-886.

Chapitre 10

Évaluation des menaces et opportunités socio-économiques de la Forêt Classée du Pic de Fon

Eduard Niesten et Léonie Bonnehin

RÉSUMÉ

Ce chapitre présente une analyse des facteurs socio-économiques contribuant aux menaces pesant sur la biodiversité dans la Forêt Classée guinéenne du Pic de Fon. La première section décrit les méthodes employées dans cette analyse, à savoir une étude de la bibliographie disponible et un atelier d'échange d'informations organisé par CI à Conakry. Le chapitre rapporte ensuite les informations générées par l'étude de la littérature et l'atelier. Suivent alors une brève discussion de l'analyse et des conclusions basées sur ces constatations, et le chapitre se termine par des recommandations générales qui forment le cadre du Plan d'Action Initial pour la Biodiversité.

INTRODUCTION ET MÉTHODOLOGIE

Une étude approfondie de la littérature grise et publiée n'a fourni que très peu d'informations socio-économiques spécifiques à la Forêt Classée du Pic de Fon. Le manque de données historiques et d'études chronologiques était encore plus décourageant, rendant presque impossible une analyse des tendances socio-économiques. D'importants travaux de terrain supplémentaires sont nécessaires pour développer une bonne compréhension quantitative des menaces et opportunités dans la Forêt Classée du Pic de Fon, d'une envergure dépassant le présent effort en terme de budget et de planification. Cela est particulièrement vrai dans le cas des données socio-économiques, qui seront capitales pour déterminer la stratégie de conservation appropriée.

Compte tenu du manque de documentation sur les tendances socio-économiques particulières à cette région, Conservation International a organisé un atelier à Conakry les 12 et 13 décembre 2002 afin d'obtenir des informations sur les menaces et les opportunités pour la conservation de la biodiversité dans la région du Pic de Fon, de la part d'un large éventail de parties prenantes et d'experts. Cet atelier avait pour but de compléter les recherches bibliographiques par des informations de la part des parties directement engagées dans la conservation, dans la région immédiate du projet, en Guinée et dans les régions frontalières. Cet exercice a aussi permis de solliciter une documentation supplémentaire qui n'avait pas été obtenue lors des recherches bibliographiques initiales. L'objectif global de cet atelier était de faciliter l'échange d'information.

L'atelier a été animé par CI (Léonie Bonnehin) en utilisant l'approche ZOPP (Zielorientierte Projektplanung, ou OOPP- Objective Oriented Project Planning – en anglais). Le ZOOP a une structure systématique appliquée lors d'un atelier pour l'identification des parties prenantes et l'analyse de la problématique. L'analyse se déroule généralement en quatre étapes :

1. Parties prenantes: une revue des personnes, des groupes et des organisations liés à un projet et leurs intérêts, attitudes et implications pour la planification du projet.

2. Menaces : énumération des principales menaces, y compris les causes, les effets et l'ordre de priorité.

3. Objectifs : une reformulation des problèmes sous forme d'objectifs réalistes et réalisables.

4. Opportunités : identification des opportunités et évaluation de la faisabilité des alternatives.

Cette méthodologie est caractérisée par l'absence de programmes d'actions ou de conclusions formulés à l'avance. C'est un processus participatif, avec la contribution de tous les participants, qui définit les résultats de l'Atelier de Conakry, qui devrait donc avoir un soutien très large.

Les participants ont été sélectionnés pour représenter un large éventail de perspectives et d'expertises vis à vis des efforts de conservation en Guinée, pour les Grands Plateaux du Mont Nimba, et la Forêt Classée du Pic de Fon. Au total, 33 participants provenant de 17 organisations dont le gouvernement national, les bailleurs multilatéraux, les ONG et les centres de recherche scientifique, et la société RTM&E (le Tableau 10.1 présente la liste des participants à l'atelier). La contribution et les perspectives de représentants des communautés installées autour de la Forêt Classée du Pic de Fon auraient apporté un atout supplémentaire à l'atelier. Le PGRR est l'institution qui jusqu'à présent a le mieux réussi à établir des liens avec ces communautés, mais cet engagement n'en est qu'au stade initial et les bases nécessaires à ce type d'échange n'ont pas été encore édifiées.

RÉSULTATS

Etude bibliographique
Menaces sur la biodiversité dans le Bloc Forestier de Haute Guinée

La région du Bloc Forestier de Haute Guinée est caractérisée par une extrême pauvreté, un accroissement rapide de la densité de population humaine, et un manque de financement pour la gestion gouvernementale de l'environnement. Les habitats naturels restants continuent à subir des dégradations liées à l'exploitation forestière et à la propagation de l'agriculture et de l'agroforesterie, tandis que la faune est menacée par une pression intense de la chasse pour la viande de brousse. Parallèlement, les troubles civils limitent le développement des capacités humaines et affaiblissent l'application des lois sur l'environnement, dans un contexte qui présente déjà un faible niveau de capacités institutionnelles. De plus, le flot de réfugiés intensifie la pression sur les ressources forestières, avec plus de 600 000 réfugiés localisés dans la seule Guinée (CI 1999). Par conséquent, la conservation dans l'ensemble du *hotspot* est confrontée à de sérieux problèmes, notamment la dépendance sur des pratiques de cultures sur brûlis, la forte pression des activités minières commerciales, l'importance de la chasse profondément ancrée pour la viande de brousse et la persistance de conflits civils (CI 1999).

Menaces pesant sur la biodiversité en Guinée
Le Plan d'Action National pour la Biodiversité en Guinée

Tableau 10.1. Participants à l'Atelier de Conakry

Organisations	Nb de participants
Ministère des Mines, de la Géologie et de l'Environnement, Guinée	2
Direction Nationale des Eaux et Forêts, Guinée	4
Fonds pour l'Environnement Mondial (GEF) Guinée	1
USAID (Agence Américaine pour le Développement International)	2
PGRR/Centre Forestier de N'Zérékoré	2
CEGEN (Centre pour l'Etude et la Gestion de l'Environnement du Nimba)	4
PNUD (Programme des Nations Unis pour le Développement) Guinée	1
UNESCO MAB (Programme Homme et Biosphère)	1
Rio Tinto Mining and Exploration Limited	1
AGIR (Programme Européen de Gestion Intégrée des Ressources)	1
Conservation International	3
Université de Würzburg (Allemagne)	1
Guinée Ecologie	3
PEGRN/USAID (Programme étendu de Gestion des Ressources Naturelles)	1
CERE (Centre d'Etude et de Recherche en Environnement, Université de Conakry)	1
Winrock International	1

souligne les critères suivants d'appauvrissement de la diversité biologique dans le pays : l'augmentation de la population humaine, la focalisation forcée sur les besoins de développement économique à court terme, la dépendance sur des technologies inappropriées, des moyens limités pour mettre en avant les richesses de la biodiversité, le manque de gestion gouvernementale des ressources communautaires, la migration humaine, l'instabilité politique et la guerre civile (MMGE 2002).

L'aspect le plus décourageant pour la planification de la conservation est le fait que des populations parmi les plus pauvres au monde vivent en Guinée, et particulièrement en Guinée Forestière. De plus, leurs moyens de subsistance et d'existence sont presque uniquement basés sur les ressources forestières. Au moins 90 % de la totalité de la consommation énergétique du pays se fait sous forme de bois et charbon (MMGE 2002). L'agriculture, qui compte pour environ 85 % des emplois dans la région, dépend de la conversion des forêts en zones agraires (MMGE 2002). On a estimé en 2002 que près de 140 000 ha de forêt sont détruits chaque année en Guinée (MMGE 2002). Juste deux années auparavant, Konomou et Zoumanigui (2000) estimaient la déforestation annuelle à 120 000 ha. Les feux de brousse endommagent de vastes étendues du pays chaque année. Au moins 17 des 190 espèces mammifères de Guinée sont « Menacé » et « Proche de l'extinction » et au moins 16 des 625 espèces d'oiseaux sont « En danger ». Au moins 36 espèces de plantes sont « En danger ». La chasse pour la viande de brousse, l'urbanisation, l'afflux de réfugiés et le taux d'analphabétisme approchant 70% constituent des obstacles aux efforts de conservation.

Les quatre piliers de la stratégie nationale pour la conservation de la biodiversité face à ces menaces sont : la création d'un système représentatif d'aires protégées, l'inclusion des communautés locales par des méchanismes de gestion participative, le développement des capacités humaines pour jouer des rôles variés dans la conservation et le renforcement des efforts de conservation aux niveaux local, régional et international (MMGE 2002).

La Forêt Classée du Pic de Fon

La Forêt Classée du Pic de Fon est entourée de 24 villages dont les habitants ont un impact direct sur la forêt en convertissant les zones boisées en terres cultivables, en provoquant des feux de brousse pour créer des pâtures et en pratiquant une exploitation minière artisanale (PGRR 2002b). Camara et Guilavogui (2001) indiquent que les villageois sont bien informés des limites de la Forêt Classée du Pic de Fon mais que cela ne les décourage pas. Les feux de brousse non contrôlés (pour la chasse, la préparation des terres pour les cultures ou les pâtures) et l'activité minière artisanale ont été identifiés comme étant les principales menaces à ce qui reste du bloc forestier (PGRR 2001, Camara et Guilavogui 2001). En conséquence, la chaîne du Simandou et en particulier la Forêt Classée du Pic de Fon, font face à une multitude de menaces d'origine anthropique. Les modifications démographiques sont probablement

les plus fondamentales parmi les nombreuses variables et tendances contribuant aux dynamiques complexes qui soustendent ces pressions.

Profil de la population

La Forêt Classée du Pic de Fon recouvre les préfectures de Beyla et Macenta. La densité de population dans ces préfectures est toujours relativement basse, avec 9 à 16 habitants recensés par km² (Hatch 1998). Pourtant, la densité de population est probablement plus élevée sur le piedmont austral de la Forêt Classée du Pic de Fon (Konomou et Zoumanigui 2000), dans la zone que le PGRR a identifié comme étant du plus haut intérêt pour les efforts de conservation et de gestion des ressources naturelles (PGRR 2001). Hatch (1998) a signalé un taux d'accroissement de la population atteignant les 2,9 % par an pour les préfectures de Beyla et Macenta, ce qui indique que ces densités de population s'accroissent rapidement, doublant en moins de 25 ans. Il faut cependant vérifier si les tendances démographiques des populations vivant à la périphérie immédiate de la Forêt Classée du Pic de Fon sont comparables à celles des préfectures plus importantes.

Plusieurs groupes ethniques sont représentés dans les trois préfectures, incluant des Konianké, Peuls, Kouronko, Toronka, Kissi, Malinké, Guerzé, Toma et Manian (Hatch 1998, Konomou et Zoumanigui 2000). Les populations indigènes des 24 villages immédiatement riverains de la Forêt Classée du Pic de Fon sont principalement constituées des groupes ethniques Toma, Manian, Konianké et Guerzé. Les Peuls et les Malinkés sont aussi impliqués dans la région dans les secteurs du commerce, du pastoralisme et de la prospection minière artisanale (Camara and Guilavogui 2001).

Konomou et Zoumanigui (2000) signalent une population de réfugiés de 629 275 personnes, soit environ 40 % de la population de Guinée Forestière (Konomou et Zoumanigui 2000). Bien sûr, l'afflux spectaculaire des réfugiés en provenance du Liberia et de la Sierra Leone explique sans doute le taux d'accroissement de la population observé dans les préfectures de Kérouané, Beyla, et Macenta

En résumé, les problèmes imminents liés à la population sont l'accroissement de la pression sur les ressources en raison de la croissance de la population et du flux continu de réfugiés et l'érosion des institutions communautaires de gestion des ressources en raison de la diversité ethnique croissante due aux migrations.

Activité économique

Les principales activités économiques dans la région de la Forêt Classée du Pic de Fon sont l'agriculture et le pastoralisme, ainsi que l'extraction minière artisanale à petite échelle de l'or, du fer et des diamants (Camara et Guilavogui 2001). Plusieurs sources, dont les participants de l'atelier de Conakry, mentionnent l'exploitation minière artisanale comme une menace pesant sur la biodiversité de la Forêt Classée du Pic de Fon. Pourtant, l'équipe du RAP n'a relevé qu'une seule observation d'activité minière artisanale (voir

Annexe 3, description pour le codage du site FONBA3), et l'étendue de ces activités reste donc à vérifier. La chasse au gibier est aussi une activité importante.

L'agriculture en Guinée Forestière s'articule autour de la culture du riz. Sur l'ensemble de la région, approximativement 120 000 ha sont semés chaque année pour le riz pluvial ; le riz occupe plus des trois quarts de la superficie totale cultivée dans la région (Konomou et Zoumanigui 2000). Les agriculteurs ont principalement recours aux méthodes de culture itinérante, dont les brûlis en guise de préparation aux semis. Dans les endroits relativement fertiles telles que les vallées et les dépressions aux alentours de la Forêt Classée du Pic de Fon, la propagation de la culture du riz est l'une des principales causes de la déforestation (Konomou et Zoumanigui 2000).

L'agriculture compte pour environ 90 % des emplois dans la région autour de la Forêt Classée du Pic de Fon. Le riz, le principal aliment de base pour la consommation locale, est la principale culture en combinaison avec un large éventail d'autres cultures (fonio, manioc, pommes de terre, igname, maïs, arachide, taro, etc.). Les zones forestières fournissent aussi des productions agro-forestières de rente telles que le café, l'huile de palme, le cacao, la banane et la noix de cola (Camara et Guilavogui 2001).

Des cultures sont présentes à l'intérieur des limites de la Forêt Classée du Pic de Fon. Sur un ensemble de 10 villages dans la seule préfecture de Beyla, le PGRR a répertorié au moins 561 ha de cultures à l'intérieur des limites de la Forêt Classée du Pic de Fon (PGRR 2002b). Le riz représente la plus grande proportion (45 %) suivi du manioc (17 %), de la banane (11 %) et du café (9 %). L'empiètement agricole dans la Forêt Classée du Pic de Fon comprend donc à la fois les cultures de rentes et de subsistance. Le PGRR (2001) estime que jusqu'à 35 à 40 % du bloc forestier original a été affecté par les empiètements agricoles.

Les activités d'élevage extensif dans la région comprennent l'élevage des bovins, des ovins et des caprins. Une minorité de communautés élèvent aussi des porcins. L'élevage est mené sur des pâtures naturelles et des feux de brousse sont utilisés pour stimuler la régénération de la pâture (Camara et Guilavogui 2001).

La chasse fait aussi partie des menaces pesant sur la biodiversité dans la région (Hatch 1998, Camara et Guilavogui 2001). Bien que la loi guinéenne autorise la chasse traditionnelle comme moyen de subsistance, des signes montrent que la chasse se pratique aussi de manière commerciale, alimentant la filière de la viande de brousse (PGRR 2001). De telles pratiques cynégétiques sont autorisées par les lois existantes pour les espèces qui ne disposent pas d'un statut de protection particulier, mais qui font l'objet de restrictions bien définies. Les chimpanzés et autres primates et mammifères « En danger » sont intégralement protégés, ce qui signifie que leur abattage ou capture est interdit par la loi. Pourtant, l'absence d'une présence coercitive rend fortement probable le fait que les diverses réglementations soient ignorées.

L'examen de la littérature n'a pas fourni d'informations directes permettant de vérifier si la viande de brousse est une composante importante du régime alimentaire dans les villages entourant la Forêt Classée du Pic de Fon. Pourtant, nous savons que des porcs sont élevés dans certains de ces villages, et les restrictions alimentaires d'origine religieuse ou culturelle à l'encontre de la viande de brousse vont généralement de pair avec des restrictions similaires sur le porc. Par conséquent, il y a peu de chance que les interdictions alimentaires sur la viande de brousse aient cours dans tous les villages riverains de la Forêt Classée du Pic de Fon. De plus, parmi les quatre principales régions de Guinée, c'est en Guinée Forestière que la consommation de viande de brousse est la plus importante. L'équipe du RAP a trouvé des collets et des pièges, entendu des tirs de fusil et trouvé des cartouches usagées, ce qui ne laisse aucun doute sur le fait que la chasse a bien lieu dans la Forêt Classée du Pic de Fon. Il reste à déterminer dans quelle mesure cette chasse est pratiquée pour satisfaire les besoins de subsistance ou alimenter la filière commerciale de la viande de brousse.

En résumé, les problèmes imminents liés aux activités économiques comprennent : la conversion continue des forêts pour l'agriculture dans la Forêt Classée du Pic de Fon, l'usage continu de feux pour la déforestation, la régénération des pâtures, et la chasse ; et une augmentation du niveau de chasse pour répondre aux besoins d'une population croissante.

L'exploitation du bois
L'exploitation commerciale du bois s'accélère dans toute la Guinée et bien sûr dans l'ensemble de l'écosystème forestier de Haute Guinée (voir Figure 10.1). Ceci illustre à la fois les efforts déployés pour générer des échanges extérieurs à travers l'exportation des grumes, et la pression d'une forte et croissante demande de bois de chauffe. Le bois de chauffe couvre environ 77 % des besoins énergétiques des ménages et le charbon 3 % (Diawara 2001). L'exploitation forestière est identifiée comme étant l'une des principales menaces sur la biodiversité dans le Plan d'Action National pour la Biodiversité (MMGE 2002). Sur les deux sites étudiés,

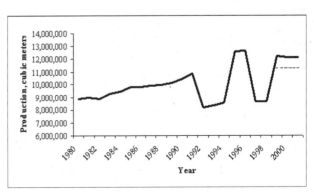

Figure 10.1. Production annuelle de bois rond, Guinée (1980-2001)

l'expédition RAP a trouvé peu d'indications sur l'étendue des impacts de l'exploitation forestière sur la Forêt Classée du Pic de Fon. Pourtant, des analyses de la FAO indiquent que pour 2002, en Guinée, la demande devrait dépasser les réserves, tant pour le bois d'œuvre que pour le bois de chauffe, ainsi que pour les produits forestiers non ligneux (PFNL) (Diawara 2001). Par conséquent, durant les 15 prochaines années, les forêts classées qui ne disposent pas de statuts légaux de protection ni d'un renforcement des capacités adéquates risquent d'être sérieusement menacées dans l'ensemble de la Guinée.

Infrastructure

Les préfectures qui recouvrent la chaîne du Simandou ne sont pas bien desservies sur le plan des infrastructures. L'accès routier à la région est généralement mauvais, et l'électricité et les infrastructures de télécommunications sont limitées et peu fiables. Les coûts de transport élevés et les liens de communication médiocres limitent le degré d'intégration de la région dans l'économie nationale. Toutefois, les investissements en infrastructure sur le point d'être réalisés vont favoriser le retour à des activités génératrices de revenus (cultures de rente, exploitation du bois et du gibier) et par conséquent encourager la conversion de la forêt dans la Forêt Classée du Pic de Fon et les zones environnantes.

Contexte institutionnel

L'Etat est pratiquement absent de la région. Officiellement, le Centre Forestier de N'Zérékoré et le CEGEN agissent sous mandats gouvernementaux, mais dépendent en fait largement de l'assistance allemande pour le premier et de l'UNESCO et de l'UNDP pour le second. Aussi, si le gouvernement doit jouer un rôle actif, un investissement substantiel dans l'accroissement des capacités sera-t-il nécessaire. Au niveau des communautés locales, les institutions n'exercent pas d'influence notable sur la gestion des ressources au sein de la Forêt Classée du Pic de Fon. Ceci n'est pas surprenant étant donné que les activités dans la Forêt Classée sont illégales et que l'insécurité de l'emploi n'incite pas à investir dans des régimes de gestion ou de coopération. Cependant, de solides recherches sur le terrain sont nécessaires pour permettre une évaluation plus complète de la présence institutionnelle, des capacités et des besoins en ce qui concerne la gestion des ressources au niveau de la communauté.

Contexte politique régional

Au cours de trois dernières années, les relations entre Lansana Conté et les autres dirigeants dans la région se sont considérablement améliorées, indiquant peut-être une diminution de la probabilité d'un conflit armé entre la Guinée et ses voisins. Cependant la Guinée et le Liberia continuent de s'accuser mutuellement de supporter les mouvements rebelles contre le pouvoir de l'autre régime. La stabilité politique en Guinée semble s'être consolidée au cours des dernières années, mais ce n'est en aucun cas une certitude. Par conséquent les afflux de réfugiés et le déplacement de Guinéens à l'intérieur du pays demeurent des éventualités non négligeables dans un futur proche. Les événements en Côte d'Ivoire illustrent cruellement à quel point une stabilité et une prospérité apparentes peuvent être illusoires. Le Liberia reste une cause de préoccupation, une insurrection d'intensité modérée continue à couver au Sénégal, et il reste à voir si la Sierra Leone peut soutenir la progression de la paix alors que les Nations Unies se retirent du pays.

Comme noté ci-dessus, Konomou et Zoumanigui (2000) mentionnent une population de réfugiés atteignant presque les 630 000 personnes en Guinée Forestière. Près de N'Zérékoré, Seredou, et Kouankan, des camps hébergent des réfugiés du Liberia et de la Sierra Leone. L'UNEP (2000) décrit les défis environnementaux qui ont accompagné la présence d'une importante population de réfugiés au sud-est de la Guinée, et tout Plan d'Action pour la Biodiversité de la région doit prendre en compte la possibilité d'autres augmentations de la population de réfugiés. Par exemple, si la situation en Côte d'Ivoire se détériore davantage, de nouveaux réfugiés risquent d'affluer dans la préfecture de Beyla (Tahirou 2002).

RÉSULTATS DE L'ATELIER

L'Atelier s'est ouvert avec une présentation des résultats initiaux obtenus par l'exercice du RAP dans la Forêt Classée du Pic de Fon, incluant une introduction au contexte général de conservation de la biodiversité dans le *hotspot* des Forêts de Haute Guinée. Les participants se sont ensuite entendus sur les objectifs de l'atelier, tels qu'énoncés ci-dessous :

1. Identifier et confirmer les principales parties prenantes dans la région.

2. Identifier et caractériser les menaces sur la biodiversité et les obstacles à la conservation dans la région.

3. Etablir un ordre de priorité initial à ces menaces.

4. Identifier les opportunités et les solutions possibles à ces menaces sur la biodiversité.

Objectif 1 de l'Atelier : Identifier et confirmer les principales parties prenantes dans la région

Les participants à l'atelier de Conakry ont identifié 14 catégories de parties prenantes relatives à la région du Pic de Fon, et ont dressé la liste des préoccupations et des attentes de la part de chacun de ces groupes de parties prenantes vis à vis de la conservation de la biodiversité. Ont été définis comme parties prenantes tous les acteurs pouvant influencer, de manière positive ou négative, la recherche de la conservation de la biodiversité dans la Forêt Classée du Pic de Fon. Le Tableau 10.2 présente la liste des parties prenantes et de leurs préoccupations et attentes telles qu'identifiées au cours de l'atelier.

Table 10.2. Préoccupations et attentes des participants.

Participants	Préoccupations	Objectifs/Attentes relatifs à la conservation
Gouvernement de la Guinée	Impact de l'exploitation des ressources naturelles sur la ressource de base Besoins des électeurs	Accroissement de la protection de la biodiversité et gestion soutenable des ressources naturelles Développement économique Accroissement des ressources financières d'appui à la conservation
CERE/ Université de Conakry		Meilleurs équipements et infrastructures de recherche Participation aux activités de recherche Soutien financier pour les équipes de recherche Financement des centres de recherche Développement de l'expertise scientifique
Donateurs (ex. USAID, Banque Mondiale)	Faible gestion financière Mauvaise utilisation des fonds	Stratégie d'investissement pour la conservation Protection de la biodiversité
Communautés riveraines de la forêt classée du Pic de Fon	Expropriation des terres Restriction des droits d'usage des ressources	Participation aux décisions de gestion des ressources Bénéfices financiers au niveau des communautés Partage équitable des bénéfices Gestion soutenable des ressources Alternatives/opportunités économiques pour remplacer les activités destructives
Rio Tinto exploration; Secteur minier;	Contraintes sur l'extraction de la ressource Limitation ou fermeture des concessions minières	Aire de protection définie sur la base de l'étude d'impact Investissements coordonnés avec le plan de protection Réconcilier l'extraction de la ressource avec la conservation Conservation de la biodiversité
Secteur de l'exploitation forestière		Méthodes d'exploitation améliorées Gestion soutenable de la ressource Protection des concessions d'exploitation forestière Filière commerciale d'exploitation forestière organisée Marchés locaux
ONG environnementales	Participation insuffisante des communautés locales Manque de financements adéquats Conflit entre les ONG et les autres institutions Non respect des accords Dispersion des fonds pour la conservation Exclusion des ONG des processus décisionnaires	Protection des espèces menacées Education environnementale Renforcement des capacités des ONG Gestion soutenable des ressources Participation à la planification, la mise en œuvre et la recherche de stratégies de conservation Accroître la mise en valeur de la biodiversité Restaurer les habitats dégradés Renforcer la protection des zones hautement prioritaires Promouvoir le respect des règles législatives
CEGEN/ Centre Forestier de N'Zérékoré		Connaissance approfondie de la biodiversité Renforcement des capacités institutionnelles Développement de partenariats Développement d'un plan de gestion régional Marquage des limites de concession minière Co-gestion des ressources avec les communautés locales
Artisans	Réglementation restrictive Accès aux ressources refusé	Meilleur accès au crédit/micro-crédit Gestion sensée des matières premières Droits d'accès à la ressource Accroître la valeur des produits artisanaux Formation et assistance pour améliorer la commercialisation Accès à de meilleures méthodes de gestion et technologies Développement des filières d'exportation
Apiculteurs		Modernisation des méthodes Commercialisation du miel

Participants	Préoccupations	Objectifs/Attentes relatifs à la conservation
Fermiers des communautés riveraines de la forêt classée du Pic de Fon	Limitation des zones agricoles Expropriation des terres	Introduction de nouvelles méthodes Amélioration des méthodes Accroître la valeur des produits Augmentation des rendements/productions Crédit agricole Sécurité financière Valeur ajoutée à la production
Pêcheurs des communautés riveraines du Pic de Fon	Réglementation restrictive Réduction de l'accès à la ressource	Techniques de pêche améliorées Protection des bassins versants Démarcation des zones de pêche Gestion collaborative Accroître la valeur des prises
Chasseurs d'oiseaux opérant dans la forêt classée du Pic de Fon	Réglementation restrictive Capture d'oiseaux interdite	Augmentation des quotas Utilisation soutenable des ressources en oiseaux
Chasseur de gibier opérant dans la forêt classée du Pic de Fon	Interdiction de la chasse Arrêt de l'utilisation de la faune sauvage Réglementation restrictive Réglementation des zones de chasse	Utilisation soutenable de la ressource faune sauvage Définition de zones de chasse Participation à la gestion de la faune sauvage

Tableau 10.3. Principales menaces sur la biodiversité du Pic de Fon

Menaces/Problèmes	Facteurs contributeurs
Déforestation	Exploitation forestière, coupes agressives Cultures sur brûlis Feux de brousse Déboisement pastoral
Pression agricole	Zones cultivées dans la Forêt Classée / demande de terres fertiles Conflits de propriété terrienne Immigration dans la région, croissance des populations humaines Propagation incontrôlée des cultures de rentes
Déclin de la faune et de la flore	Surexploitation Braconnage, pêche non réglementée Surexploitation des produits forestiers Collecte incontrôlée des plantes alimentaires et/ou médicinales
Destruction de l'écosystème	Pollution sonore des machines associées aux activités d'exploration de Rio Tinto Pollution de l'air et des cours d'eau Destruction des rivières et lits de torrents par l'exploitation artisanale de l'or et du diamant Déforestation à proximité des sources Introduction d'espèces exogènes
Faiblesses institutionnelles	Manque de respect des lois / réglementations contradictoires Faiblesse des capacités humaines / manque de suivi
Croissance de la population humaine	Mouvement de populations en réponse à l'exploitation Croissance démographique Présence d'éclaircies dans la forêt classée Migration de réfugiées
Généralisation de la pauvreté	Très bas niveau de vie Analphabétisme Infrastructures d'éducation et de santé inadéquates

Tableau 10.4. Actions de conservation requises dans la Forêt Classée du Pic de Fon

Solutions	Actions
Perfectionnement des connaissances sur la Forêt Classée du pic de Fon	Recherche scientifique sur la biodiversité Soutenir l'évaluation environnementale de la forêt classée du Pic de Fon et la réalisation d'un inventaire biologique Mise en place d'une structure de suivi écologique dans la forêt classée du Pic de Fon
Développement d'une stratégie de gestion soutenable	Adopter un modèle approprié pour l'exploitation forestière Protéger les sites exceptionnels tels que les sources chaudes Mettre en œuvre une gestion soutenable des ressources Inclure les communautés locales dans la gestion Développer un plan de gestion Promouvoir l'établissement permanent en opposition aux modes de production itinérants
Renforcement des capacités institutionnelles pour la conservation	Renforcement du statut légal de conservation de la forêt classée du Pic de Fon Appliquer la législation Implication financière de l'Etat Soutenir l'harmonisation des lois pertinentes
Prise de conscience du public	Faire prendre conscience au public de l'importance de la conservation de la biodiversité Répandre la connaissance des lois pertinentes Education environnementale
Soutien du développement des communautés locales	Soutenir les alternatives économiques compatibles avec la conservation de la biodiversité Promouvoir un tourisme basé sur la vie sauvage Répandre des techniques agricoles compatibles avec l'écologie Créer des centres communautaires et des postes de santé

Objectif 2 de l'Atelier : Identifier et caractériser les menaces sur la biodiversité et les obstacles à la conservation dans la région

Il a été demandé aux participants d'identifier et de donner des preuves des principales menaces pesant sur la biodiversité dans la Forêt Classée du Pic de Fon. Cet exercice n'établissait pas de différence explicite entre les causes proches et les tendances les plus éloignées. Les participants ont listé les menaces sur la biodiversité et les obstacles à la conservation à grande échelle, et identifié les facteurs contribuant à ces menaces et obstacles. Le Tableau 10.3 présente la liste des principales menaces sur la biodiversité dans la Forêt Classée du Pic de Fon, répertoriées à l'Atelier de Conakry. La liste reflète l'opinion bien informée d'experts soigneusement choisis, mais n'implique pas que des preuves solides ou de la documentation soient disponibles pour chaque menace. Plusieurs des menaces listées dans le Tableau 10.3 restent donc à vérifier sur le terrain et à quantifier.

Objectif 3 de l'Atelier : Etablir un ordre de priorité initial des menaces sur la biodiversité au Pic de Fon

Après avoir identifié les principales menaces sur la biodiversité, les participants à l'atelier ont tenté de parvenir à un consensus sur l'ordre initial à donner à ces menaces. Mais ils ont conclu que trop peu de données étaient disponibles pour établir un rang de priorité définitif. Par contre, l'atelier a choisi de définir la priorité des éléments d'une possible stratégie de conservation pour la Forêt Classée du Pic de Fon, discutée dans la section suivante, et ont

fortement insisté sur le fait que des recherches et collectes de données supplémentaires étaient nécessaires en tant qu'étape immédiate.

Objectif 4 de l'Atelier : Identifier les opportunités et les solutions possibles aux menaces pesant sur la biodiversité au Pic de Fon

La pénurie en données concrètes sur les tendances biologiques et socio-économiques de la Forêt Classée du Pic de Fon et de la zone environnante limite considérablement notre capacité à établir des stratégies de conservation appropriées pour la Forêt Classée du Pic de Fon. Néanmoins, sur la base de leurs expériences et connaissances en tant qu'experts de la région, les participants à l'atelier ont développé plusieurs recommandations pour faire face aux menaces identifiées dans le Tableau 10.3. Le Tableau 10.4 donne la liste des recommandations de l'atelier de Conakry pour les composantes d'une stratégie de conservation de la Forêt Classée du Pic de Fon. L'ordre de priorité de ces actions est discuté dans la section Recommandations de ce chapitre.

DISCUSSION

Les sections précédentes présentent les résultats de deux méthodes de collecte de données –l'étude de la bibliographie et l'atelier qui s'est tenu à Conakry. Sur le plan qualitatif, l'image d'ensemble qui se dégage de ces deux méthodes est cohérente. Cependant, aucun des deux exercices n'a fourni d'indications adéquates quant à l'échelle, l'intensité, et les prévisions dans le temps caractérisant les différents types de menaces sur la biodiversité dans la Forêt Classée du Pic de Fon. En l'absence de données quantitatives fiables sur ces facteurs, tout ordre de priorité donné à ces menaces et aux réponses à leur apporter demeure provisoire.

L'agriculture semble être l'une des menaces les plus pressantes sur l'intégrité de la Forêt Classée du Pic de Fon. L'empiètement agricole des communautés voisines est déjà prouvé par la présence de parcelles cultivées pour l'alimentation ou la rente. Le fait de planter des cultures arbustives pérennes comme le café requiert un investissement important de la part des fermiers, ce qui indique qu'ils savent que les cultures à long terme à l'intérieur des limites de la Forêt Classée du Pic de Fon ne risquent pas de provoquer de réaction de la part des autorités. De plus, l'utilisation de méthodes de rotation des cultures exercera forcément une pression persistante sur les habitats restants du fait que des zones sont défrichées pour créer de nouvelles parcelles.

L'étude de la littérature et l'atelier ont tous deux permis d'identifier la chasse pour la viande de brousse comme l'une des menaces majeures pesant sur la biodiversité dans la Forêt Classée du Pic de Fon. Cette conclusion est confirmée par l'équipe du RAP, qui a observé de nombreux collets et pièges, entendu des coups de feu et trouvé des cartouches laissées par des chasseurs. L'usage largement répandu des feux comme moyen de chasser le gibier dans cette région peut aussi détruire de grandes étendues de l'habitat, même si l'équipe du RAP n'a rien trouvé qui indique que cette pratique était employée dans les zones étudiées de la Forêt Classée du Pic de Fon.

L'utilisation de feux pour préparer des pâturages pour les animaux d'élevage a été identifiée, dans la bibliographie et pendant l'atelier, comme une menace générale dans la région mais il n'y a pas de référence spécifique à des empiètements par l'élevage dans la Forêt Classée du Pic de Fon. Toutefois, chacun des 24 villages entourant la Forêt Classée du Pic de Fon pratique l'élevage. Par conséquent, combinés à l'usage de feux pour défricher la forêt à des fins agricoles et pour chasser le gibier, les feux de brousse incontrôlés destinés à stimuler la régénération des pâturages demeurent une menace importante pour les habitats au sein de la Forêt Classée du Pic de Fon.

Plusieurs sources mentionnent une exploitation minière artisanale illégale à l'intérieur des limites de la Forêt Classée du Pic de Fon, et les participants à l'atelier de Conakry ont également identifié cette activité comme une menace. En particulier, la bibliographie et les participants soulignent l'impact destructeur de l'exploitation minière artisanale ou à petite échelle sur le lit des cours d'eau et leur parcours. Comme l'équipe du RAP n'a rapporté qu'une seule observation d'activité minière artisanale, elle est identifiée comme une menace possible, demandant des recherches et vérifications supplémentaires.

Bien que les données passées en revue au cours de l'étude de la bibliographie et de l'atelier de Conakry mentionnent l'exploitation forestière, l'équipe du RAP n'a rien vu qui indique que la zone ait subi une exploitation du bois à grande échelle par le passé. Toutefois, l'exploitation commerciale du bois prend de l'ampleur en Guinée et la demande en bois de chauffe dépasse rapidement les capacités de régénération de la ressource de base. Selon toute probabilité, les membres des communautés voisines de la Forêt Classée du Pic de Fon prélèvent du bois pour l'énergie combustible, la construction et d'autres activités, mais l'échelle et l'impact de telles activités restent à déterminer. L'équipe du RAP a noté que la zone qui s'étend de l'entrée de la forêt classée jusqu'au pied de la montagne près du camp 2 était dépourvue d'arbres (à part quelques squelettes), probablement à cause de l'utilisation locale du bois pour les maisons, le mobilier, le bois de chauffe etc, en plus du défrichage agricole. A ce stade, l'exploitation forestière est classée comme une menace qui peut être significative dès à présent, et sinon, risque de se préciser davantage dans le futur.

Enfin, l'extraction non contrôlée des autres ressources de la Forêt Classée du Pic de Fon, comme le poisson et les produits forestiers non ligneux (PFNL), a également été identifiée au cours de l'étude de la bibliographie et de l'atelier. Cependant, aucun des deux exercices n'a fourni d'information concernant les espèces de poissons couramment pêchées ni la gamme des PFNL généralement prélevés, sans parler des taux d'extraction ni de l'impact global. Sans informations supplémentaires sur la nature ou l'échelle d'extraction, la pêche et le prélèvement de PFNL peuvent seulement être considérés comme des menaces possibles sur la biodiversité.

Tableau 10.5: Menaces sur la biodiversité dans la forêt classée du Pic de Fon

Principales menaces à la biodiversité		
Extension de l'agriculture	Chasse non réglementée	Déboisement pastoral
Cultures sur brûlis	Feux de brousse incontrôlés	

Menaces possibles à la biodiversité		
Exploitation minière artisanale illégale	Exploitation du bois	Collecte non réglementée des PFNL
Pêche excessive	Opérations minières commerciales	

Par conséquent, l'empiètement agricole et la chasse pour la viande de brousse apparaissent comme les menaces les plus importantes pesant sur la biodiversité dans les forêts restantes du Pic de Fon. L'usage de feux incontrôlés pour ces deux activités ainsi que pour la préparation des pâturages pour les animaux d'élevage peut avoir un effet particulièrement dévastateur sur le milieu naturel. D'autres menaces comme l'exploitation minière artisanale illégale, l'exploitation excessive des ressources en poissons, l'exploitation du bois, et le prélèvement des PFNL, restent à vérifier et à quantifier. Le Tableau 10.5 résume les conclusions concernant les menaces sur la biodiversité dans la Forêt Classée du Pic de Fon.

RECOMMANDATIONS

Au cours de l'atelier qui s'est tenu à Conakry, les participants ont dressé un ordre de priorité des actions de conservation nécessaires. Ils se sont accordés sur le fait que dans un futur immédiat (à l'horizon de 1 à 2 ans), les actions urgentes requises comprennent la consolidation du renforcement des capacités, particulièrement au regard de la réglementation de la chasse, des recherches supplémentaires sur de nombreux aspects biologiques et socio-économiques de la région, une campagne d'éducation-sensibilisation des communautés locales, et l'entretien du marquage des limites de la Forêt Classée du Pic de Fon. Il a aussi été souligné que l'appui au développement des communautés locales est essentiel afin de promouvoir des modes de vies soutenables, de faire décroître les pressions sur les ressources forestières et pour la réalisation des objectifs de conservation à long terme.

Les efforts produits au Mont Nimba se présentent comme un modèle idéal pour l'ébauche d'une stratégie pour la Forêt Classée du Pic de Fon. Le programme du CEGEN, organisme responsable de la coordination des efforts de conservation dans la région du Mont Nimba, comporte la protection des zones adjacentes, complétée par des mesures qui assurent des bénéfices aux communautés locales. Ces mesures incluent l'expérimentation d'alternatives à la chasse au gibier comme source de protéines, des projets sanitaires et la fourniture de services de santé. De plus, le CEGEN a engagé des discussions avec le gouvernement de Guinée qui se sont récemment traduites par l'intégration de la chaîne du Simandou dans son mandat.

Les collectes de données socio-économiques recommandées doivent obligatoirement avoir lieu au niveau communautaire, par le biais d'études, d'interviews et d'échanges directs entre les chercheurs et les villageois. Conduire de telles études requiert d'importants efforts pour communiquer l'objet de la recherche aux villageois tout en gérant les attentes qui peuvent émerger vis-à-vis des bénéfices pour les communautés locales. Ceci demande un investissement considérable en temps et en ressources dans l'établissement de relations et un franc degré d'implication dans la poursuite des travaux une fois l'étude terminée. Une collecte de données au niveau communautaire qui n'adhérerait pas scrupuleusement à ces principes serait irréfléchie sur de nombreux plans. Comme exemple des ressources et du temps nécessaires pour un processus complet de consultation, une proposition au « Global Environment Facility » pour financer un exercice similaire au Mont Nimba, a budgétisé 560 000 $ pour en financer les activités sur une période de 12 mois (GEF 1999).

Il sera difficile de formuler une stratégie de conservation pour le Pic de Fon dans le contexte plus large des Hauts Plateaux du Nimba. L'inclusion du Pic de Fon au mandat du CEGEN est un pas important dans cette direction. De nombreuses menaces relevées pour la Forêt Classée du Pic de Fon font partie des dynamiques qui opèrent au niveau régional, dont la chasse pour la viande de brousse, l'élevage extensif, l'utilisation non contrôlée de feux, et l'afflux de réfugiés. Pour cette raison, la planification des investissements pour la conservation doit aussi être conduite sur une base régionale. Les perspectives régionales sont cruciales en tant que menaces majeures pour les grands mammifères, en particulier la fragmentation de l'habitat. Le corridor construit pour lier la population de chimpanzés de Bossou à celles du Mont Nimba est un exemple d'une telle planification et action régionale (Ham et al. Sous presse).

MATÉRIEL CONSULTÉ

[Author/institution uncertain]. 1995. Livre Blanc/ Monographie Nationale sur la Diversité Biologique. Guinea/UNBio: Conakry. (494 pp.)

Arid Lands Information Center (University of Arizona). [date uncertain]. Draft Environmental Profile – Guinea. United States National Committee for Man and the Biosphere [UNMAB]. USAID/Office of Forestry, Environment, and Natural Resources: Washington DC. (229 pp.)

Bonnehin, L. 2002. «Analyse des Menaces et Opportunités de Conservation de la Biodiversité au Pic de Fon, Massif de Simandou, Guinée». Rapport de l'Atelier ZOPP, Conakry Dec. 12-13 2002. Conservation International.

Camara, L. 2000. «Revue et Amélioration des données relatives aux produits forestiers en République de Guinée». Programme de partenariat CE-FAO (1998-2002). Food and Agriculture Organization of the United Nations: Rome.

Camara, N. 1999. "Revue des données du Bois-Energie en Guinée". Programme de partenariat CE-FAO (1998-2002). Food and Agriculture Organization of the United Nations: Rome.

Camara, W. and K. Guilavogui. 2001. "Identification des villages riverains autour des forets classes du Pic de Fon et Pic de Tibe". PGRR/Centre Forestier N'Zérékoré, Mesures Riveraines.

CEGEN. (no date). "The Nimba Mountains: balancing environment and sustainable development". Produced by Fauna & Flora International.

[CI] Conservation International. 1999. From the Forest to the Sea: Biodiversity Connections from Guinea to Togo. Conservation International: Washington, DC.

Diawara, D. 2000. "Les Données Statistiques sur les Produits Forestiers Non-Ligneux en Republique de Guinée". Programme de partenariat CE-FAO (1998-2002). Food and Agriculture Organization of the United Nations: Rome.

Diawara, D. 2001. "Document National de Prospective – Guinée". L'Etude prospective du secteur forestier en Afrique (FOSA). Food and Agriculture Organization of the United Nations: Rome.

Direction Générale des Services Economiques. 1953. Arrêté Portant Classement la forêt de "Fon".

[GEF] Global Environment Facility. 1999. «PDF-B Proposal – Conservation of biodiversity through integrated participatory management in the Nimba Mountains."

Government of the Republic of Guinea. 1999. "Global Environment Facility PDF-B Proposal – Conservation of Biodiversity through Integrated Participatory Management in the Nimba Mountains".

Government of the Republic of Guinea. 2000. "Interim Poverty Reduction Strategy Paper". International Monetary Fund: Washington DC.

Ham, R., T. Humle, D. Brugière, M. Fleury, T. Matsuzawa, Y. Sugiyama and J. Carter. In process. "Chapter 2.5: Guinea" in Status Survey and Recommendations for the Conservation of The West African Chimpanzee, (report in progress, contact CABS/CI for details).

Hatch & Associates, Inc. 1998. "Preliminary Environmental Characterisation Study: Simandou Iron Ore Project Exploration Programme, South-Eastern Guinea". Montreal, Canada.

Konomou, M. and K. Zoumanigui. 2000. «Zonage Agro-Ecologique de la Guinée Forestière». Centre de Recherche Agronomique de Seredou. Ministère de l'Agriculture, des Eaux et Forets: Guinée.

Ministère de l'Agriculture, des Eaux et Forets/Direction Nationale des Forets et de la Faune. 199?. «Exposé des Motifs: Projet de Loi portant Code de la Protection de la Faune Sauvage et réglementation de la Chasse. Guinea: Conakry.

Ministère des Travaux Publics et de l'Environnement/ Direction Nationale de l'Environnement. 1997. Monographie Nationale sur la Diversité Biologique. Guinea/PNUE: Conakry. (146 pp.)

[MMGE]. Ministry of Mines, Geology, and the Environment. 2002 (January). National Strategy and Action Plan for Biological Diversity; Volume 1: National Strategy for Conservation Regarding Biodiversity and the Sustainable Use of these Resources. Guinea/UNDP/GEF: Conakry.

PGRR. 2001. «Pic de Fon – Situation Actuelle, Analyse et Recommandations (Proposition d'axes de collaboration)». Centre Forestier N'Zérékoré.

PGRR. 2002a. «Proposition de stratégie pour la surveillance du Pic de Fon – Attente PGRF». Centre Forestier N'Zérékoré, Section Conservation – Biodiversité.

PGRR. 2002b. «Rapport d'Activités – Délimitation du versant sud-ouest de la Forêt Classée du Pic de Fon». Centre Forestier N'Zérékoré (Antenne Bero).

Rio Tinto. 2002. Règlements Concernant l'Environnement, la Santé, la Sécurité et les Relations avec les Communautés et Dispositions Concernant l'Environnement et la Sécurité Destinées aux Entrepreneurs. Rio Tinto Mining and Exploration Ltd Africa/Europe Region.

Robertson. 2001. Guinea. In: Fishpool, L.D.C. and M.I. Evans (eds.). Important Bird Areas in Africa and Associated Islands: Priority sites for conservation. Newbury and Cambridge, UK: Pisces Publications and BirdLife International. Pp 391-402.

Szczesniak, P. 2001. "The Mineral Industries of Côte d'Ivoire, Guinea, Liberia, and Sierra Leone". U.S. Geological Survey Minerals Yearbook.

Tahirou, B. 2002. "Collecte de données bibliographiques sur la Forêt Classée du Pic de Fon et du périmètre de prospection de Rio Tinto". Draft report submitted to Conservation International.

United Nations Environment Programme (UNEP, in collaboration with UNCHS and UNHCR). 2000. Environmental Impact of Refugees in Guinea. UNEP/Regional Office for Africa: Nairobi, Kenya.

USAID/Guinea. 2000. «Briefing Book for Invited Partners». Synergy Workshop.

USAID/Guinea Natural Resources Management project. 1996. «Scoping statement – Nyalama classified forest co-management initiative». (28 pp.)

U.S. Department of State [USDOS]. 2000. FY 2001 Country Commercial Guide: Guinea. U.S. & Foreign Commercial Service, U.S. Department of State: Washington DC.

World Bank. 2001. Country Assistance Strategy Progress Report for the Republic of Guinea. Report No. 22451. World Bank: Washington DC.

Wright, L. and M. Sylla. 2001. "Mining Annual Review 2001 – Guinea". The Mining Journal.

Wright, L. and M. Sylla. 2002. "Mining Annual Review 2002 – Guinea". The Mining Journal.

Ziegler, S., G. Nikolaus and R. Hutterer. 2002. „High Mammalian Diversity in the Newly Established National Park of Upper Niger, Republic of Guinea". Oryx. Vol. 36, No. 1: pp. 73-80.

Appendix/Annexe 1

Plant species recorded from three habitats of the Forêt Classée du Pic de Fon

Espèces de plantes recensées dans trois habitats de la Forêt Classée du Pic de Fon

Jean Louis Holié and Nicolas Londiah Delamou

LEGEND :
FR = Forest ; MG = Montane Grassland ; FG = Gallery Forest
Le = Liana ; He = Grass ; Abt = Shrub ; Abe = Tree
Shrub : diameter less or equal to 20 cm
Tree : diameter greater than 20 cm

IUCN Categories:
EN = Endangered; VU = Vulnerable; nt = near threatened

FR = Forêt ; MG = prairies de montagne ; FG = Forêt Galerie
Le = Liane ; He = Herbe ; Abt = Arbuste ; Abe = Arbre
Arbuste = Diamètre inférieur ou égal à 20 cm
Arbre = Diamètre supérieur à 20 cm

Catégories UICN:
EN = Menacé d'Extinction; VU = Vulnérable ; nt = Presque Menacé

		FR	MG	FG	Veg. Type	IUCN Status
Acanthaceae (8)	Adhatoda robusta	X			He	
	Brillantaisia nitens	X			He	
	Eremomastax speciosa	X			He	
	Hypoestes verticillaris	X			He	
	Justicia flava	X			He	
	Sanocrater berhautii	X			He	
	Staurogyne capitata	X			He	
	Whitfieldia bateritia	X			He	
Agavaceae (4)	Dracaena arborea	X			Abe	
	Dracaena elliotii	X			Abt	
	Dracaena mannii	X			Abe	
	Dracaena surculosa	X			Abt	
Amaranthaceae (2)	Celosia trigyna	X			He	
	Cyathula prostrata	X			He	
Anacardiaceae (4)	Antrocaryon micraster	X			Abe	VU
	Pseudospondias microcarpa	X			Abe	
	Spondias mombin	X			Abe	
	Tricoscypha longiflora	X			Abe	
Annonaceae (12)	Artabotrys velutinus	X			Le	
	Piptostigma fasciculata	X			Abe	
	Cleistopholis patens	X			Abe	
	Enantia polycarpa	X			Abe	
	Isolona campanulata	X			Abt	

		FR	MG	FG	Veg. Type	IUCN Status
	Isolona campanulata	X			Abt	
	Neostenanthera hamata	X			Abe	VU
	Pachypodanthium staudtii	X			Abe	
	Uvariopsis guineensis	X			Abt	
	Xylopia aethiopica	X			Abe	
	Xylopia parviflora	X			Abe	
	Xylopia quintasii	X			Abe	
	Xylopia villosa	X			Abe	
Apocynaceae (15)	*Alstonia boonei*	X			Abe	
	Baissea lane poolei	X			Le	
	Baissea multiflora	X			Le	
	Baissea zygodioides	X			Le	
	Funtumia elastica	X			Abe	
	Funtumia latifolia	X			Abe	
	Holarrhena floribunda	X		X	Abe	
	Landolphia dulcis	X			Le	
	Landolphia hirsuta	X			Le	
	Landolphia owariensis	X			Le	
	Rauvolfia vomitoria	X			Abt	
	Strophanthus hispidus	X			Le	
	Tabernaemontana longiflora	X			Abt	
	Voacanga africana	X			Abt	
	Voacanga bracteata	X			Abt	
Araceae (6)	*Amorphophallus dracontioides*	X			He	
	Anubias gracilis	X			He	
	Cercestis afzelii	X			Le	
	Culcasia saxatilis	X			Le	
	Culcasia scandens	X			Le	
	Raphidophora africana	X			Le	
Araliaceae (1)	*Schefflera barteri*			X	Le	
Arecaceae (5)	*Ancisthrophyllum secundiflorum*	X			Le	
	Calamus deerratus	X			Le	
	Elaeis guineensis	X			Abe	
	Eremospatha macrocarpa	X			Le	
	Raphia vinifera	X			Abe	
Asclepiadaceae (4)	*Ceropegia nigra*	X			Le	
	Gongronema latifolium	X			Le	
	Parquetina nigrescens	X			Le	
	Secamone afzelii	X			Le	
Asteraceae (13)	*Ageratum conyzoides*	X			He	
	Aspilia latifolia	X		X	He	
	Bidens pilosa	X			He	
	Blumea guineensis		X		He	
	Chromolaena odorata	X			He	
	Elephantopus sp.	X			He	

		FR	MG	FG	Veg. Type	IUCN Status
	Erigeron floribundum	X			He	
	Gynura cernua	X			He	
	Microglossa densiflora	X			Le	
	Mikania cordata	X			Le	
	Spilanthes prostrata	X			He	
	Synedrella nodiflora	X			He	
	Vernonia conferta	X			Abt	
Balanitaceae (1)	*Balanites wilsoniana*			X	Abe	
Balonophoraceae (1)	*Thonningia sanguinea*	X			He	
Begoniaceae (1)	*Begonia rostrata*	X			He	
Bignoniaceae (4)	*Markhamia tomentosa*	X		X	Abe	
	Newbouldia laevis				Abt	
	Spathodea campanulata			X	Abe	
	Stereospermum acuminatissimum			X	Abe	
Bombacaceae (2)	*Bombax buonopozense*	X			Abe	
	Ceiba pentandra	X			Abe	
Boraginaceae (1)	*Cordia platythyrsa*	X			Abe	VU
Burseraceae (2)	*Canarium schweinfurthii*	X			Abe	
	Dacryodes klaineana	X			Abe	
Caesalpiniaceae (12)	*Amphimas pterocarpoides*	X			Abe	
	Anthonotha macrophylla	X			Abe	
	Bussea occidentalis	X			Abe	
	Chidlowia sanguinea	X			Abe	
	Cryptosepalum tetraphyllum	X			Abe	VU
	Dialium aubrevillei	X			Abe	
	Dialium dinklagei	X			Abe	
	Distemonanthus benthamianus	X			Abe	
	Erythrophleum suaveolens	X			Abe	
	Gilbertiodendron limba	X			Abe	
	Mezoneuron benthamianum	X			He	
	Paramacrolobium coeruleum	X			Abe	
Campanulaceae (1)	*Cephalostigma perrottetii*	X			He	
Cannaceae (1)	*Canna indica*	X			He	
Chrysobalanaceae (2)	*Parinari excelsa*	X		X	Abe	
	Rubus sp.	X			Le	
Clusiaceae (7)	*Allanblackia floribunda*	X			Abe	
	Garcinia kola	X			Abe	VU
	Garcinia menfiensis	X			Abt	
	Garcinia polyantha	X			Abe	
	Harungana madagascariensis	X			Abt	
	Mammea africana	X			Abe	
	Vismia guineensis	X			Abt	

		FR	MG	FG	Veg. Type	IUCN Status
Combretaceae (9)	*Combretum hispidum*	X			Le	
	Combretum micronatum	X			Le	
	Combretum paniculatum	X			Le	
	Combretum platypterum	X			Le	
	Combretum racemosum	X			Le	
	Holoptelea grandis	X			Abe	
	Terminalia glaucescens		X		Abe	
	Terminalia ivorensis	X			Abe	VU
	Terminalia superba	X			Abe	
Commelinaceae (3)	*Palisota hirsuta*	X			He	
	Palisota paniculata	X			He	
	Polyspatha paniculata	X			He	
Connaraceae (5)	*Byrsocarpus coccineus*	X			Le	
	Cnestis corniculata	X			Le	
	Cnestis ferruginea	X			Le	
	Cnestis racemosa	X			Le	
	Connarus africanus	X			Le	
Convolvulaceae (3)	*Calycobolus africanus*	X			Le	
	Ipomoea involucrata	X			Le	
	Merremia pterygocaulos	X			Le	
Cucurbitaceae (1)	*Adenopus breviflorus*	X			Le	
Cyatheaceae (1)	*Cyathea manniana*	X			He	
Cyperaceae (2)	*Rhynchospora corymbosa*	X			He	
	Scleria barteri	X			He	
Dennstaedtiaceae (1)	*Pteridium aquilinum*	X			He	
Dilleniaceae (2)	*Tetracera alnifolia*	X				
	Tetracera potatoria	X			Le	
Erythroxylaceae (1)	*Erythroxylum mannii*	X			Abe	
Euphorbiaceae (26)	*Alchornea cordifolia*	X			Le	
	Alchornea hirtella	X			Le	
	Antidesma laciniatum	X			Abt	
	Antidesma venosum	X			Abt	
	Bridelia aubrevillei	X			Abe	
	Bridelia ferruginea	X			Abt	
	Bridelia micrantha	X			Abt	
	Discoglypremna caloneura	X			Abe	
	Drypetes afzelii	X			Abt	VU
	Drypetes principum	X			Abt	
	Drypetes singroboensis	X			Abt	VU
	Hymenocardia acida	X			Abt	
	Hymenocardia lyrata	X			Abt	
	Macaranga heterophylla	X			Abt	
	Macaranga rosea	X			Abt	
	Maesobotrya barteri	X			Abt	

		FR	MG	FG	Veg. Type	IUCN Status
	Manniophyton africanum	X			Le	
	Mareya micrantha	X			Abt	
	Microdesmis keayana	X			Abt	
	Phyllanthus amarus	X			He	
	Phyllanthus discoideus	X		X	Abe	
	Phyllanthus floribundus	X			Le	
	Ricinodendron heudelotii	X			Abe	
	Tetrorchidium didymostemon	X			Abt	
	Uapaca guineensis	X			Abe	
	Uapaca heudelotii	X			Abe	
Fabaceae (25)	*Abrus precatorius*	X			Le	
	Abrus stictosperma	X			Le	
	Dalbergia ecastaphyllum	X			Le	
	Dalbergia hostilis	X			Le	
	Dalbergiella welwitschii	X			Le	
	Desmodium ovalifolium	X			He	
	Desmodium salicifolium	X			He	
	Desmodium velutinum	X			He	
	Dolichos dinklagei		X		He	
	Dolichos nimbaensis		X		He	
	Droogmansia scaettaiana	X			He	
	Eriosema glomeratum	X			He	
	Erythrina senegalensis	X			Abe	
	Erythrina vogelli	X			Abe	
	Indigofera prieuriana	X			He	
	Kotschya ochreata	X			He	
	Leptoderris brachyptera	X			Le	
	Leptoderris fasciculata	X			Le	
	Lonchocarpus cyanescens	X			Le	
	Millettia rhodantha	X			Abe	
	Millettia stapfiana	X			Abt	
	Mucuna poggei	X			Le	
	Ostryoderris leucobotrya	X			Abt	
	Platysepalum hirsutum	X			Le	
	Pterocarpus erinaceus	X			Abe	
	Pterocarpus santalinoides	X			Abe	
Flacourtiaceae (2)	*Caloncoba echinata*	X			Abt	
	Lindackeria dentata	X			Abt	
Hernandiaceae (2)	*Illigera pentaphylla*	X			Le	
	Illigera vespertilio	X			Le	
Hippocrateaceae (4)	*Hippocratea pallens*	X			Le	
	Hippocratea welwitschii	X			Le	
	Salacia cornifolia	X			Le	
	Salacia erecta	X			Le	

		FR	MG	FG	Veg. Type	IUCN Status
Icacinaceae (3)	Leptaulus daphnoides	X			Abe	
	Polycephalium capitatum	X			Le	
	Rhaphiostylis beninensis	X			Le	
Lamiaceae (4)	Hoslundia opposita	X			Abt	
	Hyptis lanceolata				Le	
	Hyptis suaveolens				Le	
	Solenostemon monostachyus		X		He	
Lecythidaceae (3)	Combretodendrum africana	X			Abe	
	Napoleonaea leonensis	X			Abt	
	Napoleonaea vogelii	X			Abt	
Leeaceae (1)	Leea guineensis	X			Abt	
Liliaceae (1)	Gloriosa superba	X			Le	
Linaceae (2)	Hugonia afzelii	X			Le	
	Hugonia planchonii	X			Le	
Loganiaceae (4)	Anthocleista nobilis	X			Abe	
	Anthocleista vogelii	X		X	Abe	
	Strychnos aculeata*	X			Le	
	Strychnos spinosa	X				
Malvaceae (6)	Hibiscus rostellatus			X		
	Hibiscus surattensis			X		
	Sida alba	X				
	Sida rhombifolia	X				
	Sida urens	X				
	Urena lobata	X				
Marantaceae (8)	Halopegia azurea	X				
	Hypselodelphys poggeana	X			Le	
	Hypselodelphys violacea	X			Le	
	Marantochloa cuspidata	X			He	
	Marantochloa leucantha	X			He	
	Marantochloa purpurea	X			He	
	Sarcophrynium brachystachys	X			He	
	Thaumatococcus daniellii	X			He	
Melastomataceae (3)	Dissotis erecta		X		He	
	Dissotis phaeotricha	X			He	
	Osbeckia senegalensis	X			He	
Meliaceae (6)	Carapa procera	X			Abe	
	Entandrophragma candollei	X			Abe	VU
	Entandrophragma utile	X			Abe	VU
	Guarea cedrata	X			Abe	VU
	Khaya grandifoliola	X			Abe	VU
	Trichilia heudelotii	X			Abe	
Melianthaceae (1)	Bersama abyssinica	X		X	Abe	

		FR	MG	FG	Veg. Type	IUCN Status
Menispermaceae (3)	Kolobopetalum ovatum	X			Le	
	Stephania dinklagei	X			Le	
	Tiliacora varnekei	X			Le	
Mimosaceae (16)	Albizia ferruginea	X			Abe	VU
	Albizia sassa	X			Abe	
	Albizia zygia	X			Abe	
	Arthrosamanea altissima	X			Abe	
	Aubrevillea platycarpa	X			Abe	
	Bussea occidentalis	X			Abe	
	Calpocalyx brevibracteatus	X			Abe	
	Dichrostachys glomerata	X			Abe	
	Entada africana		X		Abt	
	Entada mannii	X			Le	
	Newtonia duparquetiana	X			Abe	
	Parkia bicolor	X			Abe	
	Parkia biglobosa		X	X	Abe	
	Pentaclethra macrophylla	X			Abe	
	Piptadeniastrum africanum	X			Abe	
	Xylia evansii	X			Abe	
Moraceae (13)	Antiaris africana	X				
	Bosqueia angolensis	X			Abe	
	Ficus barteri			X	Abt	
	Ficus capensis	X		X		
	Ficus eriobotryoides		X	X	Abt	
	Ficus exasperata	X				
	Ficus gnaphalocarpa	X		X		
	Ficus leprieurii	X			Abt	
	Milicia excelsa	X		X	Abe	nt
	Milicia regia	X			Abe	VU
	Morus mesozygia	X		X	Abe	
	Musanga cecropioides	X			Abe	
	Treculia africana	X			Abe	
Myristicaceae (1)	Pycnanthus angolensis	X			Abe	
Ochnaceae (4)	Lophira alata	X			Abe	VU
	Ochna afzelii	X			Abt	
	Ochna membranacea	X			Abt	
	Ouratea calantha	X			Abt	
Octoknemataceae (1)	Octoknema borealis	X			Abe	
Olacaceae (4)	Olax sp.	X			Abt	
	Ongokea gore	X			Abe	
	Ptychopetalum anceps	X				
	Strombosia glaucescens	X			Abe	

		FR	MG	FG	Veg. Type	IUCN Status
Orchidaceae (3)	*Angraecum* sp.	X			Le	
	Bulbophyllum recurvum	X			He	
	Bulbophyllum sp.	X			He	
Passifloraceae (4)	*Adenia cissampeloides*	X			Le	
	Adenia gracilis	X			Le	
	Adenia lobata	X			Le	
	Smeathmannia pubescens	X			Abt	
Piperaceae (2)	*Piper guineense*	X			Le	
	Piper umbellatum	X			He	
Poaceae (13)	*Bouteloua eriopoda*		X		He	
	Bouteloua gracilis		X		He	
	Hyparrhenia involucrata		X		He	
	Hyparrhenia rufa		X		He	
	Imperata cylindrica		X		He	
	Leptaspis cochleata					
	Mapania sp.	X			He	
	Olyra latifolia	X			He	
	Oplismenus hirtellus	X	X		He	
	Pennisetum purpurium		X		He	
	Setaria chevalieri	X			He	
	Setaria megaphylla	X			He	
	Streptogyna crinata	X			He	
Polygalaceae (2)	*Carpolobia alba*	X			Abt	
	Carpolobia sp.	X			Abt	
Polypodiaceae (1)	*Platycerium angolense*	X			He	
Proteaceae (1)	*Protea angolensis*	·	X		He	
Pteridaceae (1)	*Ceratopteris cornuta*	X			He	
Rhamnaceae (1)	*Gouania longipetala*	X			Le	
Rhizophoraceae (1)	*Cassipourea congensis*	X			Abt	
Rubiaceae (41)	*Bertiera racemosa*	X		X	Abt	
	Borreria radiata			X	He	
	Canthium subcordatum	X			Abt	
	Canthium vulgare	X		X	Abt	
	Cephaelis biaurita	X		X	Abt	
	Cephaelis peduncularis	X			Abt	
	Craterispermum caudatum			X	Abt	
	Craterispermum laurinum	X		X	Abt	
	Crossopteryx febrifuga		X		Abt	
	Dictyandra arborescens		X		Abt	
	Euclinia longiflora	X		X	Abt	
	Gaertnera sp.		X	X	Abt	
	Geophila hirsuta	X			He	
	Geophila neurodictyon	X			He	

		FR	MG	FG	Veg. Type	IUCN Status
	Geophila obvallata	X			He	
	Ixora laxiflora	X			Le	
	Massularia acuminata	X		X	Abt	
	Mitragyna stipulosa	X			Abe	
	Morinda geminata	X		X	Abe	
	Morinda longiflora	X			Le	
	Morinda lucida	X		X	Le	
	Mussaenda arcuata	X			Le	
	Mussaenda erythrophylla	X			Le	
	Mussaenda linderi	X			Le	
	Nauclea diderrichii	X			Abe	VU
	Nauclea latifolia	X		X	Abt	
	Oxyanthus formosus	X			Abt	
	Oxyanthus unilocularis	X			Abt	
	Pavetta platycalyx	X			Abt	
	Pavetta sp.	X			Abt	
	Psilanthus mannii	X			Abt	
	Psychotria sp.	X		X	Abt	
	Rothmannia longiflora	X			Abt	
	Rutidea parviflora	X			Le	
	Sabicea venosa	X			Le	
	Sherbournia calycina	X			Le	
	Tarenna edulis			X	Abt	
	Tarenna nutidila	X		X	Abt	
	Tarenna vignei subglabrata			X	Abt	
	Tricalysia macrophylla	X		X	Abt	
	Urophyllum afzelii	X		X	Abt	
	Vangueriopsis spinosum	X		X	Abt	
Rutaceae (3)	Fagara macrophylla	X			Abe	
	Teclea verdoorniana			X	Abe	
	Zanthoxylum zanthoxyloides	X			Abt	
Sapindaceae (8)	Allophylus africanus	X			Abt	
	Blighia sapida	X			Abe	
	Blighia welwitschii	X			Abe	
	Cardiospermum grandiflorum	X			He	
	Chytranthus carneus	X			Abt	
	Deinbollia pinnata	X			Abt	
	Lecaniodiscus cupanioides	X			Abe	
	Paullinia pinnata	X			Le	
Sapotaceae (7)	Afrosersalisia afzelii	X			Abe	
	Afrosersalisia chevalieri	X			Abe	
	Anigeria altissima	X			Abe	
	Chrysophyllum perpulchrum	X			Abe	
	Chrysophyllum pruniforme	X			Abe	

		FR	MG	FG	Veg. Type	IUCN Status
	Chrysophyllum subnudum	X			Abe	
	Neolemonniera clitandrifolia	X			Abe	EN
Selaginellaceae (3)	Selaginella kalbreyeri	X			He	
	Selaginella myosurus	X			He	
	Selaginella sp.	X			He	
Simaroubaceae (2)	Dacryodes klaineana	X			Abe	
	Harrisonia abyssinica	X		X	Abt	
Solanaceae (5)	Physalis angulata	X			He	
	Solanum incanum	X			He	
	Solanum nigrum	X			He	
	Solanum nodiflorum	X			He	
	Solanum verbascifolium	X			He	
Sterculiaceae (7)	Cola caricaefolia	X			Abt	
	Cola cordifolia	X			Abe	
	Cola millenii	X			Abt	
	Cola setigera	X			Abt	
	Mansonia altissima	X			Abe	
	Sterculia oblonga	X			Abe	
	Sterculia tragacantha	X			Abe	
Tiliaceae (3)	Glyphaea brevis	X			He	
	Grewia mollis	X			He	
	Triumfetta cordifolia	X			He	
Ulmaceae (6)	Celtis adolfi-friderici	X			Abe	
	Celtis brownii	X			Abe	
	Celtis mildbraedii	X			Abe	
	Celtis rodanta	X			Abe	
	Celtis zenkeri	X			Abe	
	Trema orientalis	X			Abt	
Vitaceae (2)	Cissus difusiflora	X			Le	
	Cissus populnea	X			Le	
Zingiberaceae (3)	Aframomum elliotii	X			He	
	Aframomum heudelotii	X			He	
	Costus afer	X			He	

Appendix/Annexe 2

Plant species recorded during the Pic de Fon RAP survey with food, medicinal or international trade value

Espèces de plantes avec une valeur alimentaire, médicinale ou commerciale, recensées durant l'etude

Jennifer McCullough

Family	Genus/species	Common name	Food	Medicine	Int. trade
Acanthaceae	*Adhatoda robusta*				unspecified
Acanthaceae	*Justicia flava*	Justicia	leaves, raw		
Amaranthaceae	*Celosia trigyna*	Cockscomb	leaves, raw		
Anacardiaceae	*Antrocaryon micraster*	Antrocaryon	kernel, dried		
Anacardiaceae	*Spondias mombin*	Mombin, yellow; hogplum; jobo	fruit pulp; fruit jam and jellies	unspecified	
Annonaceae	*Cleistopholis patens*	Avom; Awom	fruit		
Annonaceae	*Isolona campanulata*			pharmaceutical	research as anti-malarial
Annonaceae	*Xylopia aethiopica*		seed		
Apocynaceae	*Alstonia boonei*			unspecified	
Apocynaceae	*Funtumia elastica*			unspecified	
Apocynaceae	*Landolphia dulcis*	African rubber			latex
Apocynaceae	*Landolphia hirsuta*	African rubber			latex
Apocynaceae	*Landolphia owariensis*	African rubber	fruit		latex
Apocynaceae	*Rauvolfia vomitoria*	rauvolfia	unspecified	unspecified	medicine
Apocynaceae	*Strophanthus hispidus*	strophanthus			strophanthus
Apocynaceae	*Voacanga africana*	voacanga	unspecified	unspecified	medicine
Arecaceae	*Elaeis guineensis*	Oil palm	fruit; seed oil/fat; fermented leaf vegetables; animal feed		oil palm
Arecaceae	*Raphia vinifera*	Raffiapalm	fruit drink/wine; terminal bud		
Asclepiadaceae	*Parquetina nigrescens*			unspecified	
Asteraceae	*Ageratum conyzoides*			unspecified	
Asteraceae	*Bidens pilosa*	Beggarticks, railway; burr marigold; black jack; spanish needles	leaves: raw, cooked		
Balanitaceae	*Balanites wilsoniana*		oil		
Bombacaceae	*Bombax buonopozense*	Bombax; red flowered silk; cotton tree; kapokier	seed: dried; leaves: raw; calyces: raw, dried		
Bombacaceae	*Ceiba pentandra*	Ceiba; kapok	fresh leaf vegetables; seed; young leaves		silk-cotton

Family	Genus/species	Common name	Food	Medicine	Int. trade
Burseraceae	*Canarium schweinfurthii*	Papo-canarytree; incense tree; African elemi	fruit, nut oil/fat		
Caesalpiniaceae	*Bussea occidentalis*	Asamantawa	unspecified		
Chrysobalanaceae	*Parinari excelsa*	Rough-skinned-plum; Guineaplum	fruit raw		
Chrysobalanaceae	*Rubus* sp.	Blackberry; raspberry; youngberry	fruit raw		
Clusiaceae	*Allanblackia floribunda*	Tallow-tree	seeds, dried		fat (soapmaking, cosmetics)
Clusiaceae	*Garcinia kola*	Kola, bitter; false kola	seed; chewing sticks; fruit		
Clusiaceae	*Garcinia polyantha*		seed; chewing sticks	sap used to make wound dressing	
Clusiaceae	*Mammea africana*	African mammee-apple; African apricot	fruit; nut		
Combretaceae	*Combretum paniculatum*	Combretum, panicled	leaves		
Combretaceae	*Terminalia glaucescens*			unspecified	
Combretaceae	*Terminalia superba*	Limba/Afara			timber
Connaraceae	*Byrsocarpus coccineus*			pharmaceutical	pharmaceutical
Euphorbiaceae	*Alchornea cordifolia*		leaves as vegetables		
Euphorbiaceae	*Antidesma venosum*	Antidesma	fruit		
Euphorbiaceae	*Hymenocardia acida*	Hymenocardia	young shoots		
Euphorbiaceae	*Maesobotrya barteri*		fruit; jam and jellies		
Euphorbiaceae	*Mareya micrantha*			unspecified	
Euphorbiaceae	*Ricinodendron heudelotii*	Mankettinut, musodo; African woodoilnut	oil seeds; fruit		
Euphorbiaceae	*Uapaca guineensis*	Sugar-plum; red cedar	fruit		
Euphorbiaceae	*Uapaca heudelotii*		fruit		
Fabaceae	*Abrus precatorius*	Rosarypea; jequirity	leaves; raw	unspecified	
Fabaceae	*Mucuna poggei*	Velvetbean; mucuna bean	Seeds, dried		
Fabaceae	*Pterocarpus erinaceus*	Padauk	wood; animal feed; leaves		kino (resin/dye)
Fabaceae	*Pterocarpus santalinoides*		fresh leaf vegetables		
Lamiaceae	*Solenostemon monostachyus*	Solenostemon	leaves, raw		
Liliaceae	*Gloriosa superba*			unspecified	unspecified
Loganiaceae	*Strychnos spinosa*	Poisonnut, kaffirorange; monkey-ball; Dzai	leaves raw; fruit raw		
Malvaceae	*Urena lobata*	Cadillo; bun-okra; aramina plant	leaves, raw; flowers		
Marantaceae	*Marantochloa cuspidata*			unspecified	
Meliaceae	*Carapa procera*			unspecified	

Family	Genus/species	Common name	Food	Medicine	Int. trade
Meliaceae	Entandrophragma utile	Sipo/Utile			timber
Meliaceae	Trichilia heudelotii	Bitterwood	seed, dried, oil; press cake; aril, dried; fruit;		
Mimosaceae	Albizia ferruginea	Albizia	leaves, raw		
Mimosaceae	Albizia sassa	Albizia	leaves, raw		
Mimosaceae	Albizia zygia	Albizia	leaves, raw		
Mimosaceae	Bussea occidentalis	Samanta	seed		
Mimosaceae	Parkia bicolor	African locust bean; nittatree	seed spices and condiments; fresh vegetables; seeds; fruit: pulp, meal dried; pods raw		
Mimosaceae	Pentaclethra macrophylla	Owalaoiltree; atta bean; congo acacia; apara seed	seed; Oil		
Moraceae	Ficus barteri		leaves		
Moraceae	Ficus capensis	Fig, cape; bush fig	leaves; fruit: raw		
Moraceae	Ficus eriobotryoides		leaves		
Moraceae	Ficus exasperata		leaves		
Moraceae	Ficus gnaphalocarpa		leaves		
Moraceae	Morus mesozygia	Mulberry	fruit		
Moraceae	Treculia africana	African breadfruit	seed: fresh, dry, whole, flour; bark; animal feed;		
Ochnaceae	Lophira alata	Azobe; Meni-oil tree; African oak; false-shea; Lophira, winged	seed, dried		timber
Piperaceae	Piper guineense			unspecified	
Piperaceae	Piper umbellatum	Peppers	leaves, raw		
Poaceae	Pennisetum purpurium	Napiergrass; elephant grass	young shoots		
Rubiaceae	Fagara macrophylla		leaves		
Rubiaceae	Mitragyna stipulosa			unspecified	
Rubiaceae	Morinda lucida			unspecified	
Rubiaceae	Mussaenda arcuata	Mussaenda	fruit		
Rubiaceae	Nauclea latifolia	Guineapeach; African peach; country fig; doundake	fruit pulp	unspecified	
Rubiaceae	Zanthoxylum zanthoxyloides			unspecified	
Sapindaceae	Blighia sapida	Akee, Raphe	unspecified		
Sapindaceae	Cardiospermum grandiflorum	Heartseed, showy; balloon vine	vine		
Sapindaceae	Paullinia pinnata	Paullinia	Aril		
Sapotaceae	Chrysophyllum perpulchrum	Starapple	fruit		
Sapotaceae	Chrysophyllum pruniforme	Starapple	fruit		
Sapotaceae	Chrysophyllum subnudum	Starapple	fruit		

Family	Genus/species	Common name	Food	Medicine	Int. trade
Simaroubaceae	*Harrisonia abyssinica*			bilharzia, chronic wounds (roots)	
Solanaceae	*Physalis angulata*	Groundcherry; sumberry	fruit, raw		
Solanaceae	*Solanum incanum*	Tomato, bitter; jakato	fruit, raw		
Solanaceae	*Solanum nigrum*	Nightshade, black; wonderberry; msobo	leaves: raw, dried	unspecified	
Solanaceae	*Solanum nodiflorum*	nightshade, guinea; lumbush	leaves; fruit		
Sterculiaceae	*Cola millenii*		fruit		
Tiliaceae	*Glyphaea brevis*	Glyphaea	leaves, raw		
Tiliaceae	*Grewia mollis*	Grewia	young leaves, raw		
Tiliaceae	*Triumfetta cordifolia*	Burweed	leaves, raw		
Ulmaceae	*Trema orientalis*			unspecified	
Vitaceae	*Cissus populnea*	Treebine	fruit: ripe, unripe		
Zingiberaceae	*Aframomum elliotii*	Aframomum	fruit, raw		
Zingiberaceae	*Aframomum heudelotii*	Aframomum	fruit, raw		

Sources:

Cunningham, A.B. 1993. African Medicinal Plants: setting priorities at the interface between conservation and primary health care. Working paper 1. UNESCO, Paris.

[FAO] Food and Agriculture Organization of the United Nations. 1990. Appendix 3 - Commonly Consumed Forest and Farm Tree Foods in the West African Humid Zone. The Major Significance of 'Minor' Forest Products: The Local Use and Value of Forests in the West African Humid Forest Zone. FAO, Rome.

[FAO] Food and Agriculture Organization of the United Nations and the US Department of Health, Education and Welfare. 1968. Appendix 4 - Index of Scientific Names of Edible Plants. Food Composition Table for Use in Africa.

MacKinnon, J. and K. MacKinnon. 1986. Review of the proteced areas system in the Afrotropical Realm. IUCN, Gland, Switzerland, in collaboration with UNEP. 259 pp.

Marshall, N.T. 1998. Searching for a Cure: Conservation of Medicinal Wildlife Resources in East and Southern Africa, TRAFFIC International.

Safowora, A. 1982. Medicinal Plants and Traditional Medicine in Africa. John Wiley and Sons Limited, Chichester.

Appendix/Annexe 3

Code and short description of habitats investigated in the Pic de Fon classified forest

Code et brève description des habitats étudiés dans la Forêt Classée du Pic de Fon

Mark-Oliver Rödel and Mohamed Alhassane Bangoura

m-h = man-hours spent searching in an area.
m-h = Homme-heures passées à prospecter une zone donnée

Nr.	Code	Latitude (N)	Longitude (W)	Date	m-h	Description
1	Whisky1 / FONBA	08°33.546'	08°52.870'	9 November	3	near base camp of Rio Tinto, gallery forest along a small stream, surrounded by Guinea savanna
2	Whisky2 / FOND1	08°31.861'	08°54.361'	8 & 26-29 November	19	1,650 m asl, montane grassland southwest of the Pic de Fon, small pockets of gallery forest along small mountain creeks, forests directly connected to the larger high forests in lower altitude, in the grassland small patches of swampy areas with small puddles, in areas where bare rock occurs some rocky depressions are water filled
3	FOND2	08°32.943'	08°53.964'	28 November	4	1,100 m asl deep valley with near pristine high forest, fast flowing river with few more stagnant parts, eastern slope of the Pic de Fon
4	FOND3	08°33.270'	08°54.428'	30 November	0.5	1,662 m asl, montane grassland on mountain top, close to the climate station
5	FONBAN	08°33.330'	08°58.790'	7 December	6	village Banko, surrounded by secondary forest, plantations and savannah, gallery forest close to a river
6	FONBA1	08°31.499'	08°56.204'	1-4 & 6 December	36	650 m asl, second base camp, app. 6 km from village Banko in the direction to the Simandou ridge, app. 1.5 km within the forest reserve, the camp is situated close to true high forest, there are a lot of mountain streams of different size, some hills with savannah vegetation, in the direction to the village banana, cacao and coffee plantations, secondary forests with swampy areas
7	FONBA2	08°30.876'	08°56.593'	2 December	17.5	small brook, fast flowing, banks heavily vegetated, close to banana plantation
8	FONBA3	08°30.950'	08°56.561'	2 & 5 December	8	larger pond, bank with high grass, small secondary forest, cacao and banana plantations, several small brooks, secondary forest in good state with swampy valley, several stagnant ponds, artisan gold mining
9	FONBA4	08°32.053'	08°58.424'	5 December	14	very swampy area close to the village, between a cacao plantation and savannah, in the rainy season larger ponds with very dense shrubby vegetation at the banks, during the investigations only two smaller puddles remained water filled; way to base camp through plantations and secondary forest, partly savannah, several small rivers

Nr.	Code	Latitude (N)	Longitude (W)	Date	m-h	Description
10	FONBA5	no GPS data	no GPS data	2 December	3	small deep pond with large leaved herbs in swampy areas in secondary forest with open understorey (close to FONDBA2), secondary forest close to banana and cacao plantation
11	NIONSO	no GPS data	no GPS data	2 December	2.25	river Mea, situated at the limit between the Pic de Fon reserve and the village of Nionsomorido, savannah
12	MORIBA	no GPS data	no GPS data	5-6 November	4.25	river Dofa, close to village Morigbadougou, northeast flank of Pic de Fon, savannah and small gallery forest along the river
13	TRAORE	no GPS data	no GPS data	3-4 November	10.5	Tiyeko near Traorela, village situated in the North-east of Pic de Fon, gallery forest at the mountain base (Ouleba forest), flowing rocky river, savannah close to the village
14	Mount Focon	no GPS data	no GPS data	29 November	6	South-west flank of Pic de Fon steep mountain slope, humid forest, swampy areas, lower storey comparatively open

Nr.	Code	Latitude (N)	Longitude (W)	Date	m-h	Description
1	Whisky1 / FONBA	08°33.546'	08°52.870'	9 Novembre	3	proche du camp de base de Rio Tinto, forêt galerie le long d'un petit torrent/rivière, entouré de savane guinéenne
2	Whisky2 / FOND1	08°31.861'	08°54.361'	8 & 26-29 Novembre	19	1650 m d'altitude, prairie de montagne au sud ouest du Pic de Fon, petites poches de forêt galerie le long de petits ruisseaux de montagne, forêts directement connectées au plus grandes forêts de moindre altitude, dans la prairie de petites zones marécageuses avec de petites flaques, dans les zones ou la roche est à nue, des dépressions rocheuses sont remplies d'eau
3	FOND2	08°32.943'	08°53.964'	28 Novembre	4	1100 m d'altitude, vallée profonde avec forêt presque primaire, rivière à débit rapide avec quelques parties stagnantes, versant oriental du Pic de Fon
4	FOND3	08°33.270'	08°54.428'	30 Novembre	0.5	1662 m d'altitude, prairie d'altitude au sommet de la montagne, à proximité de la station climatique
5	FONBAN	08°33.330'	08°58.790'	7 Décembre	6	village de Banko, entouré de forêt secondaire, de plantations et de savanes, forêt galerie proche d'une rivière
6	FONBA1	08°31.499'	08°56.204'	1-4 & 6 Décembre	36	650 m d'altitude, second camp de base distant d'environ 6km du village de Banko dans la direction de la chaîne du Simandou, à environ 1,5 km de la forêt classée, le camp est situé à proximité de la vraie forêt, il y a de nombreux torrents de montagne de différentes tailles, quelques collines avec une végétation savanicole, en direction du village, des plantations de bananes, de cacao, de café, forêt secondaire avec des zones marécageuses
7	FONBA2	08°30.876'	08°56.593'	2 Décembre	17.5	petits ruisseaux, courant rapide, berges à végétation dense, proche de plantation de banane

Nr.	Code	Latitude (N)	Longitude (W)	Date	m-h	Description
8	FONBA3	08°30.950'	08°56.561'	2 & 5 Décembre	8	grande mare, berge avec de hautes herbes, petite forêt secondaire, plantations de banane et de cacao, plusieurs petits ruisseaux, forêt secondaire en bon état avec vallées marécageuses, nombreuses mares stagnantes, extraction artisanale de l'or
9	FONBA4	08°32.053'	08°58.424'	5 Décembre	14	zone très marécageuse proche du village, entre plantation de cacao et savane, en saison des pluies, mares plus grandes avec une végétation buissonnante très dense sur les berges, durant l'expertise seulement deux plus petites flaques d'eau subsistaient ; chemin du camp de base entre plantations et forêt secondaire, en partie de la savane, nombreuses petites rivières
10	FONBA5	Pas de relevé GPS	Pas de relevé GPS	2 Décembre	3	petite mare profonde avec plantes feuillues dans les zones marécageuses de la forêt secondaire avec un sous bois ouvert (proche de FONDBA2), forêt secondaire proche des plantations de banane et de cacao
11	NIONSO	Pas de relevé GPS	Pas de relevé GPS	2 Décembre	2.25	rivière Mea, située à la limite entre la Forêt Classée du Pic de Fon et le village de Nionsomorido, savane
12	MORIBA	Pas de relevé GPS	Pas de relevé GPS	5-6 Décembre	4.25	rivière Dofa, proche du village de Morigbadougou, versant nord-est du Pic de Fon, savane et petite forêt galerie le long de la rivière
13	TRAORE	Pas de relevé GPS	Pas de relevé GPS	3-4 Décembre	10.5	Tiyeko proche de Traorela, village situé au Nord –Est du Pic de Fon, forêt galerie sur le piedmont (forêt de Ouléba), torrent rocailleux, savane proche du village
14	Mount Focon	Pas de relevé GPS	Pas de relevé GPS	29 Décembre	6	Versant escarpé au Sud-Ouest du Pic de Fon, forêt humide, zones marécageuses, étage inférieur comparativement ouvert

Appendix/Annexe 4

Amphibian and reptile tissue samples and voucher specimens collected during the Pic de Fon RAP survey

Echantillons de tissus d'amphibiens et reptiles ainsi que spécimens de référence collectés durant l'expertise RAP du Pic de Fon

Mark-Oliver Rödel and Mohamed Alhassane Bangoura

List of tissue samples (DNS) and voucher specimens. MOR = collection RODEL, MAB = collection BANGOURA; B = preserved in formaldehyde, F0 = preserved in ethanol with corresponding tissue sample, S = preserved in ethanol without corresponding tissue sample.

Liste des échantillons de tissus (DNS) et spécimens de référence. MOR = Collection de Rodel, MAB = Collection de Bangoura ; B = Conservé dans le formaldéhyde, FO = Conservé dans l'éthanol avec les échantillons de tissu correspondant, S = Conservé dans l'éthanol sans les échantillons de tissu correspondant.

Species	DNS	MOR	MAB
Bufo maculatus	F080	B33, F080	B34
B. togoensis	F057, 75-78	F075-76, 78, S4-5	F057
Bufo (?) sp.		B25	
Hoplobatrachus occipitalis		B24	B1, 20
Conraua sp.	F05-7, 34, 47	B11, F05-6, 34 + tadpole	F07, 47
Amnirana albolabris	F083-64	B15, F083-84	B18
Amnirana sp.		B13-14	
Petropedetes natator	F01-4, 16, 85	F01, 3-4, 16, 85 + tadpole	B4, F02
Phrynobatrachus accreansis	F027-29, 48, 50, 55-56, 73	B2, 40, 47-48, F055, 73	B3, 38-39, 41, F027-29, 56, S1
Phrynobatrachus alleni	F054	F054	
Phrynobatrachus alticola	F017-24, 33, 41-44, 61, 64	B7, 36, 52, F024, 33, 41-44, 61, 64	B8-10, 37, 53, F017-23
Phrynobatrachus fraterculus	F051-53, 67-68, 91	F067-68	F051-53, 91
Phrynobatrachus liberiensis	F058-60, 65	F059-60	F058, 65
Phrynobatrachus phyllophilus	F093	F093	S6
Phrynobatrachus cf. *maculiventris*		B49-51	
Ptychadena aequiplicata			S7
Ptychadena sp. aff. *aequiplicata*	F081	F081	

Species	DNS	MOR	MAB
Ptychadena sp. 1	F08, 32, 37-39, 94-95, 106-109	B12, F032, 39, 94-95, 108 + tadpoles	F08, 37-38, 106-107, 109
Arthroleptis sp.	F013-14, 24, 26, 31, 62-63, 105	F013, 24, 31, 62, 105	B5-6, F014, 26, 63
Cardioglossa leucomystax	F082, 88-90	F088, 90	F082, 89
Astylosternus occidentalis		B19	
Leptopelis hyloides	F09, 30, 45-46, 66	B43, F09, F066	B31-32, 42
Hyperolius chlorosteus	F071-72, 86-87	B16, F086	B17, F071-72, 87, S3
Hyperolius fusciventris	F0100-101	F0101	F0100
Hyperolius concolor	F069-70, 99, 102, 104	F069-70, 99, 104	F0102
Hyperolius picturatus	F010-12, F040	B26, 28-30, 45, F010-11, 40, 3 froglets, 1 tadpole	B27, 44, 46, F012
Afrixalus dorsalis	F096-97	F096, S8	F097
Afrixalus fulvovittatus	F0110	F0110-111+ tadpole	2 without #
Kassina cochranae	F035-36, 79	F035-36, 79 + 6 tadpoles	

Appendix/Annexe 5

Reptile species encountered during the RAP
and records obtained by Rio Tinto staff (RT)
and villagers (V)

Espèces de reptiles rencontrées durant le
RAP et recensements obtenus par l'équipe de
Rio Tinto (RT) et les villageois (V)

Mark-Oliver Rödel and Mohamed Alhassane Bangoura

Family/Famille	Genus/Genre	Species/Espèces	Seen	Record	CITES
Gekkonidae	*Hemidactylus*	*muriceus*	x	-	-
Scincidae	*Mabuya*	*affinis*	x	-	-
	Gophioscincopus	sp.	x	-	-
Agamidae	*Agama*	*agama*	x	-	-
Chamaeleonidae	*Chamaeleo*	*senegalensis*	x	-	2
Varanidae	*Varanus*	*ornatus*	x	-	2
Pythonidea	*Python*	*sebae*	-	RT	2
Colubridae	*Philothamnus*	*irregularis*	-	RT	-
	Afronatrix	*anoscopus*	x	-	-
Viperidae	*Bitis*	*arietans*	x	-	-
		gabonica	-	RT	-
Elapidae	*Naja*	*melanoleuca*	-	RT	-
		nigricollis	-	RT	-
Crocodylidae	*Osteolaemus*	*tetraspis*	-	V	2

Appendix/Annexe 6

List of bird species recorded in the Forêt Classée du Pic de Fon

Liste des espèces d'oiseaux recensées dans la Forêt Classée du Pic de Fon

Ron Demey and Hugo J. Rainey

Species/Espèces	Highland/Haute altitude		Lowland/Basse altitude		Threat/ Statut de conservation	Endemic/ Endémisme	Biome	Habitat
	Abundance/ Abondance	Breeding/ Signes de reproduction	Abundance/ Abondance	Breeding/ Signes de reproduction				
Ciconiidae (1)								
Ciconia episcopus			R					a
Threskiornithidae (1)								
Bostrychia rara/olivacea			R					l
Accipitridae (15)								
Pernis apivorus			R					a
Gypohierax angolensis			U					a
Gyps africanus			R					a
Polyboroides typus			C					c,a
Circus aeruginosus	U							g
Accipiter tachiro			U					c,a
Accipiter melanoleucus			R					a
Urotriorchis macrourus			U				GC	m
Kaupifalco monogrammicus	U		R					e,l
Buteo auguralis			R					a
Hieraeetus ayresii			R					a
Lophaetus occipitalis	R		R					a
Spizaetus africanus			R				GC	a
Stephanoaetus coronatus	U							a
Polemaetus bellicosus	U							a
Falconidae (1)			R					
Falco biarmicus								a
Phasianidae (4)								
Ptilopachus petrosus	C							g
Francolinus lathami			U				GC	l
Francolinus ahantensis	C		C				GC	l,e
Francolinus bicalcaratus	R							g
Rallidae (1)								

Species/Espèces	Highland/Haute altitude		Lowland/Basse altitude		Threat/ Statut de conservation	Endemic/ Endémisme	Biome	Habitat
	Abundance/ Abondance	Breeding/ Signes de reproduction	Abundance/ Abondance	Breeding/ Signes de reproduction				
Sarothrura pulchra	C		F				GC	l,r
Columbidae (8)								
Treron calva	U		C					c
Turtur brehmeri			C				GC	l
Turtur tympanistria	U		C					l
Turtur afer			F					s
Columba iriditorques	R		C				GC	c
Aplopelia larvata			R					l
Streptopelia semitorquata			F					s
Streptopelia vinacea	R							s
Musophagidae (2)								
Corythaeola cristata	R		C					c,m
Tauraco persa	C		C				GC	c,m
Cuculidae (8)								
Cuculus clamosus			R					s
Cercococcyx mechowi			F				GC	m
Cercococcyx olivinus			C				GC	m
Chrysococcyx cupreus	C		C					c
Chrysococcyx klaas	F		F					c
Ceuthmochares aereus	F		F					m,l
Centropus leucogaster			F					l
Centropus senegalensis	U		F	b				g,s
Tytonidae (1)								
Tyto alba			R					s
Strigidae (3)								
Bubo africanus	R							s
Bubo poensis			R				GC	c,m
Strix woodfordii			F					m
Caprimulgidae (1)								
Macrodipteryx longipennis	R							s
Apodidae (2)								
Rhaphidura sabini			F				GC	a
Neafrapus cassini			F				GC	a
Trogonidae (1)								
Apaloderma narina			U					m
Alcedinidae (6)								
Halcyon badia			U				GC	m
Halcyon leucocephala	R		R					e,s
Halcyon malimbica	R		F					m,l
Ceyx pictus			U					e
Alcedo leucogaster			R				GC	l,r
Alcedo quadribrachys			R					r

Species/Espèces	Highland/Haute altitude		Lowland/Basse altitude		Threat/ Statut de conservation	Endemic/ Endémisme	Biome	Habitat
	Abundance/ Abondance	Breeding/ Signes de reproduction	Abundance/ Abondance	Breeding/ Signes de reproduction				
Meropidae (4)								
Merops muelleri			R				GC	m,e
Merops gularis	R						GC	e
Merops albicollis	C		C					a,g,s
Merops apiaster			U					s
Coraciidae (1)								
Eurystomus glaucurus			R					e
Phoeniculidae (1)								
Phoeniculus bollei			R					e
Bucerotidae (2)								
Tockus fasciatus			C				GC	c,e,s
Ceratogymna elata			F		nt		GC	c,e
Capitonidae (9)								
Gymnobucco calvus	C		C				GC	c,e
Pogoniulus scolopaceus			C				GC	e
Pogoniulus atroflavus	C		C				GC	c
Pogoniulus subsulphureus	U		C				GC	c,m,e
Pogoniulus bilineatus	C		C					c,m,e
Buccanodon duchaillui	R		C				GC	c
Tricholaema hirsuta			C				GC	c,m,e
Lybius vieilloti	U		R					s
Trachylaemus purpuratus	R						GC	m,l
Indicatoridae (4)								
Melichneutes robustus			R				GC	c,e
Indicator maculatus			R				GC	m,l
Indicator conirostris	R		R					c,m
Indicator willcocksi			U				GC	c,m
Picidae (5)								
Campethera maculosa			U				GC	m
Campethera nivosa			F				GC	m,l
Dendropicos gabonensis	F		F				GC	c,m,e
Dendropicos fuscescens			U					e,s
Dendropicos pyrrhogaster			F					e
Eurylaimidae (1)								
Smithornis capensis			F					l
Alaudidae (1)								
Mirafra africana	F							g
Hirundinidae (6)								
Psalidoprocne nitens	R		C				GC	a,e,s
Psalidoprocne obscura			C				GC	a,e,s
Riparia riparia			U					a,s
Hirundo abyssinica			R					a,s

Species/Espèces	Highland/Haute altitude		Lowland/Basse altitude		Threat/ Statut de conservation	Endemic/ Endémisme	Biome	Habitat
	Abundance/ Abondance	Breeding/ Signes de reproduction	Abundance/ Abondance	Breeding/ Signes de reproduction				
Hirundo preussi			C					a,s
Hirundo rustica	C		C					a,s
Motacillidae (5)								
Motacilla flava	U							g
Motacilla clara			R					r
Anthus similis	F							g
Anthus trivialis	U							e,g
Anthus cervinus	C							g
Campephagidae (2)								
Campephaga quiscalina	C		F					c
Lobotos lobatus			R		VU	UG	GC	c,m
Pycnonotidae (19)								
Andropadus virens	C		C					l,e
Andropadus gracilis			C				GC	c,e
Andropadus ansorgei	R		C				GC	c,e
Andropadus curvirostris			U				GC	l,e
Andropadus gracilirostris	F		C					c,e
Andropadus latirostris	C		C	b				l,e
Baeopogon indicator	F		C				GC	c,e
Chlorocichla simplex			C				GC	e
Thescelocichla leucopleura			F				GC	e,r
Pyrrhurus scandens	C		C				GC	c,m,e
Phyllastrephus baumanni			R		DD		GC	e
Phyllastrephus icterinus	U		F				GC	m,l
Bleda syndactyla			U				GC	l
Bleda canicapilla			F				GC	l
Criniger barbatus	F		C				GC	m,l
Criniger calurus	C		C				GC	m,l
Criniger olivaceus			R		VU	UG	GC	m
Pycnonotus barbatus	C		C					e,s
Nicator chloris	F		F				GC	m,e
Turdidae (14)								
Stiphrornis erythrothorax	F		F				GC	l
Sheppardia cyornithopsis	R		U				GC	l
Luscinia megarhynchos	F		F					e,s
Cossypha polioptera	R		R					l
Cossypha niveicapilla	U							l
Cossypha albicapilla			R	b			SG	e
Alethe diademata	F		U				GC	l
Alethe poliocephala	F		U					l
Neocossyphus poensis	F		F				GC	l
Stizorhina finschi	F		C				GC	m,l
Cercotrichas leucosticta			R				GC	l
Saxicola torquata	F							g

Species/Espèces	Highland/Haute altitude Abundance/ Abondance	Highland/Haute altitude Breeding/ Signes de reproduction	Lowland/Basse altitude Abundance/ Abondance	Lowland/Basse altitude Breeding/ Signes de reproduction	Threat/ Statut de conservation	Endemic/ Endémisme	Biome	Habitat
Saxicola rubetra	C		F					g,s
Turdus pelios	U		U					e
Sylviidae (27)								
Bathmocercus cerviniventris			F		nt	UG	GC	e,r
Melocichla mentalis	F		U					g,s
Acrocephalus scirpaceus	R							e,g
Hippolais polyglotta	R		U					e,s
Cisticola erythrops	U		U					g,s
Cisticola cantans	R		U					g,s
Cisticola lateralis			F					s
Cisticola natalensis	R							g
Cisticola brachypterus	F							g
Prinia subflava	C		C					e,g,s
Schistolais leontica	F				VU	UG	GC	e
Apalis nigriceps	C		C				GC	c
Apalis sharpii	C		C			UG	GC	c
Camaroptera brachyura	F		F					e
Camaroptera superciliaris			U				GC	e
Camaroptera chloronota	C		C				GC	l
Macrosphenus kempi	U						GC	e
Macrosphenus concolor	C		F				GC	m,l
Eremomela badiceps	C	b	C				GC	c
Sylvietta virens			F				GC	e
Sylvietta denti	C		C				GC	c,m
Phylloscopus trochilus	C		F					c,e,s
Phylloscopus sibilatrix			F					c,e
Sylvia borin	R		R					e,s
Sylvia atricapilla	U		R					e,s
Hyliota violacea			U				GC	c
Hylia prasina	C		C				GC	m,l,e
Muscicapidae (5)								
Fraseria ocreata			R				GC	c
Fraseria cinerascens			R				GC	e,r
Muscicapa epulata			R				GC	e
Muscicapa ussheri	F		U				GC	c,e
Ficedula hypoleuca	C		F					e,s
Monarchidae (4)								
Erythrocercus mccallii	R		F				GC	c,m
Elminia nigromitrata	F		C				GC	l
Trochocercus nitens			C				GC	l
Terpsiphone rufiventer	F		C	b			GC	m,l
Platysteiridae (6)								

Species/Espèces	Highland/Haute altitude		Lowland/Basse altitude		Threat/ Statut de conservation	Endemic/ Endémisme	Biome	Habitat
	Abundance/ Abondance	Breeding/ Signes de reproduction	Abundance/ Abondance	Breeding/ Signes de reproduction				
Megabyas flammulatus	F		F				GC	c
Bias musicus			U					c,e
Dyaphorophyia castanea	R		U				GC	m,l
Dyaphorophyia blissetti	R		F				GC	m,l
Dyaphorophyia concreta	R		F					l
Platysteira cyanea	C							l,e
Timaliidae (6)								
Illadopsis rufipennis			F					l
Illadopsis fulvescens	R		R				GC	l
Illadopsis cleaveri			F				GC	l
Illadopsis rufescens	R		F		nt	UG	GC	l
Illadopsis puveli	C		U				GC	l
Phyllanthus atripennis	U						GC	m,l
Paridae (1)								
Parus funereus			R				GC	c
Remizidae (1)								
Pholidornis rushiae			R				GC	c,m
Nectariniidae (12)								
Anthreptes rectirostris			C				GC	c,m
Cyanomitra verticalis	R							e
Cyanomitra cyanolaema	F		C				GC	c
Cyanomitra obscura	C		C	b				m,l
Chalcomitra senegalensis	F		U					e
Hedydipna collaris			C					e
Hedydipna platura	R							e
Cinnyris chloropygius			C					e
Cinnyris venustus	C		U					e,s
Cinnyris johannae			R				GC	c,e
Cinnyris superbus			U				GC	c,e
Cinnyris cupreus	F	b						e
Zosteropidae (1)								
Zosterops senegalensis	F		C					e,s
Malaconotidae (5)								
Malaconotus cruentus			U				GC	m
Malaconotus multicolor	C		F					c,m
Tchagra australis			U					e
Dryoscopus gambensis	U							c,m
Laniarius aethiopicus	R							e
Oriolidae (2)								
Oriolus nigripennis			U				GC	c
Oriolus brachyrhynchus	C		C				GC	c
Dicruridae (3)								

Species/Espèces	Highland/Haute altitude		Lowland/Basse altitude		Threat/ Statut de conservation	Endemic/ Endémisme	Biome	Habitat
	Abundance/ Abondance	Breeding/ Signes de reproduction	Abundance/ Abondance	Breeding/ Signes de reproduction				
Dicrurus ludwigii	C		C					m
Dicrurus atripennis			R				GC	m,l
Dricurus modestus			U					c,e
Sturnidae (4)								
Onychognathus fulgidus			U				GC	c
Lamprotornis chloropterus	R		R					s
Lamprotornis iris	R		R		DD		SG	s
Cinnyricinclus leucogaster			U					s
Passeridae (1)								
Petronia dentata	R						SG	s
Ploceidae (11)								
Ploceus nigricollis			F					e
Ploceus nigerrimus			R				GC	e
Ploceus cucullatus			C					e,s
Ploceus superciliosus			R	b				s
Ploceus preussi	R						GC	c
Malimbus nitens			F				GC	m,l,e
Malimbus malimbicus	F		U				GC	m
Malimbus scutatus			F				GC	c,e
Quelea erythrops			R					s
Euplectes hordeaceus			F					s
Euplectes ardens	U		F					g,s
Estrilididae (11)								
Nigrita canicapilla	C		C					c
Nigrita bicolor			F				GC	c
Nigrita fusconota			R				GC	c
Spermophaga haematina	R		C				GC	e
Mandingoa nitidula	R		R					e
Euschistospiza dybowskii	U						SG	e,g
Lagonosticta rubricata	U		F					g,s
Estrilda melpoda	F		U					g,s
Estrilda astrild	U		U	b				g,s
Lonchura cucullata	R		R					g,s
Lonchura bicolor			F	b				e
Viduidae (1)								
Vidua camerunensis	R		R					e,g,s
Fringillidae (1)								
Serinus mozambicus	C							e,g
Emberizidae (2)								
Emberiza hortulana	R							g
Emberiza tahapisi	C							g
Species total: 233	131		198		UG = 6		GC = 104	
							SG = 4	

Abundance:
C - Common: encountered daily, either singly or in significant numbers
F - Fairly common: encountered on most days
U - Uncommon: irregularly encountered and not on the majority of days
R - Rare: rarely encountered, one or two records of single individuals

Breeding:
b - evidence of breeding observed

Threat:
DD - Data Deficient
VU - Vulnerable
nt - Near Threatened

Endemic:
UG - endemic to Upper Guinea Forest block

Biome:
GC - restricted to Guinea-Congo Forests biome
SG - restricted to Sudan-Guinea Savannah biome

Habitat:
c - forest canopy
m - mid forest storey
l - lower forest storey and ground
e - forest edge
r - rivers, streams and ponds
a - aerial and flying overhead
g - highland grassland
s - lowland savannah

Abondance:
C - Commune: observée quotidiennement, seule ou en nombre conséquent
F - Assez commune: observée presque chaque jour
U - Peu commune: irrégulièrement observée et pas tous les jours
R - Rare: rarement observée, une ou deux observations d'individus solitaires

Reproduction:
b - preuve de reproduction observée

Statut de conservation:
VU - Vulnérable
DD - Insuffisamment documenté
nt - Quasi-menacé

Endémisme:
UG - endémique au bloc forestier de Haute Guinée

Biome:
GC - confinée au biome des forêts guinéo-congolaises
SG - confinée au biome de la savane soudano-guinéenne

Habitat:
c - canopée
m - strate moyenne de la forêt
l - strate inférieure de la forêt et sol
e - lisière
r - cours d'eau et mares
a - dans les airs et survolant le site
g - prairies de montagne
s - savane de plaine

Appendix/Annexe 7

List of birds trapped in mist-nets in the
Forêt Classée du Pic de Fon

Liste des oiseaux capturés au filet dans
la Forêt Classée du Pic de Fon

Ron Demey and Hugo J. Rainey

Species/Espèces		Number/Nombre	
		Highland/ Haute altitude	Lowland/ Basse altitude
Turtur tympanistria	Tambourine Dove	1	
Aplopelia larvata	Lemon Dove		1
Halcyon malimbica	Blue-breasted Kingfisher	1	
Alcedo leuocogaster	White-bellied Kingfisher		2
Merops muelleri	Blue-headed Bee-eater		1
Pogoniulus scolopaceus	Speckled Tinkerbird		2
Pogoniulus subsulphureus	Yellow-throated Tinkerbird		2
Campethera nivosa	Buff-spotted Woodpecker		1
Motacilla clara	Mountain Wagtail		2
Andropadus virens	Little Greenbul	1	8
Andropadus latirostris	Yellow-whiskered Greenbul		41
Baeopogon indicator	Honeyguide Greenbul		1
Chlorocichla simplex	Simple Leaflove		1
Phyllastrephus icterinus	Icterine Greenbul		2
Bleda syndactyla	Red-tailed Bristlebill		2
Bleda canicapilla	Grey-headed Bristlebill		4
Criniger barbatus	Western Bearded Greenbul		8
Criniger calurus	Red-tailed Greenbul		1
Stiphrornis erythrothorax	Forest Robin		3
Sheppardia cyornithopsis	Lowland Akalat	1	2
Cossypha polioptera	Grey-winged Robin Chat	1	
Cossypha niveicapilla	Snowy-crowned Robin Chat	1	
Cossypha albicapilla	White-crowned Robin Chat		1
Alethe diademata	Fire-crested Alethe	1	1
Alethe poliocephala	Brown-chested Alethe	3	2
Neocossyphus poensis	White-tailed Ant Thrush	1	
Cercotrichas leucosticta	Forest Scrub Robin		2
Bathmocercus cerviniventris	Black-headed Rufous Warbler		1
Acrocephalus scirpaceus	European Reed Warbler	1	
Schistolais leontica	Sierra Leone Prinia	2	

Species/Espèces		Number/Nombre	
		Highland/ Haute altitude	Lowland/ Basse altitude
Camaroptera brachyura	Grey-backed Camaroptera	3	
Camaroptera chloronota	Olive-green Camaroptera		4
Phylloscopus trochilus	Willow Warbler	7	
Sylvia borin	Garden Warbler	1	
Hylia prasina	Green Hylia		3
Ficedula hypoleuca	Pied Flycatcher	1	
Elminia nigromitrata	Dusky Crested Flycatcher		4
Terpsiphone rufiventer	Red-bellied Paradise Flycatcher	1	5
Dyaphorophyia blissetti	Red-cheeked Wattle-eye		1
Dyaphorophyia concreta	Yellow-bellied Wattle-eye		2
Platysteira cyanea	Common Wattle-eye	3	
Illadopsis rufipennis	Pale-breasted Illadopsis		6
Illadopsis puveli	Puvel's Illadopsis		2
Cyanomitra obscura	Western Olive Sunbird	1	7
Chalcomitra senegalensis	Scarlet-chested Sunbird	1	
Cinnyris venustus	Variable Sunbird	2	
Cinnyris cupreus	Copper Sunbird	1	
Zosterops senegalensis	Yellow White-eye	1	
Dicrurus atripennis	Shining Drongo		1
Malimbus nitens	Blue-billed Malimbe		4
Euplectes ardens	Red-collared Widowbird	2	
Nigrita bicolor	Chestnut-breasted Negrofinch		2
Spermophaga haematina	Western Bluebill		5
Mandingoa nitidula	Green Twinspot	1	1
Euschistospiza dybowskii	Dybowski's Twinspot	3	
Lagonosticta rubricata	Blue-billed Firefinch	1	
	Totals	43	138
	Grand total of 183 individuals of 56 species.		

Appendix/Annexe 8

Ecological characterization and comparison of the presence of terrestrial small mammals from Pic de Fon (PF) compared with data from Mount Nimba (N), the Ziama range at Sérédou (Z), the National Park of Upper Niger in Guinea (UN), and from Parc National du Mont Péko in Côte d'Ivoire (MP).

Caractérisation écologique et comparaison de la présence des petits mammifères terrestres du Pic de Fon (PF) avec les données du Mont Nimba (N), de la chaîne du Ziama à Sérédou (Z), du Parc National de Haute Niger (UN), et du Parc National du Mont Péko en Côte d'Ivoire.

Jan Decher

F - forest/forêt
S - savannah/savane
M - mixed strategy/stratégie mixte
PF - this study; cette étude
N - Coe 1975, Gautun et al. 1986, Heim de Balsac et Lamotte 1958, Verschuren and Meester 1977
Z - Roche 1971
UN - Ziegler, Nikolaus, and Hutterer 2002
MP - Lim and Van Coeverden de Groot 1997

Common names follow Wilson and Cole (2000).
Les noms communs correspondent à Wilson and Cole (2000).

Group/Genus/Species	Common name	Ecolog. Charact.			Localities				
Groupe/Genre/Espèces	Nom commun	F	M	S	PF	N	Z	UN	MP
Insectivores									
Atelerix albiventris	African hedgehog			1				+	
Crocidura bottegi	Bottego's Shrew			1		+			
Crocidura crossei	Crosse's Shrew	1						+	
Crocidura cf. *denti*	Dent's Shrew	1			+			+	
Crocidura foxi	Fox's Shrew			1		+		+	
Crocidura fuscomurina	Tiny Musk Shrew			1		+		+	
Crocidura douceti	Doucet's Musk Shrew		1			+			
Crocidura flavescens	Greater Red Musk Shrew		1			+			+
Crocidura grandiceps	Large-headed Shrew	1			+				
Crocidura jouvenetae	Jouvenete's Shrew	1				+			
Crocidura lamottei	Lamotte's Shrew			1					
Crocidura muricauda	Mouse-tailed Shrew	1		1		+		+	

Group/Genus/Species	Common name	Ecolog. Charact.			Localities				
Groupe/Genre/Espèces	Nom commun	F	M	S	PF	N	Z	UN	MP
Crocidura nanilla	Tiny White-toothed Shrew			1				+	
Crocidura nimbae	Nimba Shrew	1				+			
Crocidura olivieri	Olivier's Shrew		1			+	+	+	
Crocidura poensis	Fraser's Musk Shrew		1			+	+		
Crocidura theresae	Therese's Shrew			1			+		
Sylvisorex megalura	Climbing Shrew		1			+	+	+	
Micropotamogale lamottei	Nimba Otter Shrew	1				+	+		
Rodents									
Anomalurops sp.	Scaly tailed flying squirrel	1						+	
Anomalurus beecrofti	Beecroft's Scaly-tailed Squirrel	1				+	+		
Anomalurus derbianus	Lord Derby's Scaly-tailed Squirrel	1				+	+		
Idiurus macrotis	Pygmy Scaly-tailed flying Squirrel	1				+			
Epixerus ebii	Western Palm Squirrel	1				+			
Funisciurus pyrrhopus	Fire-footed Rope Squirrel	1				+	+	+	+
Heliosciurus gambianus	Gambian Sun Squirrel	1			+	+	+	+	
Heliosciurus rufobrachium	Red-legged Sun Squirrel	1			+		+	+	
Paraxerus poensis	Green Bush Squirrel	1			+	+	+		
Protoxerus stangeri	African Giant Squirrel	1			+	+	+		
Protoxerus aubinnii	Slender-tailed Squirrel	1			+	+			
Xerus erythropus	Striped Ground Squirrel			1	+	+	+	+	
Arvicanthis sp.	African Grass Rat			1				+	
Cricetomys sp.	Gambian or Giant Pouched Rat		1		+	+	+	+	
Dasymys rufulus	African Marsh Rat		1			+	+	+	
Dephomys defua	Defua Rat	1				+	+		
Dendromus melanotis	Climbing Mouse	1					+		
Grammomys buntingi	Bunting's Thicket Rat	1					+		
Grammomys rutilans	Shining Thicket Rat		1			+			
Hybomys trivirgatus	Temminck's Striped Mouse	1			+	+	+		
Hybomys planifrons	Miller's Striped Mouse	1			+	+		+	
Hylomyscus alleni	Allen's Wood Mouse	1			+	+	+		
Lemniscomys bellieri	Bellier's Striped Grass Mouse			1				+	
Lemniscomys striatus	Typical Striped Mouse			1		+	+	+	
Lophuromys sikapusi	Rusty-bellied Brush-furred rat	1				+	+		
Malacomys edwardsi	Edward's Swamp Rat	1			+	+	+		+
Mastomys erythroleucus	Multimammate Rat		1			+	+	+	
Mastomys hildebrandtii	Hildebrandt's Multimammate Rat		1					+	
Mastomys natalensis	Natal Multimammate Rat		1					+	
Myomyscus (Myomys) daltoni	Dalton's mouse		1					+	

Group/Genus/Species	Common name	Ecolog. Charact.			Localities				
		F	M	S	PF	N	Z	UN	MP
Myomyscus (Myomys) derooi	Deroo's Mouse		1					+	
Mus musculoides	Temminck's Mouse		1		+	+	+	+	
Mus setulosus	Peter's Mouse			1	+	+	+		
Oenomys ornatus	Ghana Rufous-nosed Rat	1				+	+		
Praomys rostratus	Forest Soft-furred Mouse	1			+	+		+	+
Praomys tullbergi	Tullberg's Soft-furred Mouse	1				+	+		
Rattus rattus	House Rat		1			+	+		+
Tatera guineae	Guinea Gerbil			1				+	
Tatera kempi	Kemp's Gerbil			1		+	+		
Uranomys ruddi	Rudd's Mouse			1		+		+	
Graphiurus nagtglasii (=hueti)	Huet's Dormouse	1				+	+		
Graphiurus lorraineus	Lorrain Dormouse	1				+	+		
Graphiurus crassicaudatus	Jentink's Dormouse	1				+			
Hystrix cristata	Crested Porcupine		1					+	
Atherurus africanus	Brush-tailed Porcupine		1		+	+		+	
Thryonomys swinderianus	Greater Cane Rat			1		+	+	+	